Biological control by natural enemies

Biological control by natural enemies

PAUL DEBACH

Emeritus Professor of Biological Control
Department of Entomology, University of California, Riverside, USA

DAVID ROSEN

Professor of Entomology
The Hebrew University, Faculty of Agriculture, Rehovot, Israel

Second edition

The right of the
University of Cambridge
to print and sell
all manner of books
was granted by
Henry VIII in 1534.
The University has printed
and published continuously
since 1584.

CAMBRIDGE UNIVERSITY PRESS
Cambridge
New York · Port Chester · Melbourne · Sydney

Published by the Press Syndicate of the University of Cambridge
The Pitt Building, Trumpington Street, Cambridge CB2 1RP
40 West 20th Street, New York, NY 10011-4211, USA
10 Stamford Road, Oakleigh, Melbourne 3166, Australia

First published 1974
Second edition 1991

Printed in Great Britain at the University Press, Cambridge

British Library cataloguing in publication data
DeBach, Paul
Biological control by natural enemies. – 2nd ed.
1. Pests. Biological control
I. Title   II. Rosen, David
628.96

Library of Congress cataloguing in publication data
DeBach, Paul.
Biological control by natural enemies / Paul DeBach, David Rosen.
– 2nd ed.
p. cm.
Includes bibliographical references (p. 386) and index.
ISBN 0-521-39191-1
1. Insect pests–Biological control. 2. Weeds–Biological
control. 3. Agricultural pests–Biological control. I. Rosen,
David, 1936- . II. Title.
SB933.3.D43 1990 |991|
632'.96–dc20 90-2388 CIP

ISBN 0 521 39191 1 hardback

SE

CONTENTS

PREFACE

TO THE SECOND EDITION

This second edition is part of a long continuing collaboration in biological control research and application between the two authors, beginning in 1966. Although nominally working half a world apart, we have during the past 25 years worked together during sabbaticals etc., for more than five years. This collaborative research has resulted in 16 joint publications, including one book. Our views and philosophy of biological control, while very similar, nevertheless are not always the same, hence much reflection, discussion and some compromise have necessarily gone into this revision. We believe this has resulted in a much improved book.

During the 16 years following the publication of the first edition, research and application in biological control have expanded significantly. We have had personal experience with a lot of this, have had first-hand accounts of much more from colleagues, and have thoroughly reviewed the literature up to the late 1980s. The great expansion of biological control has been brought up to date in this book, and numerous pertinent recent citations have been added. We have also added many illustrations of natural enemies, expanded the discussions of their biology and ecology, and included descriptions of several recent successes of biological and integrated control.

When the first edition of this book was published in 1974, chemical pest control was at its zenith. Although public and professional attitudes towards pesticides have changed markedly since then, and integrated pest management, or IPM, has almost become a household word, in practice the overuse and misuse of pesticides continue to increase (see Chapters 1 and 10). Too many so-called IPM programs are still based on chemical spray schedules and lack any real biological control component. The need for biological control – as an excellent solution to pest and environmental problems and as the mainstay of integrated pest management programs – is therefore now greater than ever.

The original preface to the first edition is still largely apropos, so we have left it unchanged.

PAUL DEBACH
DAVID ROSEN

PREFACE

TO THE FIRST EDITION

Years ago, when I was a 19-year-old sophomore student at the University of California at Los Angeles taking a course in Economic Entomology from Professor A. M. Boyce, Professor Harry S. Smith, the late world authority on biological control, was invited over from the then University of California's Citrus Experiment Station at Riverside to give a series of lectures on biological control. This was the turning point in my professional career. Biological control was so intellectually satisfying, so biologically intriguing and so ecologically rational a means of pest control that I immediately opted to become a specialist in this field and to do my PhD research when the time came under Professor Smith at Riverside. This I did and I have always been happy with my choice. It is also why I am writing this book.

The main purpose of this book is to outline the workings and potentialities of biological control of pests for the general reader and student interested in environmental phenomena. It is not primarily to discuss the drawbacks and dangers of pesticide usage. The only reason for bringing in any discussion of pesticides is that these two approaches to pest control have been virtually incompatible since the widespread unilateral use of pesticides became so greatly emphasized in recent years. Biological control cannot operate effectively, if at all, in the face of modern chemical pest control as it is generally practiced.

The insecticide crisis is a multifaceted one affecting society and the environment in diverse ways. The adverse effects have furnished the source for numerous scientific papers, reports and books. Some pertinent ones are cited in this book but I shall not go into more detail here.

It is ironic that the adverse effects of pesticide chemicals on biological control have received relatively little public attention as compared to, say, widely publicized effects on fish or birds, yet these effects on biological control hold the main key to the whole chain of events leading to the massive proliferation of insecticide usage. This will be illustrated in Chapter 1. If we can maintain, improve and, when necessary, increase biological control, pesticide chemical usage can be very greatly reduced overall and at times, places and on certain crops dispensed with entirely. It is as simple as that, and technologically much more readily achievable than is generally realized.

The term biological control has come to be used in a variety of ways. Professor H. S. Smith coined the term in 1919 to apply to the control or regulation of pest populations by natural enemies, i.e. by predators, parasites or pathogens and this is briefly what biological control means in this book.

Other biological methods of pest control exist and are discussed in Chapter 9. All differ basically in philosophical, biological and ecological principles from biological control and this is why I maintain the distinction in terms.

This book advocates that the first and most important step in pest control (particularly in agriculture where most pest control is required) should be aimed at the achievement of classical biological control by the discovery, importation and colonization of new natural enemies from abroad. But at the same time it stresses as always being indispensable, the conservation or protection from adverse factors of established natural enemies and further shows that at times, and as necessary, natural enemy activity can be purposefully aided by man as, for instance, by insectary mass production and periodic field colonization.

I fully recognize that all pests are not equally susceptible to classical biological control – although we have to learn this by trying – and that the method may not be of any practical consequence at all for a few pests, so that other non-chemical or *selective* chemical methods will continue to be required and should be developed or utilized as dictated by sound bio-ecological research in each major habitat of every agro-ecosystem. Such research on biological and integrated control is still in its infancy and until it grows up the massive use of chemical pesticides is not likely to become much reduced.

PAUL DEBACH

ACKNOWLEDGMENTS

Many of the friends and colleagues who helped with the first edition of this book, plus numerous others, provided invaluable assistance with the revised edition. We are especially grateful to Eric C. G. Bedford, René Le Berre, Shirley A. Briggs, Martha Brown, H. Denis Burges, José L. Carrillo S., Yau-i Chu, Jack R. Coulson, Irma Crouzel, Everett J. Dietrick, Henry S. Dybas, Louis A. Falcon, André Fougeroux, K. van Frankenhuyzer, Raymond E. Frisbie, Dan Gerling, Uri Gerson, Itamar Glaser, Richard D. Goeden, David J. Greathead, W. D. Guthrie, Marvin K. Harris, Sherif A. Hassan, J. H. Hatchett, Hans R. Herren, Marjorie A. Hoy, Carl B. Huffaker, A. R. Jutsum, Michael G. Klein, P. Koppert, E. Fred Legner, Li Li-ying, Robert F. Luck, Karl Maramorosch, Fowden G. Maxwell, John S. Noyes, Asher K. Ota, Mike Rose, Jay A. Rosenheim, Sara S. Rosenthal, Yoram Rössler, Arnon Shani, Mine Soydanbay, Frederick W. Stehr, Allen L. Steinhauer, M. Tanaka, Louis Tejada, Joop C. van Lenteren, Jeff K. Waage, Stanley G. Wellso, Robert A. Wharton, Max J. Whitten, Brian J. Wood and Ruth Ann Yonah.

Illustrations have been graciously made by, furnished by, or their permission to use granted by the following individuals and organizations: Max Badgley, Leopoldo E. Caltagirone, James R. Carey, Jack K. Clark, Harold Compere, Donald L. Flaherty, T. W. Fisher, Gordon Gordh, Kenneth S. Hagen, Robert M. Hendrickson, Marjorie A. Hoy, Carl B. Huffaker, Charles A. Kennett, Kenneth Middleham, Mike Rose, Paul H. Rosenfeld, Frank E. Skinner, Vernon M. Stern and Robert van den Bosch, all of the University of California; Jean R. Adams, Robert C. Bjork, Gerald R. Carner, Sam R. Dutky, Robert M. Faust, R. H. Foote, Frank D. Parker and R. L. Ridgway, all of the United States Department of Agriculture; E. A. Heinrichs, Michiel C. Rombach and B. Merle

Shepard of the International Rice Research Institute, Los Baños, Philippines; John Green, Douglas F. Waterhouse and Max J. Whitten of the CSIRO, Canberra, Australia; Chris Prior and Harry C. Evans of the CAB International Institute of Biological Control; David P. Annecke and Rami Kfir, Plant Protection Research Institute, Pretoria, South Africa; Margaret K. Arnold, NERC Institute of Virology, Oxford; Ronald Chatham, Bryan, Texas; Dan Gerling, Tel Aviv University; W. H. Haseler, The Alan Fletcher Research Station, Queensland, Australia; Karl Maramorosch, Rutgers University; Peter Neuenschwander, the Africa-wide Biological Control Program, IITA, Ibadan, Nigeria; J. R. Norris, Cadbury Schweppes Ltd., The University, Reading; Asher K. Ota, Hawaiian Sugar Planters Experiment Station, Honolulu; Petr Starý, Czechoslovakian Academy, Prague; Joop C. van Lenteren, Agricultural University, Wageningen; Academic Press, London; CAB International, Wallingford, Oxon., England (formerly Commonwealth Agricultural Bureaux); and the Plenum Publishing Corporation, New York. Plates of eucalyptus snout-beetle by F. G. C. Tooke, as originally published in *Entomology Memoir 3*, reproduced under Copyright Authority 4672 of 6.3.1972 of the Government Printer of the Republic of South Africa.

To the memory of
HAROLD COMPERE
Specialist, University of California, Riverside

A long-time colleague, a dear friend and a continuous source of inspiration.

Foreign explorer *par excellence*, internationally known taxonomist in the parasitic Chalcidoidea, expert biologist and biological control historian. His personal experience, tying back through his father George Compere's career, spanned in just two lifetimes all of the developments in biological control since the first great success occurred in Los Angeles a century ago.

(For photographs see Fig. 5.1, pp. 138–139.)

1

Pests, pesticides and biological control

The continuing population explosion has confronted Mankind with many problems, including the major ones of imminent starvation on the one hand and the threat of worldwide environmental pollution on the other. It is imperative that whatever we do to alleviate one problem does not seriously aggravate another. Thus, in order to feed the ever-increasing population of the world and avoid starvation, we must devise effective means to control the various pests that take such a heavy toll of our agricultural crops, but we should endeavor to do so by employing methods that reduce, rather than increase, environmental pollution. Biological control of pests is an effective means of doing just this, and that is what this book is about.

Ever since the dawn of history, man has been plagued with numerous pest species – some devastating his crops, causing injury to his domestic animals or destroying his property, others attacking him directly, sucking his blood, causing annoyance or transmitting disease. Most of our pests are insects and related arthropods, such as mites, but it does not follow that most insects and other arthropods are pests. To the contrary, numerous economically harmless arthropod species are found feeding on our agricultural crops, forest trees and ornamental plants, by comparison with only a few that are ever considered pests. Obviously, only those organisms that cause significant economic injury are regarded as pests.

What causes an arthropod population to attain pest status and inflict economic injury? As pointed out by our colleagues Vernon Stern *et al.* (1959) and others, the factors may be many and varied, but they can be grouped into three main categories: invasions, ecological changes and socio-economic changes (see also Horn, 1988).

Invasions are, unfortunately, all too common. Intercontinental travel and commerce during the last several centuries, and especially the deliberate transfer of numerous crop plants and ornamentals, often bearing cryptic infestations of insects and mites, have enabled many species to disperse and become established as pests in new habitats. (See Elton's classical treatise, 1958, *The Ecology of Invasions by Animals and Plants.*) Indeed, a large proportion of the important pests in modern agro-ecosystems are species of foreign origin. Even in this sophisticated day and age, with many quarantine restrictions, invasions repeatedly occur because of the rapidity and volume of air transport and other forms of travel. Regulatory control in the form of diligent quarantines is therefore of primary importance in preventing pest problems.

Sailer (1983) has listed 837 insect and related arthropod species, including some 80 important pests, that have invaded the United States during the period 1920–1980. He estimates that every year, 11 potential pests gain entry into the country, seven of which are likely to be injurious, and that every third year or so a major pest invades the United States. In 1986, nearly 50 000 interceptions of exotic organisms were made at US ports of entry by the US Department of Agriculture. Recent invasions have included the Mediterranean fruit fly (*Ceratitis capitata*), which has now been successfully eradicated for the eleventh time since 1929, the Mexican rice borer (*Eoreuma loftini*), the citrus blackfly (*Aleurocanthus woglumi*), the woolly whitefly (*Aleurothrixus floccosus*), the bayberry whitefly (*Parabemisia myricae*), the black parlatoria scale (*Parlatoria zizyphi*) and the Russian wheat aphid (*Diuraphis noxia*). The latter species alone has caused annual crop losses estimated at US $100 million in 11 US states and several Canadian provinces.

The same is true for other parts of the world. England, for instance, was invaded by the grape phylloxera (*Daktulosphaira vitifolii*) in 1985 and by the western flower thrips (*Frankliniella occidentalis*) in 1986. Other notorious examples include the American serpentine leafminer (*Liriomyza trifolii*) in the Netherlands in 1975, the mango mealybug (*Rastrococcus invadens*) in Africa in 1982, and the leucaena psyllid (*Heteropsylla cubana*) in Hawaii in 1984. Also many important weed species in the United States and elsewhere are foreign invaders.

Ecological changes are another major cause of pest outbreaks. The history of agriculture has been the history of constant ecological change. By creating monocultures, by selecting high-yielding plant cultivars, by eliminating competitors or natural enemies through various agrotechnical practices, etc., man has often inadvertently created conditions favorable to certain species and has thus induced a manyfold increase in their populations. Even the 'Green Revolution', an unprecedented step forward in Mankind's efforts to alleviate food and natural fiber shortages, may have only aggravated the pest situation in this respect.

Finally, *socio-economic changes* are often as important in affecting pest status as are actual changes in the physical or biotic environment of the agro-ecosystem. Economic thresholds – that is, the density levels at which pest populations should be controlled to prevent them from causing economic injury – are determined by such factors as the market value of the crop and the cost of control measures, as well as by consumer habits and taste. Changes in public attitude, for instance, towards 'cosmetically' pest-damaged agricultural produce may drastically lower the thresholds and thus cause a hitherto insignificant organism to be considered a serious pest, although its actual population density is low and the effect on yield is of no consequence. The following examples may illustrate this point. The opening of the Japanese market, with its stringent quarantine restrictions, to Israeli citrus fruit has brought about a general lowering of economic thresholds and a resulting general intensification of pest control operations on citrus in Israel. On the other hand, a severe freeze in Spain or Morocco, resulting in a reduced supply of citrus fruit in European markets, would enable Israeli growers to raise the culling threshold for, say, scale-insect presence on fruit exported to the European Community.

One way or another, serious pest problems are a prominent fact of life in modern agriculture, and new ones appear regularly as just discussed. World crop losses are estimated at 35 percent – about 12 percent due to insects and mites, 12 percent to plant pathogens, 10 percent to weeds and 1 percent to mammals and birds – and are worth about US $400 billion (i.e. US billion = thousand million) annually (Pimentel, 1986). Additional losses occur after harvest. Effective control of pests remains an absolute necessity in any modern agro-ecosystem, and thus far has been achieved by an impressive array of chemical, biological, cultural, mechanical and autocidal techniques. (We shall discuss many of them in Chapter 9.) More and more, however, since World War II, *chemical control*, or the use of toxic pesticides, has become the dominant strategy for pest control, and this has been the cause of mounting concern in recent years.

Pesticides: a powerful weapon that may misfire

The modern era of chemical pest control began with the advent of synthetic organic insecticides in the early 1940s. First came DDT and other chlorinated hydrocarbons, with their broad spectrum of pesticidal activity and long residual effects. These were soon followed – and eventually largely replaced — by the organophosphorus and carbamate pesticides, with shorter residual effects but with considerably higher toxicity to all living things, and more recently by the new synthetic pyrethroids. By and large, these modern pesticides have provided us with potent means for suppressing arthropod pests and other noxious organisms. For one thing, they have certainly been responsible for great gains made in man's never-ending war against arthropod-borne diseases. As pointed out by Luck, van den Bosch & Garcia (1977), chemical control of the mosquito vectors of malaria, although far from achieving its global eradication goals, has saved millions of lives and reduced suffering in many parts of the world. The incidence of various other diseases has also been greatly reduced by chemical control of their arthropod vectors. In agriculture, too, which accounts for more than half the total use of pesticides worldwide, chemical pest control has been instrumental in achieving great increases in crop yields.

However, as is now widely recognized, this has been a mixed blessing. Not only do chemical pesticides, at best, afford only temporary relief from pest problems, but their massive overuse and frequent misuse have resulted in grave problems. Ever-increasing cost has been just one of these problems, with disastrous consequences as to the profitability of certain agricultural crops, especially in developing countries.

Pesticidal toxicity has become a major environmental problem. Many modern pesticides are rather general biocides, toxic to humans as well as to many other non-target organisms. Between 400 000 and 2 million pesticide poisonings occur worldwide each year, most of them among farmers in developing countries, and 10 000 to 40 000 of them result in death (Postel, 1987). In the United States alone, some 45 000 cases of human pesticide poisoning occur annually, resulting in nearly 3000 hospitalizations and in 200 deaths (Pimentel *et al.*, 1980). In addition there are the risks of cancer, sterility, and other long-term effects on human health. Producers, users, and the general public are liable to be affected. Other toxicity hazards involve domestic animals, honey bees and other pollinators, wildlife and natural enemies of pests. Non-biodegradable pesticides may contaminate soils, water systems and food chains, and have become a major component of environmental pollution. It would be wise to remember, in this respect, that the world's biosphere is one finite, integral system. When pesticides that are banned in so-called developed countries continue to be manufactured in those countries in large quantities and exported to developing countries, the environmental repercussions of their continued use may be felt anywhere in the world.

Last but not least is pesticidal phytotoxicity, or the poisoning of plants. Here again, besides obvious instances of injury to susceptible plants, which in severe cases may amount to defoliation, loss of crop or even death of plants, there may also be subtle, often unnoticed, effects on plant physiology. Thus, many pesticides may reduce the photosynthetic rate of various plants, and this may have adverse effects on crop yields. Jones *et al.* (1986) have recently reviewed these cryptic effects in some detail.

Some general references broadly covering nearly all aspects of the subject of adverse environmental effects of pesticides include Brown (1978), Carson (1962), Croft (1990), Delucchi (1987), Doutt (1964*a*), Graham (1970), Hallenbeck & Cunningham-Burns (1985), Harmer (1971), Jepson (1989), Mellanby (1967), Metcalf (1980), National Research Council (1987), Newsom (1967), Pimentel & Perkins (1980), Postel (1987), Ragsdale & Kuhr (1987), Report of the Environmental Pollution Panel (1965), Report of the Secretary's Commission on Pesticides (1969), Rudd (1964, 1971), Sheets & Pimentel (1979) and van den Bosch (1978). The Worldwatch Institute, 1776 Massachusetts Avenue NW, Washington, DC 20036, issues annual '*State of the World*' reports and various other publications covering these and many other environmental issues.

Pimentel *et al.* (1981) estimate that the annual cost of the environmental and social problems caused by pesticides is at least $3 billion in the United States alone. Indeed, as pointed out by Luck, van den Bosch & Garcia (1977), chemical control has become 'a troubled pest management strategy'.

Fostering pests through misuse of pesticides

The vast majority of objections to the application of chemical pesticides do not come from the people who use them most – the farmers. By and large, the man who depends for his livelihood on crop production is convinced that he must use pesticides regularly or perish. Such conviction has been actively and effectively fostered by the pesticide industry. All sorts of data and discussions in scientific and lay journals have emphasized the apparent need for, and the great good derived by the farmer from, the faithful use of pesticides.

Although these claims are certainly justified in particular cases, they cannot be accepted as general broad truths, as many entomologists and farmers have learned to their sorrow. We shall return to this controversy somewhat later. What we wish to emphasize in this section has been largely ignored or overlooked by the sellers and users of chemical pesticides. It is that pesticides often *cause* actual pest increases rather than controlling them, and more and more frequently are creating new pests of species that formerly were innocuous rarities. Although the idea may at first sound counter-intuitive, it can easily be demonstrated experimentally in the field.

Some recipes for fostering pests

Our main area of field research has been with citrus pests, and we can provide recipes for the use of several insecticides that are perfect for raising citrus pest insects in tremendous numbers in the field. So perfect in fact, that anytime we wished, we could use DDT (previous to its banning*) to cause such great increases in certain scale insects and mites on citrus trees that complete tree defoliation, even death, would result. Some of our colleagues have done the same thing with various insecticides on other crops. We did not do this in an endeavor to find out how much good chemicals do, rather the opposite, to find out how much good natural enemies do. Thus we used chemicals purposely to decimate natural enemies on certain trees, and by comparing pest trends on these with trees in the same grove on which biological control was undisturbed, the extent of biological control was revealed (see Chapter 7 for further discussion of this and other methods of evaluation). The major requirement is that the chemical must kill a greater proportion of natural enemies than it does pests, as is quite often the case anyway. Of course, if one can apply insecticides purposely to cause explosions of pest insects, it is obvious that such explosions can occur accidentally when the purpose of application is the opposite, i.e. to control pests.

There follow several recipes for the use of insecticides which in our experience, and that of others, generally have produced exceptional pest increases. Of course, such demonstrations must be done in plots where natural enemies have not already been recently killed by insecticides.

The California red scale (*Aonidiella aurantii*) can nearly always be increased dramatically on citrus as follows: take 500 g of 50% wettable DDT* and mix into 100 L of water (approximately 4 lb in 100 gallons, i.e. US gallon = 3.785 liters). Using a 12-liter (or three-gallon) garden sprayer, spray one or two liters (about one or two quarts) of this mixture

* The use of DDT is no longer legal in many countries.

Table 1.1. *Adverse effects of DDT, southern California.*
Relative California red scale population densities in various biological control plots (untreated) and in plots in which natural enemy activity had been suppressed by prior applications of a light DDT spray.

Property and location	No. of DDT applications	Final scale population density‡		Fold increase, DDT/ untreated
		Untreated	DDT-sprayed	
Bothin, Santa Barbara	11*	3	425	142
Sullivan, Santa Barbara	11*	16	575	36
Beemer, Pauma Valley	9†	1	580	580
Irvine Company, Irvine	29*	6	1336	223
Rancho Sespe, Fillmore	54*	1	1250	1250
Ehrler, Riverside	10†	4	390	98
Sinaloa Ranch, Simi	55*	3	1015	338
Stow Ranch, Goleta	37*	8	850	112
Hugh Walker Grove, Orange County	47*	1	463	463

*Monthly applications.
†Quarterly applications.
‡Initial scale population densities were similar in all plots.

lightly over the tree, repeating at monthly intervals until the tree is defoliated or dying, as a result of the increase in red scale. This may require from about 6 to 12 applications, depending upon the degree of initial infestation; the higher the numbers initially, the faster the explosion occurs. Most growers do not like this sort of test and have usually insisted that we stop before the trees were killed, but it is a wonderful way to raise red scale. Table 1.1 shows results from such tests obtained in a variety of citrus groves in southern California over a period of years.

From these figures it is readily seen that, as used, DDT caused pest increases of anywhere from 36-fold to over 1200-fold in from one to several years. A typical effect of such DDT-caused outbreaks is seen in Fig. 1.1, which shows a lemon tree nearing complete defoliation, as compared to a nearby unsprayed tree. Fig. 1.2 shows graphically, month by month, the progress of a DDT-induced outbreak. DDT caused the economic injury level to be significantly exceeded within a matter of months.

Similar results have been obtained with the California red scale in various other countries. In Israel, one of the authors (DR) observed a coincident explosion of the Florida wax scale (*Ceroplastes floridensis*) on the same DDT-treated trees.

The same formulation and application of DDT used for the California red scale caused similar outbreaks of the yellow scale (*Aonidiella citrina*) on orange trees in California. In fact, this is an easier way to raise yellow scale than it is in the insectary.

DDT again was an excellent material to use to produce massive increases in cottony-cushion scale (*Icerya purchasi*) in the field. In fact, this was one of the first cases in which it was found that DDT could produce such remarkable results. A reliable formulation consisted of 180 g of 50% wettable DDT powder, 2 L of kerosene and 15 g of blood albumen spreader mixed in 100 L of water and applied as a complete coverage spray with a commercial spray rig. For best results, applications were made in early May and repeated again in July or early August. When light to moderate cottony-cushion scale populations were present initially, tree defoliation, even death, would be attained by the massive scale increase occurring within about one year. With lighter infestations to start with, four applications over two years would be required for the

Fig.1.1. A lemon tree (right) nearing complete defoliation from massive increases of the California red scale caused by previous DDT sprays. A nearby unsprayed tree (left) showing dense foliage is under complete biological control.

same spectacular results. The greater the acreage treated, the more effective will be the population explosion.

Our colleagues Carl Huffaker, Charles Kennett and Glen Finney (1962) deliberately used light dosages of DDT on olive trees in the field to provoke massive 'upsets' of the olive scale (*Parlatoria oleae*). They repeated the procedure over a period of several years in many olive groves, so its reliability is undisputed. A résumé of some of their typical results is shown in Table 1.2. The relative population increase due to DDT over unsprayed trees is truly amazing, ranging from 75-fold to nearly 1000-fold within the period of two years.

Endrin and dieldrin, both chlorinated hydrocarbon insecticides of the cyclodiene group, have been used to produce massive infestations of the purple scale (*Lepidosaphes beckii*) on citrus. The formulations and dosage

Fig.1.2. Increases in California red scale infestation caused by light monthly applications of DDT spray as compared to nearby untreated trees under biological control in the same grove. (From DeBach, Rosen & Kennett, 1971.)

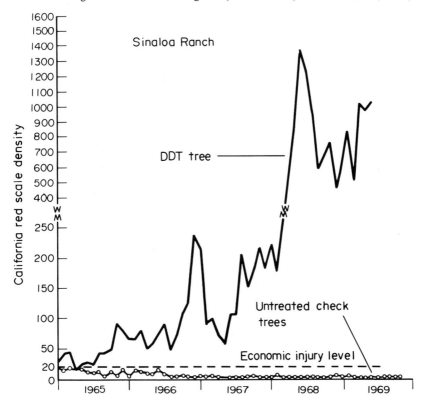

Table 1.2. Relative population density increases of the olive scale on unsprayed trees and on DDT-sprayed trees.

Location	1958 Fall		1959 Spring		1959 Fall		1960 Spring		1960 Fall		Relative fold increase, DDT-sprayed to unsprayed
	DDT-sprayed*	Un-sprayed*	DDT-sprayed	Un-sprayed	DDT-sprayed	Un-sprayed	DDT-sprayed	Un-sprayed	DDT-sprayed	Un-sprayed	
Lindsay	0.0	0.0	0.3	0.3	5.5	1.2	25.5	1.9	67.6	0.9	75:1
Seville	2.0	1.6	4.2	9.0	2.7	0.0	12.6	0.0	29.8	0.03	993:1
Hills Valley	0.8	1.3	10.6	0.1	12.1	0.4	25.6	0.2	90.7	0.1	907:1
Clovis	1.2	3.0	12.4	0.2	43.9	0.5	55.6	1.9	287.8	2.5	140:1
Herndon	2.0	3.8	38.1	3.2	134.0	0.7	60.4	0.0	169.8	1.5	113:1
Madera	16.9	11.6	23.9	3.4	43.5	0.4	120.3	0.6	204.2	0.5	408:1

*Pretreatment counts.
(Adapted from Huffaker, Kennett & Finney, 1962.)

used for both are as follows: take a 50% wettable powder and mix with water at the rate of 240 g per 100 L (2 lb per 100 US gallons). Apply 2.5 L (two-thirds of a US gallon) per tree monthly with a Hudson sprayer. This results in a light, scattered coverage. These materials did not cause quite such rapid outbreaks as DDT did with California red scale but persistence was rewarded, so that after about thirty-six months the purple scale population density index remained a low 158 on the untreated trees as compared to massive increases to 7402 on the dieldrin-sprayed trees (about a 47-fold increase) and to 20 615 on the endrin-sprayed trees (a 130-fold increase). The latter infestation index represents a tree virtually encrusted with purple scales and suffering from dead and dying leaves, twigs and small limbs.

In Malaysia, too, Wood (1971) was able to deliberately increase bagworm populations (*Metisa plana*) by applying low-volume dieldrin sprays to two plots of oil palms within larger blocks of trees. Population trends in these plots were compared to untreated trees extending some 450 m away from the centrally treated plot. There was approximately a 5- to 10-fold increase in bagworm larvae in the treated plots as compared to populations unaffected by the spray. This was the result of just one spraying.

The citrus red mite (*Panonychus citri*), and probably mites in general, are the easiest pests to increase by the application of a variety of pesticidal chemicals. The formulation of DDT used for the California red scale also works wonders in causing enormous citrus red mite infestations to develop. In some of the first tests, two and a half months following a single application of DDT, the citrus red mite population index on the DDT-treated trees was 2303 as compared to only 377 on untreated trees in the same grove. We also found that cryolite and zinc sulfate (a nutritional spray) caused significant increases in mite populations. The use of DDT on citrus in southern California was abandoned early in the game by citrus growers because of its obvious effect in causing such mite increases.

However, not only chlorinated hydrocarbons can be used to foster pest outbreaks. Other pesticides may be just as effective. Following field observations that commercial use of the organophosphorus insecticide parathion was associated with subsequent increases in cyclamen mite (*Steneotarsonemus pallidus*) populations on strawberries, Huffaker & Spitzer (1951) and Huffaker & Kennett (1956) deliberately used parathion dust in paired-plot comparisons to evaluate this phenomenon. In the first work cited, they dusted certain strawberry plots with

Table 1.3. *Relative population indices of the cyclamen mite on strawberry plots treated with parathion dust and on untreated plots. Counts made 4–6 weeks after the treatment.*

Plot no.	Untreated	Treated	Approximate % increase, treated over untreated
1	7	23	230
2	17	37	120
3	13.5	40	195
4	13.5	34	150

(Adapted from Huffaker & Spitzer, 1951.)

Fig.1.3. Population increase of Pacific mite in carbaryl-sprayed plots in a previously unsprayed vineyard. (Redrawn after Flaherty & Huffaker, 1970.)

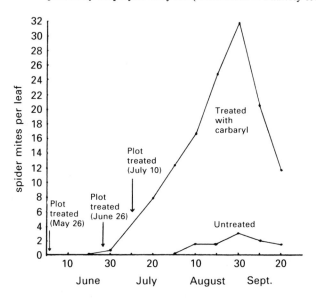

parathion in the second half of July and left other plots untreated. Cyclamen mite population indices obtained at about the peak of infestation, 4–6 weeks following the parathion application, are shown in Table 1.3. It is clear that the use of parathion dust caused increases in cyclamen mite populations of about 120 to 230 percent within a month to six weeks, thus showing that parathion is another chemical one can use to create marked increases in pest populations in the field.

Similarly, carbaryl, a widely-used carbamate insecticide, has been shown experimentally to cause marked increases of the Pacific mite (*Tetranychus pacificus*) on grapes in the San Joaquin Valley of California when used in a mixture of 2.25 kg of 50% wettable powder per 1400 L of water per ha (2 lb per 150 US gallons per acre) (Flaherty & Huffaker, 1970). Fig. 1.3 shows graphically a typical population trend resulting from carbaryl treatment in a vineyard with no previous pesticide history, and Fig. 1.4 shows the actual mite damage to the vines as a consequence of carbaryl treatment.

Carbaryl can do marvels also with other pests. Fig. 1.5 shows the results of a test carried out in a cassava field in Nigeria, experimentally infested

Fig.1.4. Severe Pacific mite damage of upper foliage on Thompson Seedless grape vines as a result of treatment with carbaryl. (From Flaherty & Huffaker, 1970.)

with the cassava mealybug (*Phenacoccus manihoti*). In the sprayed plot, a beneficial parasite was decimated and the mealybug population exploded to 200 per growing tip, whereas in the untreated plot there were only about 10 mealybugs per tip and parasitism was considerably higher (Neuenschwander & Herren, 1988).

The organophosphorus insecticide TEPP has been used experimentally to promote substantial increases of the bollworm (*Heliothis zea*) on cotton, as shown in Fig. 1.6. Most evidence indicates that other commercially used insecticides produce similar results. These insecticides are used primarily against lygus bugs, but the bollworm usually increases following their application.

Another organophosphorus insecticide that can be used to advantage is Abate (temephos). In the words of Bedford (1976), 'Abate is in fact so good at causing pest repercussions that it is a very valuable tool to the researcher wishing to induce infestations of various citrus pests by eliminating natural enemies'. A single spray of Abate in five orchards caused such outbreaks of the California red scale that an average of 29.1 percent of the crop had to be culled, as compared to only 1.7 percent in the 'control' halves of the same orchards.

The modern synthetic pyrethroids are especially useful in inducing heavy outbreaks of planthoppers and spider mites. Heinrichs *et al.* (1982)

Fig.1.5. Population increases of cassava mealybug in an experimentally infested cassava field in Nigeria, sprayed with carbaryl, as compared to an untreated plot. The columns show mean rates of parasitism (by the parasite *Epidinocarsis lopezi*) for weeks 3–12 after infestation. (Redrawn after Neuenschwander & Herren, 1988.)

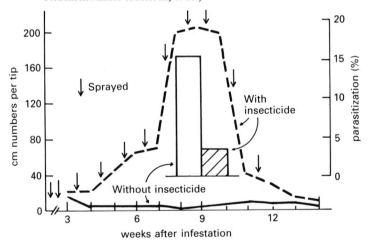

induced tremendous outbreaks of the rice brown planthopper (*Nilaparvata lugens*) in the Philippines by spraying decamethrin at 25 g active ingredient per hectare. Fig. 1.7 shows a 1000-fold increase produced by four sprays. Fewer sprays resulted in smaller outbreaks. Plaut & Mansour (1981) treated apple trees in Israel with cypermethrin, deltamethrin and permethrin and produced 36- to 60-fold increases in the two-spotted spider mite (*Tetranychus urticae*) within four weeks. Gerson & Cohen (1989) have reviewed numerous similar effects of pyrethroids on a variety of mite and insect species.

Applying pesticides to the soil can be as effective as spraying them upon the plants. Morrison, Bradley & Van Duyn (1979) have shown that aldicarb, a systemic carbamate insecticide, can produce an impressive, eight-fold increase in corn earworm (*Heliothis zea*) populations on soybeans when applied as granules to the soil.

The foregoing are but a few of the striking instances where chemicals can be used, either purposely or accidentally, to cause great increases in pest insect or mite populations. Such population explosions are generally

Fig.1.6. Increase in bollworms induced by spraying cotton plants with TEPP. (Redrawn after van den Bosch *et al.*, 1971.)

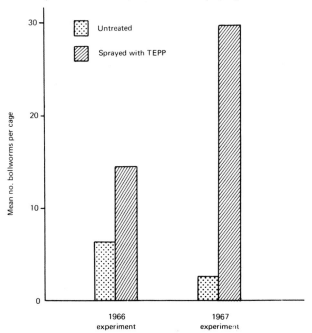

referred to as 'upsets' or 'upsets in natural balance' when non-target organisms are affected, and as 'resurgences' when target pests increase very rapidly following chemical treatments, often becoming more serious than they were originally.

Upsets of natural balance by chemicals

There was a period, following the widespread adoption of the post-World War II chlorinated hydrocarbon and organophosphorus pesticides, when the first documented reports of striking upsets in crops generally met with disbelief or disavowal on the part of both pesticide-

Fig.1.7. Population increase of rice brown planthopper induced by four sprays of decamethrin, as compared to an untreated plot. Arrows refer to insecticide application dates. Insect counts were made by sampling 40 hills per plot with a D-Vac^R suction machine. (Redrawn after Heinrichs *et al.*, 1982.)

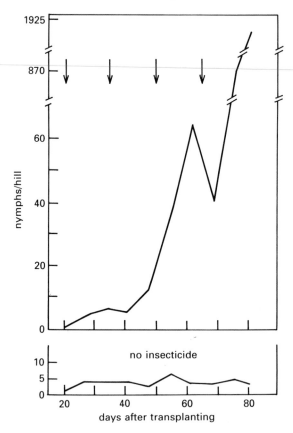

oriented entomologists and the industry. Today few, if any, knowledge-able professionals deny such effects to be rather commonplace. They are in fact extremely commonplace, as can readily be ascertained from the entomological literature. Yet, in the years following the publication of Rachel Carson's *Silent Spring* (1962), little general attention, publicity and recognition was directed toward this very broad and basic problem, as compared to the effects of pesticides on some other groups of organisms such as fish or birds. Recently the emphasis has become more balanced, as evident in the reviews of Metcalf (1980), Pimentel *et al.* (1980) and others, who have stressed the role of pesticides in inducing pest upsets. However, the main point here is that if pesticide usage is significantly reduced by greater use and conservation of natural enemies and other non-chemical methods, the problems with non-target organisms and of rapid rebound of target pests will be automatically alleviated. Not that these side problems are unimportant but, as will soon be discussed, the production of upsets by chemicals in agro-ecosystems, where more than 50 percent of the pesticides are used, leads to ever-increasing chemical usage. It follows that a major solution to reduction in the use of chemical pesticides lies in minimizing upsets and resurgences.

It would take a large book to adequately cover the recorded literature concerning upsets and resurgences. We can only refer to a few, better studied, major cases here and indicate as well some general references. In our experience, in that of various of our colleagues and from numerous literature reports, it seems clearly evident that in almost all intensively treated crops today, chemical applications are being made against upset pests (i.e. ones that formerly were innocuous) nearly as frequently as against the few key pests actually lacking established effective natural enemies. It is probably safe to generalize that in intensively treated crops it is no longer possible to distinguish between real pests (i.e. ones actually lacking established effective natural enemies) and man-made upset pests, unless long-term basic ecological studies are conducted.

The citrus red mite has increased over the years, since the chlorinated hydrocarbons and organophosphorus chemicals were introduced, from a minor or non-pest status to the number one pest of citrus in California. In fact, similar events seem to have commonly occurred worldwide with mites on many crops, as has been reviewed by Gerson & Cohen (1989), Huffaker, van den Vrie & McMurtry (1969, 1970) and McMurtry, Huffaker & van den Vrie (1970).

An early review of the adverse effects of pesticides on insect and mite pests and their natural enemies, was made by Ripper (1956). He covered upsets and resurgences involving some 50 pest species occurring in a wide variety of taxonomic groups. Diverse chemicals were responsible, such as the chlorinated hydrocarbons, the organophosphorus compounds, sulfur, lime-sulfur, copper carbonate, calcium arsenate, derris, zinc sulfate and thiuram, and even carrier materials and diluents such as talc, used to mix with the toxicants. DDT was then the worst offender. Other comprehensive reviews of upsets and resurgences among pest insects have appeared more recently (Newsom, 1967; Croft & Brown, 1975). Some other important references which furnish a picture of the magnitude of the problem, which deal with the basic biological and ecological phenomena involved, and which in themselves cite many other cases include: Conway (1969), Doutt (1964a) Flint & van den Bosch (1981), Heinrichs & Mochida (1984), Huffaker (1971a), Massee (1954), Pickett & Patterson (1953, plus a long continuing series of papers from the same laboratory), Reynolds (1971), Smith (1971), Smith & Reynolds (1972), van den Bosch (1971a, b; 1978), van den Bosch et al. (1971) and Wood (1971). Only a few of the more impressive cases covered by these authors and others can be mentioned.

Upsets are not new phenomena; some were recognized and reported in the 1920s and 1930s, but it was the widespread use of DDT and its relatives, beginning about 1945, and of parathion and related compounds a little later, that opened Pandora's box. One of us (PD) studied one of the first great DDT-induced upsets, that of the cottony-cushion scale (*Icerya purchasi*) in California in 1946 and 1947. As will be related in Chapter 5, this scale insect was the first pest to be subjugated by importation of a natural enemy, and the result literally was the saving of the citrus industry in California. The scale remained a rare insect on citrus from 1890 until DDT was widely used on citrus and other crops in the Central Valley in the 1940s. This caused the virtual extinction of the controlling predatory vedalia beetle and an unbelievable population explosion of the scale, extending from Bakersfield in the south to Hamilton City in the north. Many groves actually looked as if they were covered with snow because of the density of the cottony egg masses. Some trees were killed and many groves defoliated with subsequent loss of crop in the short two to three years before the use of DDT was voluntarily dropped or drastically modified. In order to reattain control as rapidly as possible, growers paid $1.00 each for vedalia beetles, which were still locally present in southern California. This recolonization, along with a modification of the DDT program, rapidly resulted in a return to the normal biological control. Fig 1.8 is a schematic representation of this course of events.

Local upsets of the cottony-cushion scale have recently been observed also in Israel, where they are believed to have been induced by applications of organophosphorus and carbamate insecticides against other citrus pests.

Another scale insect, the brown soft scale (*Coccus hesperidum*), became a major pest of citrus in the Lower Rio Grande Valley of Texas in 1959 when its natural enemies were killed as a result of parathion drift from adjacent cotton. A related pesticide, methyl parathion, applied to cotton also increased the fecundity of the scale. This scale is under biological control by parasites worldwide, unless they are decimated by pesticides. When the upsetting treatments were discontinued, it ceased to be a major pest in Texas and is now believed to be under complete biological control (Dean, French & Meyerdirk, 1983). Parathion-induced outbreaks have been recorded in California and South Africa. In the latter country, DDT and parathion sprays on citrus caused 'phenomenally heavy outbreaks' of the brown soft scale, which were soon followed by severe outbreaks of several other citrus scale insects and aphids (Bedford, 1976).

Fig.1.8. Cottony cushion scale on citrus in California: effective biological control by the introduced vedalia ladybeetle and the upset due to its decimation by DDT. (Redrawn from Stern *et al.*, 1959.)

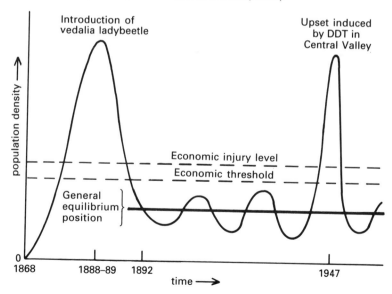

Another major upset that was induced by the spraying of pesticides on adjacent cotton was that of the long-tailed mealybug (*Pseudococcus longispinus*) on avocado in Israel in the late 1960s and early 1970s. Terrific outbreaks resulted from the drift of aerial sprays into avocado groves, and actually threatened the entire crop (Swirski *et al.*, 1980). Part of the solution was prohibiting aerial sprays with broad-spectrum pesticides at a distance of 200 m (600 ft) from the groves.

Perhaps the most amazing chronicle involving pesticide-induced upsets and resurgences has occurred with cotton, which until recently was recipient of nearly half of the agricultural insecticides used in the United States, and is still heavily treated in many countries. Reports have come from virtually all areas of the world where insecticide usage has been intensive. The effects, along with the development of pest resistance to insecticides, have been so severe as to result in the curtailment of cotton production in some major areas and the increasingly real threat of this happening in others, such as the Rio Grande Valley of Texas, the Imperial and Central Valleys of California and various Central American countries.

In the Matamoros-Reynosa area of northeast Mexico, cotton areas declined from more than 284 000 ha in 1960 to some 480 ha in 1970. As this was happening, production was moved to the new Tampico-Mante area about 300 kilometers south, where about 200 000 ha were planted to cotton in 1966. A similar but more rapid decline occurred there. By 1970 only 480 ha of cotton were left. This is attributed to the increase in intensity of pests, particularly the tobacco budworm (*Heliothis virescens*), which is an upset species, and the inability to cope with them any longer by the unilateral use of chemical pesticides in spite of attempts to increase dosages and frequency of application. Economic ruin and loss of employment have been widespread. As Adkisson (1971) has said, 'The seeds of the destruction of the Mexican cotton industry were planted when the producers decided that their insect pest problems could be best solved by the unilateral use of regularly scheduled applications of broad-spectrum insecticides.' Similar disasters have been recorded in Peru, El Salvador, Nicaragua and Australia (Bottrell & Adkisson, 1977).

In the Rio Grande Valley, cotton pests, mainly the boll weevil (*Anthonomus grandis*), were adequately controlled with chlorinated hydrocarbon insecticides for nearly 15 years, until the boll weevil became resistant to them in the late 1950s. The organophosphorus compounds, especially methyl parathion, were then used for boll weevil control, and as a result DDT became necessary to control the upset pests, the bollworm (*Heliothis zea*) and tobacco budworm (*Heliothis virescens*). By 1962, the latter two became resistant to DDT, to other chlorinated hydrocarbons and to carbamate insecticides, but could still be killed by heavy dosages of methyl parathion. This control began to become unsatisfactory by 1968, when some growers treated 15–18 times in a growing season and failed to obtain satisfactory results. Now the pest control situation was suddenly reversed as the bollworm and tobacco budworm, two pests formerly of minor importance, became more serious than the boll weevil, which faded to insignificance. The growers and entomologists were faced with a real dilemma in pest control, and ironically the worst pest species, the tobacco budworm, which no available insecticide would adequately control, had been brought to its serious pest status by the use of insecticides.

Similar insecticide-induced problems have occurred in California. Here the boll weevil does not exist, the main key pest requiring chemical control being the lygus bug (*Lygus hesperus*). In the Imperial Valley, organic insecticides such as DDT, toxaphene and endrin gave outstanding results during the early 1950s. Then they became less effective; resurgences and upsets began to occur, and other pesticides were substituted. The same pattern was repeated, with formerly minor pests or innocuous insects such as the cotton leaf perforator (*Bucculatrix thurberiella*), the cabbage looper (*Trichoplusia ni*), the beet armyworm (*Spodoptera exigua*), the salt marsh caterpillar (*Estigmene acrea*), the omnivorous leaf roller (*Platynota stultana*) and spider mites becoming more serious. Finally, it became urgently obvious that changes in pest control were necessary. Ecological research by University personnel, stressing selective insecticides and treatment based on actual need as determined by field surveys, greatly reduced the use of pesticides and alleviated the problem until 1965, when the serious pink bollworm (*Pectinophora gossypiella*) invaded the Valley from Arizona and literally exploded in this new area. Applications of broad-spectrum insecticides were made at six-day intervals in an ill-advised eradication attempt, and soon the other potential pest species already named again became so serious that continued cotton production was threatened as cotton yields dropped to the lowest level since World War II. Area-wide side effects on other crops were also noticed; for instance, vegetable crops and sugar-beets had the most severe pest infestations ever seen on these crops.

In the Central Valley, serious target-pest resurgence and upsets on cotton were experimentally induced and studied in large-scale field experiments by University personnel. In 1964 it was found that azodrin used to control the bollworm actually did not reduce the bollworm as compared to the untreated checks; in fact, the azodrin-treated plots suffered significantly more damaged bolls than the untreated checks. Again in 1965, after three applications of azodrin, more bollworms and more damage were recorded from the azodrin plots than in the untreated checks. In 1966, it was demonstrated that bidrin-treated plots sustained a cabbage looper infestation about three times as great as the checks, with considerably greater leaf damage. Three toxaphene-DDT treatment programs for lygus bug control were compared with untreated check plots in 1966. The tests covered one square mile. Beet armyworm populations were increased in all treated plots, and in two of them were 16 and 18 times higher than in checks. Cygon for lygus bug control was used in similar tests in 1970. Where July treatments were used, outbreaks of both beet armyworm and cabbage looper occurred. The overall conclusions have been that serious lepidopterous pest problems in the Central Valley involving bollworm, beet armyworm and cabbage looper are largely pesticide-induced upsets, that the cotton industry is actually losing more money with the current insecticide programs than if no insecticides were used and that, while some growers would suffer severe damage if they were to abandon all use of insecticides, most would realize little or no loss, and others would even obtain greater yields.

The sad tale of pesticide-induced problems on cotton in the Cañete Valley of Peru is best told in the words of Smith and van den Bosch (1967).*

> 'The Cañete Valley is one of the most technologically advanced of these Peruvian agro-ecosystems. The growers in it are organized into an association (La Asociación de Hacendados de Cañete) and they have their own experimental station (La Estación Experimental Agricola de Cañete) to serve their 22 000 hectares of cultivated land. Originally the valley was largely planted to sugar-cane, but in the 1920s a shift was made to the cultivation of cotton. Currently about two thirds of the valley (approximately 15 000 hectares) is planted to that crop. The Cañete agricultural operations are the most highly mechanized in Peru. Other advanced agricultural, agronomic, horticultural, and plant-protection techniques have also been adapted to crop production in the valley.

* Copyright © 1967, Academic Press, New York. Reprinted by permission.

It was the adoption of modern synthetic organic insecticides for cotton pest control which led the valley to the brink of economic disaster and ultimately to its salvation in integrated control.

The Cañete story stands as a classic example of the problems that can beset pest control which ignores ecology and relies on unilateral use of broad-spectrum insecticides. Once this approach to pest control was initiated, the Cañete was doomed to disaster because of certain factors peculiar to it. The most important of these was its status as an oasis in an otherwise arid and biologically impoverished land. Thus when synthetic organic insecticides were brought into use they were essentially applied as a blanket over the entire ecosystem. As a result, literally the entire biota of the valley was repeatedly exposed to these materials. This had two critical effects: (1) decimation of the parasite and predator fauna, which in turn relieved the key pest species as well as potential pests from biotic repression, and (2) rapid selection for insecticide resistance in the pest species.

Both developments occurred with great rapidity and with devastating effect. For example, though use of synthetic organic insecticides, principally DDT, BHC, and toxaphene, was not initiated until 1949, resistance to these materials had already developed by 1955. Where 5% DDT, 3% BHC, and a 3:5:40% BHC-DDT-sulfur mixture had proved effective in 1949, 15% DDT, 10% BHC, and a 5:10:40% BHC-DDT-sulfur mixture often applied at elevated dosages, were necessary for control by 1955. Furthermore, the interval between treatments was progressively shortened from a range of 8–15 days down to 3 days. Meanwhile, a whole complex of previously innocuous insects had risen to serious pest status. Included were *Argyrotaenia sphaleropa*, *Platynota* sp., *Pseudoplusia rogationis*, *Pococera atramenalis*, *Planococcus citri*, and *Bucculatrix thurberiella*.

Substitution of new insecticides, including organophosphorus compounds, for the older chlorinated hydrocarbons finally became necessary and then even these failed. Finally, as a result of pest resistance to the insecticides and the burgeoning complex of injurious species, the cotton yield plummeted alarmingly; the average yield per hectare in 1956 was the lowest in more than a decade.'

The solution to this crisis will be detailed in Chapter 10.

Striking upsets or resurgences have been carefully documented on oil palms and cocoa in Malaysia. In general, these developed during the late 1950s and early 1960s following the use of insecticides, particularly DDT, endrin and dieldrin, having long-term residues. On oil palms several generally innocuous species of bagworms and nettle caterpillars occasionally reached damaging levels, or other species appeared to threaten, so the use of DDT, endrin or dieldrin was initiated. Initial results were good but populations sooner or later resurged to very high densities necessitating more treatments. This led eventually to the insects becoming serious continuous pests rather than occasional ones. The successful solution to this particular paradox and certain other chemically induced pest problems will be also discussed in Chapter 10.

Cocoa was a new crop in Sabah (northern Borneo) in 1956, planted in cleared forest areas. The acreage increased rapidly and soon some indigenous insects were observed attacking cocoa trees. The first serious pests were three species of lepidopterous borers, which led to spraying with high concentrations of DDT or dieldrin in 1959. That year, several other potential pests, including leaf-eating caterpillars, aphids and mealybugs, were noticed and considered to be dangerous, so in 1960 general spraying was carried out as a prophylactic measure, using endrin, dieldrin, DDT, BHC, lead arsenate and a white oil, alone or in combination. This was repeated in 1961. Meanwhile, the pest situation was becoming worse. Branch borers became abundant by early 1961. Within months, outbreaks of four other pests occurred, two leaf-eating caterpillars, one nettle caterpillar and one homopterous planthopper. All became extremely abundant, the latter to the extent that when disturbed they rose in large clouds from the branches. Then, in July 1961, the most serious outbreak of all occurred, involving several species of bagworms, resulting in large numbers of defoliated and dying trees. Continued spraying had no effect on the bagworms. Eventually, field studies and experiments showed that virtually all of these outbreaks were insecticide-induced with the main exception of one species of ring bark borer.

The situation on tea in Sri Lanka provides yet another example of induced upsets. For years, a shot-hole borer (*Xyleborus fornicatus*) has been a serious pest there. Cultural control was largely used, until dieldrin was tried in the 1950s with good results. This subsequently resulted in upsets of the tea tortrix (*Homona coffearia*), which earlier had been brought under complete biological control by the importation of a parasitic wasp. These tortrix outbreaks, however, could be dealt with by DDT applications, so because of the importance of shot-hole borer damage, dieldrin was recommended for shot-hole borer control with follow-up DDT sprays to be made whenever tortrix upsets were noted. This policy was commonly followed with good results for several years, although DDT appeared to induce mite attacks. Then upsets of the tea tortrix became more severe and persistent, and major outbreaks of two other leaf-eating caterpillars occurred. One of these had never been previously recorded and the other had been reported as only an occasional minor pest of tea in Sri Lanka. Both of the latter caused greater damage than the shot-hole borer or the tea tortrix ever had. Subsequent studies and observations showed all of these outbreaks to be the result of insecticide usage.

An upset case of major proportions has been that of the rice brown planthopper (*Nilaparvata lugens*) in south and southeast Asia. A minor pest of tropical rice in the past, this insect increased dramatically in the early 1970s, when pesticides used against lepidopterous stem borers decimated its natural enemies. Field tests have shown massive planthopper upsets to be induced by organophosphorus, carbamate and pyrethroid insecticides alike and to be heaviest on the high-yielding, but hopper-susceptible, rice varieties introduced by the green revolution. Planthopper feeding damage, known as 'hopperburn', became increasingly extensive throughout the region, and the problem was exacerbated by hopper-transmitted virus diseases. The more pesticides were applied the more severe were the upsets, and when the chemicals were used to control the planthoppers themselves even heavier resurgences occurred. The brown planthopper became the number one pest of rice in tropical Asia, and it became evident that only a regional pesticide management program could solve the problem. We shall return to this in Chapter 10.

Two recent instances of pesticide-induced upsets, recorded in California, will serve as our last examples. Both are related to chemical eradication campaigns, carried out against newly-established pests. When a new invader is first detected, and is found to be present in a limited area, an all-out chemical campaign for its eradication may be well justified. However, the environmental repercussions of such a program may be far-reaching.

When the Japanese beetle (*Popillia japonica*) was found in San Diego in 1973, an eradication program involving foliar carbaryl sprays and ground chlordane applications was carried out in an area comprising about 100 residential blocks. A massive outbreak of the citrus red mite soon followed, causing severe defoliation to citrus trees in the eradication zone. When chemical treatments were resumed in 1974, another outbreak of the mite occurred, causing even more severe defoliation. After the treatments ended, both the purple scale and the woolly whitefly (*Aleurothrixus floccosus*), which had been under complete biological control before the eradication program began (see Chapter 6), and which for a while were held in check by the defoliation of host trees, exploded in the eradication zone. The ratio of live purple scale in the eradication zone, as compared to the surrounding untreated areas, was 1200:1 (DeBach & Rose, 1977).

A recent case is documented by Ehler & Endicott (1984). Following the eradication program carried out against the Mediterranean fruit fly (*Ceratitis capitata*) in the San Joaquin Valley of California, which involved 19 aerial malathion-bait sprays over a period of seven months in 1982 and 1983, they detected significant increases in the populations of several scale insects and aphids on olive, citrus and walnut in the treated area. This is all the more interesting, inasmuch as the bait sprays are usually considered rather selective. If their intensive use over a limited period of time was sufficient to cause pest upsets, imagine the damage that prolonged usage could do to the agro-ecosystem. Indeed, in Israel, the routine application of malathion-bait sprays against the Mediterranean fruit fly at ultra-low volume has been implicated in the gradual increase of various scale insects on citrus.

Another interesting aspect of this study was that the increase of the Mediterranean black scale (*Saissetia oleae*) on citrus was first noticed only five months after the last bait-spray application. This is not surprising, as the scale develops only one generation a year in that area, but it indicates that many upsets, when noticed, may not even be associated with the chemical treatments that had induced them.

Most of the upsets and resurgences described have been correlated with, or proven by experimental tests to be due to, adverse effects of pesticides on natural enemies. In some cases, particularly with mites, it appears that the pesticide also may act to cause increases in the reproductive or survival capacity of the pest organism. This may result from an indirect physiological effect on the host plant or from making the physical substrate more suitable. Such an effect, of course, can only add to the adverse effect of pesticides on natural enemies to cause more rapid and severe resurgences or upsets. For recent reviews see Hussey & Huffaker (1976) and Jones *et al.* (1986).

The foregoing, rather harsh, indictment of pesticides still does not present the whole problem within agro-ecosystems or of other environmental, social and economic problems derived from the use of chemicals in crops. However, as mentioned earlier, we are stressing here the adverse effects as they relate to biological control. In this regard, there is one additional major drawback within the agro-ecosystem to pesticide usage, the development of resistance to them by pests. This is not only an extremely serious disadvantage to the use of pesticides *per se* but, as they begin to fail, dosages and number of applications are increased, which magnifies the adverse effects on natural enemies.

More than 440, possibly 500, pest species are now resistant to pesticides, many to several pesticides and some to virtually all that are available for use. Resistance has been increasing in a geometric ratio since about 1950, when less than 20 species were resistant. For anyone advocating strong reliance on chemicals, the resistance picture is truly frightening, especially since no group of modern pesticides appears to be immune to it. Very good reviews of the resistance problem have been made by Georghiou & Saito (1983), National Research Council (1986) and Roush & McKenzie (1987).

Insecticides: the ecological narcotics

A very realistic comparison can be drawn between the pattern of effects from use of hard narcotics by humans and the pattern resulting from unilateral use of 'hard' pesticides in crops. At first the method of use is simple and feasible and at an early stage of usage the effects appear to be salubrious and desirable. The results that are sought are generally obtained. However, sooner or later, dosages and frequency of use have to be increased as tolerance or resistance to the chemicals develops. Applications become more frequent, and often different materials are resorted to in attempts to reproduce the original results. Ultimately, usage becomes habit-forming in that reliance on the materials becomes essentially absolute. Meanwhile, severe physiological and psychological problems (in humans) or ecological problems (in crops) develop, which may disrupt the whole system. Recognition of this state does not immediately solve the problem, because withdrawal is slow, difficult and painful, and reattainment of normality a lengthy process both in humans and in agro-ecosystems. How pesticide addiction can be alleviated will be discussed in Chapter 10.

In spite of all the foregoing discussion of the adverse effects of pesticides, for practical reasons one cannot advocate their immediate general deletion from agro-ecosystems. The problem is not that simple. To do so would result in heavy pest damage and financial loss in the short term in many cases, although in some other cases elimination of pesticides could be achieved immediately. We do stongly urge that our ultimate *goal* be the complete elimination of pesticides, and that the strongest research efforts possible be directed toward this end. This means concentration on biological control and other ecologically non-disruptive, non-pollutive methods of control.

Why the great pesticide boom?

The question naturally arises as to why, if there are such severe adverse effects, pesticide usage came to be so great and, in fact, is still increasing markedly. In US agriculture alone, the annual use of pesticides nearly tripled between 1965 and 1985 to 390 000 tons, or 2.8 kg per ha planted (about 2.5 lb per acre) (Postel, 1987).

Part of the reason was just explained. There is a reaction akin to the narcotic effects in humans that tends to make pesticides increasingly self-perpetuating, and this phenomenon has been generally recognized too late. However, there is more to the story.

Until the past few centuries, most potential pests were natives and generally had effective natural enemies that evolved with them. As mentioned earlier, as commerce and trade increased, pests were translocated without their enemies and created new, more serious problems. The reasons for these outbreaks were often not understood and led to the use of chemical poisons as attempted solutions. Insecticides began to be fairly commonly used in the 1800s as chemistry improved, as large-scale commercial farming developed, as monoculture increased and, especially as mentioned, as new invading pests from abroad increased to damaging proportions and immediate remedies were sought. What could seem more logical than poisons to kill them? Only the glimmerings of other methods of control were then being perceived. There is also a certain satisfaction, perhaps the psychology of achievement, in applying a poison and seeing the pests die rapidly as a result.

Such ready acceptance by the farmer rapidly led to the development of commercial pesticides, and eventually to the huge pesticide industry of today. As in any business, the profit motive is king, and advertising is the key to profits. We think it safe to say that over the past years very much more information on pest control has reached farmers around the world from salesmen's contacts, magazines, pamphlets, newsletters, calendars, and highway signs telling only the pesticide story, than from all the governmental, university, or other sources combined, which present a more balanced approach. Promotion of pesticide sales is big business, and has tended toward the motto used by one company for a time on TV, that 'the only good bug is a dead bug'. We have no doubt that most pesticide companies and individuals associated with them have acted ethically, honestly and in good faith in promoting their products. We think they generally believe what they say and advertise. However, they have overstressed the need for pesticides to eliminate the last 'cosmetic' thrips scar or mite russeting on fruit, when no actual damage either in quality or quantity is involved. The treatment recommendations are often based on unreal or undetermined economic injury levels (state and federal government entomologists share the blame here) and crop losses from failure to spray are commonly magnified – unintentionally, we are sure – by the inadequate method generally used by those testing insecticides to determine losses by pest insects and mites. That is, a crop area of variable size is selected and divided into subplots, generally all of which are replicated and treated with different materials, dosages, and intervals of applications, except one subplot which is 'untreated'. Such a plot often is surrounded by treated plots and receives drift chemicals, plus it is part of a larger field or orchard which has been treated previously, hence has natural enemies decimated. In no sense is it an adequate check of the potential natural or biological control that might be attained in the complete absence of treatment (see Chapter 3 on the evaluation of natural enemies). Naturally the pests tend to explode in such a plot, and by comparison with such a check plot the pesticide plots generally look much better and the amount of damage prevented by treatment appears correspondingly greater. Rudd (1964, Chapter 4) covers the pros and cons of loss from pests and the costs compared to the gains resulting from pesticide use. In referring to some figures provided by the National Agricultural Chemicals Association in 1957, documenting gains derived from the use of pesticides, he states that:

> 'Caution must be observed with such sets of figures. The means of arriving at such values are not very refined and are often unreliable. Moreover, they tell only part of the story. Too frequently, they reflect an unrealistic "optimism" not in keeping with their source.'

'Insurance' or 'by the clock' treatments are commonly pushed by pesticide salesmen and are readily practiced by farmers who expect to avoid risk of loss thereby. They are especially bad from an ecological standpoint. Such treatments ignore real need or scientific determination of economic treatment thresholds. The farmer is told something like, treat for thrips at petal-fall or spray for aphids on May 15, when it may not even be known that such insects are in fact present at that time or, if present, constitute any real hazard. A good illustration is related by Rudd:

> 'One may wonder about the accuracy of specific loss figures. In 1953 the rice leaf miner outbreak was credited with ruining 10 to 20 percent of the California rice crop. Judged by my personal damage appraisal at the time of the outbreak, this range was certainly correct for many of the rice fields. Even at that time, however, by far the greater part of the rice acreage was not hit at all. To the presumed loss of yield must be added a cost of some $1.25 million for pesticides (and a very large wildlife loss). Curiously, however, rice yields that year were among the highest on record to that date. Equally inexplicable was the reduced yield per acre and consequently reduced total production that occurred with much greater acreages in 1954, the year following the outbreak and treatment. The price of rice was also substantially lower in the year following the outbreak. I conclude, therefore, that growers extrapolated single-field losses to the total acreage, and the "insurance" spraying was the rule. Economically, insect damage seems relatively much less important than other factors. One may perhaps be permitted to wonder if perspective as well as accuracy were not somehow awry in reporting of the "threat" to the rice crop.'

It will be recalled from various instances of upsets related earlier, that in the early phases of pesticide use in crops, perhaps for quite a few years, the results may be quite satisfactory, even spectacular. At the same time, the adverse effect of pesticides on biological control makes the need for chemicals seem greater and the effectiveness of natural enemies increasingly poorer. However, we must also keep sight of the fact that in most crops there are one or more pests for which other methods of control have not as yet been discovered or perfected, hence some pesticide use is to be expected.

Some sobering statistics are offered by Pimentel *et al.* (1981). They estimate that if pesticides were completely banned from agriculture in the United States, damage from all pests – insects, diseases and weeds – would increase by nine percent over current levels, and this would probably increase retail food prices in the United States by about 12 percent. Furthermore, in spite of a more then ten-fold increase in the use of insecticides in American agriculture during the past 30 years or so, crop losses due to insects in the United States have nearly doubled – from 7 percent in the 1940s to about 13 percent today. As they point out, this loss has been effectively offset by higher crop yields. Such data makes one wonder about some of the more extravagant claims made by the chemical industry and some entomologists.

Government policy, especially in developing countries, has been a major factor inducing farmers to overuse pesticides. In an attempt to 'modernize' agriculture and increase yields, many governments have encouraged farmers to use pesticides by distributing them free of charge or by heavily subsidizing their prices. Interesting accounts of the cultural and political causes underlying the current pesticide crisis can be found in Perkins (1982) and van den Bosch (1978).

The philosophy of pest control by chemicals has been to achieve the highest kill possible, and percent mortality has been the main yardstick in the early screening of new chemicals in the laboratory. Such an objective, the highest kill possible, combined with ignorance of, or disregard for, non-target insects and mites, is guaranteed to be the quickest road to upsets, resurgences and the development of resistance by the pests. The pesticide industry has something of a parallel in the cigarette industry, in that it took many years to recognize and prove the magnitude of the adverse effects but there are still many users and advocates and the industries still push their products vigorously. Fortunately, there is a change in scientific understanding and in public attitude, which portends the development of more biologically and ecologically sound practices in the future.

The need for safer, less costly alternatives to chemical control is evident from the preceding discussion. Several such alternatives are indeed available in our arsenal for use in an integrated, interdisciplinary approach to pest management. Some, such as various cultural methods, are probably as old as agriculture itself, whereas others are relatively new. We shall discuss them in Chapter 9. However, of all the available strategies, biological control by natural enemies has been by far the most successful, and is the most promising, alternative to the unilateral use of chemical pesticides.

Biological control: definition and scope

Biological control may be variously defined as an applied field of endeavor or as a natural phenomenon. In the applied sense, it may be defined simply as the *utilization of natural enemies to reduce the damage caused by noxious organisms* to tolerable levels. On the other hand, from a more scientific standpoint this term may be used to denote one of the major ecological forces of nature, the *regulation of plant and animal numbers by natural enemies* (more on that in Chapter 3). Practically all living species are attacked by natural enemies – parasites, predators or pathogens – which feed on them in one way or another and in many cases regulate their population densities. Indeed, many potentially injurious pests are kept at very low levels and never reach economic pest proportions due to the effective action of naturally-occurring natural enemies, without deliberate intervention by man. We refer to this as *natural* biological control and discuss it in Chapter 3.

Some have attempted to broaden the applied definition of biological control to include certain other biologically-based forms of pest control, such as utilization of pheromones, sterilized males or resistant plants. All of these are, of course, important and highly desirable aspects of integrated pest management, but inasmuch as they are based on entirely different principles and employ entirely different methodologies, their inclusion in the definition of biological control would only obscure matters. Leaving semantics aside, the term 'biological control' should be restricted to the control or regulation of pest populations by natural enemies, as first proposed by Harry S. Smith in 1919.

Natural enemies can be utilized in three major ways: (1) *importation* of exotic species and their establishment in a new habitat; (2) *augmentation* of established species through direct manipulation of their populations, as by insectary mass production and periodic colonization; and (3) their *conservation* through manipulation of the environment. Although importation has been by far the most successful to date, any of these approaches may lead to effective biological control. We discuss them in Chapter 7.

When successful, the utilization of natural enemies is an inexpensive, non-hazardous means of reducing pest populations and maintaining them, often permanently, well below economic injury levels. In fact the economic threshold concept, as described earlier in this Chapter, does not apply to pest populations that are kept under biological control. The same density level that in the absence of effective natural enemies would call for chemical control measures, usually does not require such action in the presence of natural enemies that are capable of overcoming the pest and reducing it before economic injury is caused. (See Chapter 10 for further discussion.)

Success in biological control is often dependent on a thorough understanding of the organisms involved, both injurious and beneficial, and their intricate interactions. Basic studies of the systematics, biology and ecology of pests and their natural enemies are therefore an integral part of the field of biological control. More on that in Chapter 7.

The modern history of biological control can be dated from the spectacular control of the cottony cushion scale (*Icerya purchasi*) by introduced natural enemies on citrus in California in 1888 (see Chapter 5). Ever since then, hundreds of biological control projects have been successfully carried out in many parts of the world. Several of them are described in Chapters 5 and 6.

Although biological control was first practiced against insect pests, it is by no means restricted to any particular group of noxious organisms. It is applicable, and has indeed been successfully attempted, against insects and other arthropod pests, other animals as diverse as snails and rabbits, weeds and plant pathogens. In spite of the opinions of some, that biological control may only be successful in certain favorable situations, e.g. oceanic or ecological islands, salubrious climates, etc., biological control has proved very successful under very diverse circumstances: in continental areas as well as on islands, in temperate as well as tropical climates, on annual as well as perennial crops and in forests, and against practically all groups of injurious arthropods, including species of medical and veterinary importance. In fact, although the prospects may seem better in certain situations than in others, as a general rule success in biological control has been directly correlated with the amount of research and importation work carried out. Further efforts are very likely to yield many further successes.

The following are a few selected major references, dealing with all aspects of biological control: Clausen (1978), Cook & Baker (1983), DeBach (1964a), Franz (1986), Graham (1985), Hoy, Cunningham & Knutson (1983), Hoy & Herzog (1985), Huffaker (1971b), Huffaker & Messenger (1976), Mackauer, Ehler & Roland (1990), Maxwell & Harris (1974), Mukerji & Garg (1988), Ordish (1967), Papavizas (1981), Rosen (1985), Samways (1981), van den Bosch, Messenger & Gutierrez (1982), and Wood & Way (1988).

The recent trend towards integrated pest management (the fashionable catchword nowadays is IPM) should not obscure the paramount importance of biological control. Any integrated, multidisciplinary management program should definitely include active biological control projects as a major component. Unfortunately, this has not always been the case. In the words of our colleague, Carl Huffaker (1985) who during 1972–8 was coordinator of a nation-wide program to promote IPM in the United States known as the 'Huffaker Project':

> 'Despite the influx of more than $20 million to develop IPM programs for some major US crops, only a small amount has gone to develop the use of parasites, predators, pathogens and antagonists. In fact, the so-called "Huffaker" project, shortly after it was well under way, was stripped of funds for this specific area.'

We shall return to this in Chapter 10.

2

The natural enemies

It is a rare organism that has no natural enemies, if indeed there are any such. Natural enemy populations have the unique quality of being able to interact with the prey or host populations and to regulate them at lower levels than would occur otherwise. Some are effective at extremely low prey levels, others only at higher levels. Biological control operates within and often below the levels permitted by the physical factors of the environment such as temperature, moisture, light and others. This will be discussed in more detail in the next chapter.

In order to appreciate the biological workings and ecological basis of biological control, it is first desirable to have an idea of different pest groups and their major characteristic natural enemies. Although this book will stress natural enemies of insect pests and weeds because it is with these groups that nearly all applied biological control has been practiced, there are pests in nearly all groups of organisms and each is attacked by some other kind of organism, ranging from parasitic microbes to predatory mammals.

Pest mammals have their predators, parasites and pathogens, as do birds, fish, reptiles, nematodes, molluscs, weedy plants and so on through the lists of plant and animal groups. Some well-known cases of impact of natural enemies among mammals include applied biological control of pest rabbits in Australia by the imported myxoma virus and natural biological control of deer in Arizona by mountain lions.

Many natural enemies operate unsuspected, and it is amazing how complex the bio-ecological interactions may be when detailed studies of any given organism are made. Recent evidence of a deer explosion in Texas is linked to the control of the screw-worm fly of cattle by the mass release of sterile male flies. This fly not only attacks cattle but also deer, and evidently was having an unrealized but decided impact on the deer population previous to fly control. At a lower level of the phylogenetic scale, the crown of thorns starfish has been indicted as a pest of living coral reefs, and considerable alarm has been created by its reported damage in some areas. In this case the starfish is a predator, apparently tending to produce biological control, but it is also a pest because it is attacking a wanted organism – the coral. Similarly, coyotes are considered pests by some in the West, but they might be considered quite useful elsewhere in rabbit control. Space prevents any attempt to even briefly cover all major phyla; suffice it to say that biological control projects have included species as beneficial agents, i.e. natural enemies, in such diverse groups as mammals, birds, amphibians, fish, snails and nematodes, in addition to the predaceous and parasitic insects, the predatory mites and the pathogenic micro-organisms that are emphasized in this book and have been used predominantly in the biological control of insects and other arthropods.

The reasons for such emphasis lie in the fact that around 80 percent of all known animals are insects, and they include the vast majority of pests of man. They seriously affect his crops, his domestic stock and his forests and structures, as well as man himself. Efforts to control them with chemicals are responsible for a vast amount of environmental pollution and degradation. By far the greatest amount of poisonous chemicals are used against insects and mites, and over 50 percent are used for agricultural pests. Weeds are also great pests, and in recent years the use of herbicidal poisons has increased tremendously. Biological control was conceived and developed by entomologists against agricultural insect pests as well as against weeds, and most biological control projects have been directed against these groups. Hence it is only logical that we largely restrict coverage to natural enemies of insect pests (including also mites) and weeds. This is not meant to de-emphasize possibilities for biological control in other groups. There is much potential, but relatively little research emphasis has been expended in this regard.

Natural enemies of insects

There is no accurate estimate available of the total number of insect natural enemies of other insects, but there are probably as many, perhaps more, entomophagous insects as there are prey or hosts. It is commonly estimated that there are several million species of insects, including the insect natural enemies. It has also been estimated that only about 15 percent of the insect natural enemies in existence have been discovered and named. Based on the many detailed studies of natural enemies of major insect pests in their native homes, it would seem that most plant-feeding insects have more than one, and generally several to many, natural enemies. Some pest species have well over 100 recorded enemies. Many natural enemies attack several different host or prey species, but others are rather specific. Considering everything, we would hazard an estimate that there are probably as many insects that are entomophagous as there are prey or host insect species, i.e. several million natural enemy species. Obviously, even of the ones known, we can only treat broad groups. Among the best general references to these groups and their biologies are the books *Parasitic Insects* by Askew (1971), *Entomophagous Insects* by Clausen (1962), originally published in 1940 and still a classic, *Biological Control of Insect Pests and Weeds*, ed. DeBach (1964a, Chapters 6 and 7), *Theory and Practice of Biological Control*, ed. Huffaker & Messenger (1976, Chapters 5 and 6), *Insect Parasitoids*, ed. Waage & Greathead (1986), and *Beneficial Insects* by Swan (1964). The latter book will be especially helpful to the non-specialist.

Any organism that feeds on another organism is, by definition, a natural enemy. In biological control parlance, natural enemies are referred to as parasites, predators or pathogens. The first two are termed *entomophagous*, the latter 'entomogenous' or, more accurately, *entomopathogenic*. Effective biological control may be achieved through utilization of any of these groups, but parasites have been the most important to date, with predators ranking second in importance.

Both parasites and predators are animals that feed on other animals, but a *parasite* completes its development on a single host, whereas a *predator* consumes several (or many) prey individuals during its development. However, like so many other biological definitions, this convenient distinction is not very clear-cut, many species of natural enemies exhibiting characteristics of both parasite and predator. The tiny wasp *Scutellista cyanea*, shown in Fig. 2.1, is a good example. This common natural enemy of soft scale insects usually develops as an egg predator, its larva consuming hundreds of eggs underneath the body of a scale, but in their absence may also develop as a parasite of the scale insect itself. On the other hand many wasps, developing as parasites, feed as adults on the body fluids of their hosts, which they obtain by puncturing the host with their ovipositor, and this 'predatory host-feeding' may account for greater host mortality than does parasitism by these natural enemies. Fig. 2.2 shows two instances of such behavior; see Jervis & Kidd (1986) for an

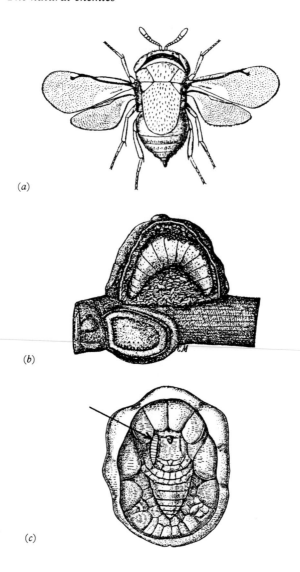

(a)

(b)

(c)

Fig.2.1. Predator or parasite? The chalcidoid wasp, *Scutellista cyanea* (family Pteromalidae): (*a*) adult female wasp; (*b*) larva developing as an egg predator in egg chamber underneath mature fig wax scale; (*c*) small larva (indicated by an arrow) developing as an ectoparasite on underside of immature scale. (From Silvestri & Martelli, 1908.)

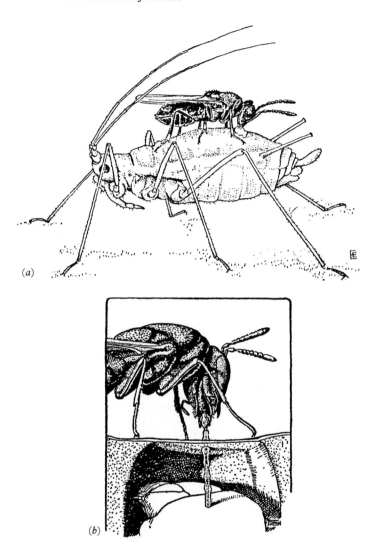

Fig.2.2. Predatory host-feeding by adult chalcidoid parasites. (*a*)
Aphidencyrtus aphidivorus (family Encyrtidae) feeding at an ovipositor
puncture it made in the abdomen of an aphid host. (From Griswold, 1926.)
(*b*) *Habrocytus cerealellae* (family Pteromalidae) feeding on a caterpillar host
through a tube it constructed from secretions of the accessory ovipositor
glands. (From Fulton, 1933.)

extensive review. *Pathogens* are micro-organisms that cause disease. We shall return to them somewhat later.

Parasitic insects

Biological characteristics and definitions

Most insects parasitic upon other insects are *protelean* parasites, i.e. they are parasitic only in their immature (larval) stages and lead free lives as adults. They usually consume all or most of the host's body and then pupate, either within or external to the host. The adult parasite emerges from the pupa and starts the next generation anew by actively searching for hosts in which to oviposit. Most adult parasites require food such as honeydew, nectar or pollen and many feed on their host's body fluids, as mentioned earlier. Others require free water as adults.

The insect parasites of insects are often termed *parasitoids*, to distinguish them from 'true' parasites, such as vertebrate-inhabiting worms, from which they differ mainly in that parasitoids usually kill their host during their own development. Although this distinction has some merit – in order to be effective in biological control, a parasite should be

Fig.2.3. Larvae of the chalcidoid *Euplectrus* (family Eulophidae), a gregarious larval ectoparasite, developing on an armyworm larva. (Photo courtesy James R. Carey.)

capable of killing the pest – we shall use the simpler and more commonly accepted term 'parasite' in this book.

Parasites may have one generation (univoltine) to one generation of the host or two or more generations (multivoltine) to one of the host. Life cycles are commonly short, ranging from 10 days to 4 weeks or so in midsummer but correspondingly longer in cold weather. However, some require a year or more, if they have hosts with but a single generation per year. In general, they all have great *potential* rates of increase.

Parasites may be categorized in many ways. A common distinction is between *ectoparasites*, feeding externally upon the host, as shown in Figs 2.3, 2.4 and 2.5, and *endoparasites*, developing internally within the host (Figs 2.6, 2.7 and 2.8), although this is certainly not absolute. Certain species may start life as endoparasites and later emerge from their host to continue feeding on it externally, whereas others may start as ectoparasites and then bore into the host.

Ectoparasites most frequently occur in hosts that live in some protected site – a larva in a leaf mine or a burrow, a pupa in a cocoon, an armored scale insect under a wax shield, etc. – where they are less likely to be dislodged and lose their host. Fig. 2.5 shows an adult ectoparasite in the act of laying its eggs upon a host larva hidden in a cotton boll. Many of them also sting and paralyze the host prior to oviposition. Endoparasites, on the other hand, are usually well protected within the host, but respiration in a liquid or semi-liquid medium may involve special adaptations in them. Certain endoparasitic larvae may directly obtain oxygen from the host's body fluids, either through their entire integument or through a posterior vesicle. Others may breathe the external air by attaching their spiracles to the host's tracheae, as do many flies of the family Tachinidae, or to their own protruding egg stalk, as do most wasps of the large family Encyrtidae. These adaptations are shown in Fig. 2.9. Many endoparasites 'mummify' their hosts upon completion of their own larval development. This is especially noticeable in parasitized aphids, in which the integument becomes distended, hard and straw-colored.

Both ectoparasites and endoparasites can be either *solitary* (when only one larva develops per individual host, as in Figs 2.7 and 2.8) or *gregarious* (when more than one larva develops normally upon a single host, as in Figs 2.3 and 2.4), but this again is not always clear-cut, as many species may develop facultatively as solitary upon a small host and gregariously upon a larger one. An extreme case of obligatory gregarious development is that of certain *polyembryonic* species, in which a single egg may give rise to several, or even hundreds, of 'identical twins' (see Fig. 2.10). Comprehensive discussions of this peculiar mode of development can be found in Doutt (1947) and Silvestri (1937).

Normal gregarious development should be distinguished from cases of *superparasitism*, in which an individual host is attacked by more parasites of a given species than can normally develop upon it (more than one in obligatorily solitary species). In such cases, intraspecific competition usually results in death of the supernumerary parasites, or in emergence of small, malformed adults, or in death of all competitors. Effective parasites are capable of discriminating between healthy and parasitized hosts, and usually avoid – at least to some extent – ovipositing upon the latter. This may involve marking a parasitized host by various mechanical

Fig.2.4. An ectoparasite of armored scale insects. This is the minute chalcidoid *Aphytis melinus* (family Aphelinidae), the most effective natural enemy of the California red scale. (*a*) Adult female ovipositing in red scale, piercing the host's armor and laying her eggs upon the soft body that lies underneath. (Photo by Jack K. Clark.) (*b*) Two larvae developing externally on body of oleander scale (scale cover removed). (From Rosen & DeBach, 1976.)

(*a*)

or chemical means. Recent research has shown that the ability to avoid superparasitism may be learned, and is improved as the female parasite gains experience (van Lenteren & Bakker, 1975).

Similarly, *multiparasitism* – when an individual host is attacked by parasites belonging to different species – usually results in the death of some or all competitors. The winner in such interspecific competition is commonly termed 'intrinsically superior'.

All the immature stages of insects are subject to attack by parasites. Accordingly, some species are egg parasites, as in Fig. 2.8, others are larval parasites, as in Fig. 2.3 or 2.11, and still others are pupal parasites,

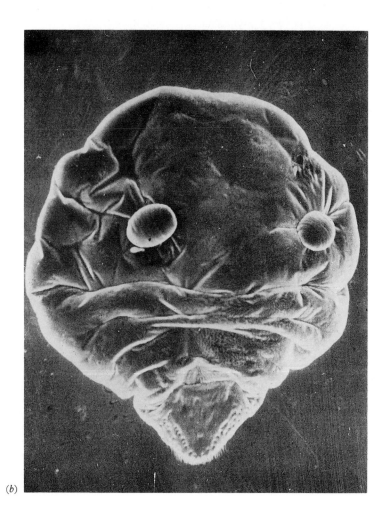

(b)

Fig.2.5. An ectoparasite, the ichneumonoid *Bracon kirkpatricki* (family Braconidae), ovipositing upon pink bollworm larva hidden inside cotton boll.

Fig.2.6. A small endoparasite (i.e. having an endoparasitic larval stage), in the act of ovipositing in a young Mediterranean black scale. This is the chalcidoid *Metaphycus helvolus* (family Encyrtidae), which is responsible for complete biological control of the black scale on citrus in many areas.

developing in the pupa of their host (Fig. 2.12). Certain species occupy intermediate positions between these categories. Some are egg-larval parasites, ovipositing in the egg of their host but completing their development in the larva, whereas others are larval-pupal or egg-pupal parasites. An egg-larval parasite is shown in Fig. 2.13. Adult insects are much less frequently parasitized, although certain wasps and flies do attack adult beetles, bugs, ants, bees, etc. Fig. 2.14 shows a tachinid fly laying its eggs on an adult beetle.

Parasitism may occur at any trophic level of a food chain. Thus, many *primary* parasites, developing upon a non-parasitic host, are in their own turn attacked by *hyperparasites*, the latter being either secondary (developing upon a primary parasite) or tertiary (developing upon a secondary), or even quaternary, etc. Secondary parasites may be *direct*, attacking their primary host (i.e. the primary parasite) within or without the secondary host (i.e. the non-parasitic host), or *indirect*, attacking the secondary host regardless of the presence or absence of a primary parasite in it but capable of developing only if one becomes available. Secondary parasites of injurious pests may seriously hamper the beneficial

Fig.2.7. A solitary endoparasite (i.e. having an endoparasitic larval stage) ovipositing in an aphid. This is the ichneumonoid *Aphidius ervi* (family Aphidiidae). (Photo courtesy Petr Starý.)

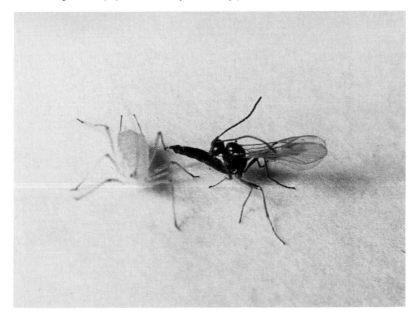

performance of primary parasites, and are therefore carefully screened out in quarantine rooms in all biological control importation projects. For recent reviews of various aspects of hyperparasitism see Rosen (1981) and Sullivan (1987).

Parasites may also be classified according to the sites at which they place their eggs, or – more rarely – larvae. Most endoparasites deposit their eggs inside the host, either freely in its body cavity or in a particular organ such as the intestine, a certain muscle or nerve ganglion, or attached to the

Fig.2.8. *Trichogramma*, a chalcidoid endoparasite of insect eggs (family Trichogrammatidae): (*a*) adult female ovipositing in *Heliothis* egg. (Photo by Gerald R. Carner, published in *Agricultural Research*, May 1983, Courtesy US Deparment of Agriculture.) (*b*) Life history: (A) female ovipositing in bollworm egg; (B) egg within host egg (dorsal view); (C, D) stages in larval development; (E) pupa; (F) adult wasp emerging from host egg. (From van den Bosch & Hagen, 1966.)

(*a*)

inner body wall, etc. Other endoparasites and most ectoparasites place their eggs externally upon the host, sometimes at a characteristic site on its body. When the host is exposed, the parasite egg may be glued to it or partly embedded in its integument, whereas when the host is concealed in a mine, a gallery, a cocoon, etc., the parasite may simply lay its eggs on or near it, often paralyzing it in the process.

Then there are parasites that deposit their eggs or larvae apart from their hosts. Certain wasps and some tachinid flies lay numerous, fully incubated 'microtype' eggs upon the food plants of their hosts. When

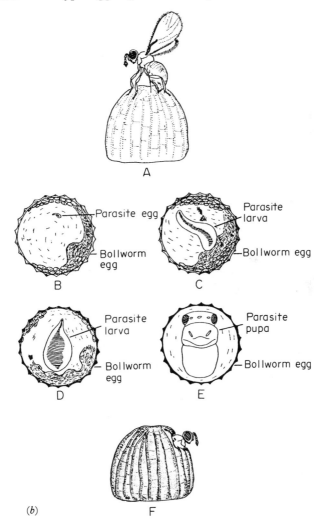

Fig.2.9. Respiratory adaptations of endoparasites. (*a*) Larva of a tachinid fly, *Prosena siberita*, an endoparasite of Japanese beetle larvae, attached to the host's trachea. (From Clausen, King & Teranishi, 1927.) (*b*) Young larva of an encyrtid wasp, *Blastothrix sericea*, an endoparasite of soft scale insects, attached to the egg-shell with the egg-stalk protruding through the host's integument. (From Maple, 1947.)

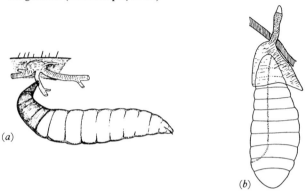

(*a*)

(*b*)

Fig.2.10. Polyembryony: (*a*) parasitized (mummified) navel orangeworm larva, each cell containing a pupa of the chalcidoid *Pentalitomastix plethoricus* (family Encyrtidae), a polyembryonic endoparasite. (*b*) A total of 789 adult wasps have emerged from this host. (From Caltagirone, Shea & Finney, 1964.)

(*a*)

(*b*)

such an egg is ingested by the host, it hatches in the digestive tract and the larva then bores through the gut into the body cavity of the host, where it may continue to develop as a primary endoparasite or, in certain species, even as an indirect hyperparasite. But the most complex life cycles are exhibited by those parasites in which it is the first instar larva, not the adult female, that is charged with seeking the host. The eggs of such

Fig.2.11. *Hyposoter* (family Ichneumonidae), a solitary larval endoparasite. (*a*) Female ovipositing in a young armyworm larva. (*b*) The parasite's cocoon attached to skin of dead host. (Photos by Jack K. Clark.)

(*a*)

(*b*)

Fig.2.12. A pupal parasite. This is the chalcidoid *Pteromalus puparum* (family Pteromalidae), ovipositing in a pupa of the imported cabbageworm. (From Doten, 1911; printed from original glass negative by Frank E. Skinner, 1960.)

Fig.2.13. An egg-larval parasite, the ichneumonid *Phanerotoma flavitestacea* (family Braconidae), ovipositing in an egg of the navel orangeworm (indicated by an arrow). The larva of this parasite completes its development only after the host larva has constructed its cocoon. (From Caltagirone, Shea & Finney, 1964.)

species are deposited and hatch upon plants, and the larvae then await, or actively search for, a suitable host. Again, certain indirect hyperparasites also develop in this manner. Obviously, most of the progeny may die without reaching a suitable host, but this is usually compensated for by an enormous fecundity, up to several thousand eggs per female.

In most parasitic species, both sexes develop in a similar manner. However, in certain wasps of the family Aphelinidae, the life history of the male differs markedly from that of the female. Whereas the females invariably develop as primary endoparasites of scale insects, mealybugs or whiteflies, the males may develop as primary ectoparasites of the same hosts, or – in other species – as primary endoparasites of lepidopterous eggs, or as hyperparasites. An extreme case is that of *adelphoparasitic* (or autoparasitic) species, in which the male may develop as a secondary parasite of an immature female of its own species. (For recent reviews see Viggiani, 1981; Walter, 1983.)

Finally, there are the *cleptoparasites*, which require a paralyzed host for the development of their progeny but are incapable of paralyzing it

Fig.2.14. A parasite of adult insects, the tachinid fly *Hyperecteina aldrichi*, ovipositing upon an adult Japanese beetle. (From Clausen, King & Teranishi, 1927.)

themselves, so they utilize hosts paralyzed by other species, after eliminating the latter. Like hyperparasites, species with marked clepto-parasitic habits should be avoided in biological control projects.

We shall now turn briefly to some of the major groups of parasitic insects.

Major groups of parasites

The main groups of parasites utilized in biological control of insect pests are the Hymenoptera (mostly wasps of the superfamilies Chalcidoidea, Ichneumonoidea and Proctotrupoidea) and Diptera (flies, especially of the family Tachinidae).

HYMENOPTERA. This is the dominent order among all entomophagous insects, both numerically and as regards their successful use in biological control. Over two-thirds of the cases of successful biological control of pest species have been achieved by hymenopterous parasites.

There are many extremely interesting biological adaptations among the Hymenoptera that are discussed in some detail in the publications previously cited. One is the ovipositor, which is a specialized egg-laying organ composed of long interlocking chitinous stylets through which the egg passes. The ovipositor acts as a drill to pierce the host or the material surrounding the host, and in many cases it also serves as a hypodermic to inject paralyzing venom into the host. It is the same organ as the stinger in bees and large wasps. In some species it is used to secrete and form a feeding tube through which the adult parasite sucks the host body fluid, much as through a straw, in order to obtain protein for continued egg production (see Fig. 2.2.*b*). The ovipositor lacks muscles except at the base, but it is supplied with nerves extending its length to the tip, which bears highly sensitive sense organs that can discern by chemical stimuli whether a host is suitable or not, and in fact whether it already contains a parasite egg. In spite of the fact that the ovipositor itself is non-muscular, it can be manipulated by the muscles attached to the base to rotate back and forth like a drill and to curve and bend in any given direction as it is used to explore either the surface or the interior of a host.

The well-known reproductive method of female production from fertilized eggs and male production from unfertilized eggs (arrhenotoky), characteristic of biparental Hymenoptera, is supplanted in some species by obligatory parthenogenesis or uniparental reproduction (thelytoky), wherein females give gise to females and males have become lost or occur only rarely. Polyembryonic reproduction in insects, mentioned earlier, occurs only in Hymenoptera. It is known to occur in at least four parasitic superfamilies. Phoresy is another interesting habit of certain parasites, especially of egg parasites that attack the freshly-laid eggs of the host. The phoretic adult parasites are small and attach themselves to the females of the host insects, then, when that female lays eggs, the parasite female leaves her and proceeds to oviposit in the newly deposited host eggs. In some pupal parasites, such phoretic adults may attach themselves to the full-grown host larva. In other species, the first-instar larvae are phoretic and attach themselves to an adult host in order to reach the host larvae in the nest or colony. Comprehensive reviews of phoresy and other aspects of parasite behavior are given by Clausen (1976) and Vinson (1985).

The most important and numerous hymenopterous parasites occur in the major groups commonly known as chalcidoids, braconids, ichneumonids and proctotrupoids.

The chalcidoids. This term refers to a large superfamily, the Chalcidoidea, that includes some thirty-odd families and thousands of species. The chalcidoids attack species in nearly all orders of insects, the preferred ones being Coleoptera (beetles), Diptera (flies and midges), Homoptera (scale insects, mealybugs, aphids, whiteflies, leafhoppers, etc.) and Lepidoptera (moths and butterflies), which also happen to include the bulk of our chief crop pests. The immature stages of chalcidoids predominantly develop in host eggs or larvae but they also commonly attack pupae, and some develop in adults of Homoptera. The adults are free-living winged 'wasps', ranging in size from the smallest of insects to ones 12 mm ($\frac{1}{2}$ inch) or more in length. They generally require honeydew or nectar as food, and some construct the feeding-tube, already mentioned, in order to feed on host body fluids. They have been responsible for more successful cases of applied biological control than all other groups of natural enemies combined. We do not consider this to be especially due to inherent superiority as much as to the fact that they include the families that specialize in attacking scale insects, mealybugs, aphids and whiteflies, and that these pest groups have received a great deal of research emphasis in biological control.

Among the chalcidoids, two families, the mymarids and the trichogrammatids, consist entirely of species that are egg parasites and consequently are usually very minute, some being less than 0.25 mm in length. It is difficult to collect many insects eggs in the field and hold them in a rearing container without obtaining numerous individuals of one egg parasite species or another. The encyrtids and aphelinids include many that specialize in attacking scale insects, mealybugs, whiteflies and related forms and have been outstandingly successful in applied biological control. The biology of aphelinids, perhaps the most interesting of all chalcidoids, has recently been reviewed by Viggiani (1984).

Figs 2.1–2.4, 2.6, 2.8, 2.10 and 2.12 show examples of various chalcidoid parasites. Another species, of the family Eupelmidae, is shown in Fig. 2.15, and a mymarid is depicted in Fig. 5.11 (p. 180).

The braconids. This is a major large family (Braconidae) of parasitic wasps in the superfamily Ichneumonoidea. They have been commonly used in applied biological control and include many species effective in natural control. Only an occasional rare species is hyperparasitic, hence the family is nearly entirely beneficial. Their general biology and life cycle is similar to the chalcidoids. Parasitism may be external or internal, solitary or gregarious. Their host preferences are principally lepidopterous larvae and coleopterous larvae but also commonly Diptera, including the damaging fruit flies. When females that have larvae which develop externally on the host oviposit, they nearly always permanently paralyze the host with a venom. This prevents the mobile host larva from dislodging the parasite eggs. The endoparasitic species seldom paralyze

Fig.2.15. An adult chalcidoid parasite of the family Eupelmidae.

the host, but some species have been found to inject a symbiotic virus into it along with their eggs. The viral particles then invade the host's tissues and may inhibit its cellular immune system. (For a discussion of this phenomenon, which is still very little understood, see Greany, Vinson & Lewis, 1984.) The free-living adults require food, and some make feeding tubes, as do the chalcidoids. Fig. 2.5 and 2.13 show examples of two adult braconids. Another species is shown in Fig. 2.16.

Various braconid subfamilies and genera specialize in distinctly particular host groups; some (Braconinae) are primarily gregarious ectoparasites on lepidopterous larvae, of which nearly all have the common habit of occurring in a burrow, cell, cocoon or within a web. Thus the adult parasites tend to search for and choose particular sites for oviposition as much as, or more so than, the individuals of the host species attacked. This is true of many other braconids. Others (Cheloninae) are principally solitary egg-larval endoparasites of lepidopterous hosts. The Microgasterinae also specialize predominantly in lepidopterous larvae, many of which are free-living. The Opiinae include the important genera

Fig.2.16. An adult braconid, *Meteorus vulgaris*. (From van den Bosch & Hagen, 1966.)

Opius and *Biosteres*, species of which mainly attack dipterous fruit fly larvae. The biology of braconids is reviewed by Matthews (1974).

The aphidiids (family Aphidiidae), until recently considered a subfamily of braconids but now regarded as a separate family, are a small group of some 300 known species worldwide. They are exclusively solitary endoparasites of aphids, and include species that have been used successfully in biological control. A comprehensive review of their biology was made by Starý (1970). Fig. 2.7 shows an example of an adult aphidiid female ovipositing in an aphid host.

The ichneumonids. This is a very large family (Ichneumonidae) numbering thousands of species, also of the superfamily Ichneumonoidea, and has great potential in biological control, judging by the many studies showing their important natural regulatory effects in crop pest populations. According to Townes (1971), 'There are more species of Ichneumonidae than all vertebrate animals combined.' In other words, more than all the mammals, birds, reptiles, amphibians and fish. The Ichneumonidae comprise about 20 percent of all parasitic insects. Surprisingly, thus far only a few have produced appreciable results when imported in applied biological control projects. Most are primary parasites, hence the family is predominantly beneficial. Some of the largest and most conspicuous parasite species occur in this family. The ovipositor especially is often noticeable and may be extremely long in relation to the size of the body (see Fig. 2.17).

The host preferences of the family are quite varied, but many subfamilies or genera tend to be fairly uniform with respect to parasitism of particular host groups. For example, the Joppinae occur only as internal parasites of larvae and pupae of Lepidoptera. Most of the Ichneumoninae probably parasitize wood- and stem-boring larvae in the Lepidoptera, Coleoptera or Hymenoptera, but many attack lepidopterous pupae and the remainder have quite varied habits. Some species have a very wide host range, tending, as in many braconids, to choose situations, such as borings, rather than particular host species for oviposition. Most of the Tryphoninae are solitary parasites of sawflies. The Ophioninae are predominantly internal parasites of lepidopterous larvae, although a few develop in the grubs of root-feeding scarabs in the soil.

This is such a large family that the biological phenomena cover a tremendously wide range of interesting and curious occurrences. Only some generalized or interesting patterns can be mentioned. Both external and internal parasitism are common, and most species oviposit on or in the host stage upon which the parasite larva is to complete development. The ectoparasites commonly oviposit on hosts enclosed in cocoons, pupal cases or burrows, and may permanently paralyze the host. This causes the host to remain in a fresh condition, yet not develop past the preferred stage, for weeks or even months. Some drill deep into bark or wood with the ovipositor, and the very long stalked eggs pass down the narrow channel and into the hidden host. The adults live for appreciable periods, often 6–8 weeks, which is advantageous when hosts are scarce. Many require free water and depend for food on nectar or pollen from particular plant species. This may affect their occurrence or efficiency in certain localities. Oviposition is keyed to the habits of the host. Free-living host larvae provide a direct stimulus via the antennae and the ovipositor for egg deposition, but host larvae enclosed in cocoons, burrows, etc. cannot be reached except by the ovipositor. Initial attraction in the latter case is probably by odor or tactile responses, and final selection of the host by sense organs on the ovipositor. The adult females of many species feed upon the body fluid of the stage they parasitize as it exudes from the ovipositor puncture, others go so far as to consume an entire host with the

Fig.2.17. A large ichneumonid parasite with an especially long ovipositor. (Photo by Kenneth Middleham.)

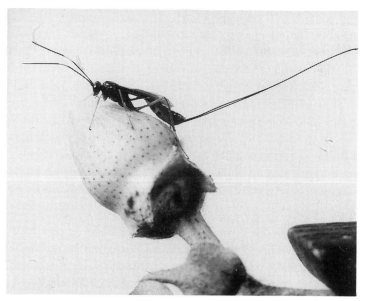

mandibles, hence become true predators, but none constructs tubes for host-feeding as many chalcidoids and braconids do.

The proctotrupoids. This superfamily, the Proctotrupoidea (also known as the Serphoidea), contains several families, the most important of which are the Platygasteridae and the Scelionidae. All are parasitic on the immature stages of other insects and they are predominantly primary internal parasites.

The species of platygasterids differ relatively little in appearance, habits and host preferences. They are endoparasites of cecidomyiid midge larvae and of homopterous nymphs, some of which are serious crop pests. The adults are necessarily small in size because their hosts are all small insects, ranging from about one to several millimeters in length, and most of them are black. The polyembryonic reproduction which occurs in the genus *Platygaster* has occasioned much biological interest. Only a few species have been utilized in biological control; one is reported to have had a substantial controlling effect on the pear midge (*Dasineura pyri*) in New Zealand, and two imported species of *Amitus* have been important in the biological control of pest whiteflies. Species of *Allotropa* are important parasites of mealybugs. Fig. 2.18 shows an adult of *Amitus spiniferus*, an important parasite of the woolly whitefly of citrus (*Aleurothrixus floccosus*). It is only about 1.5 mm in length.

The scelionids comprise a fairly large family numerically; all are small, again because they all develop in small hosts – insect eggs exclusively in this case. For the most part they parasitize eggs of Lepidoptera, Hemiptera and Orthoptera, as well as of certain flies and spiders. They generally develop as solitary individuals. Several species have been successfully utilized in biological control importation projects. In addition to interesting cases of phoresy in this group, the females of certain species exhibit the remarkable adaptation of marking the host egg with the ovipositor, much as with a pencil, following oviposition. This informs subsequent females that the host egg already contains a parasite egg, and consequently they refrain from oviposition in it. An adult of *Trissolcus basalis* in the act of marking an egg of the green vegetable bug is shown in Fig. 2.19.

DIPTERA. In addition to the tachinids, which comprise the most important dipterous family in biological control, this order contains several parasitic families, including the Cyrtidae, Nemestrinidae, Pipunculidae, Conopidae and Pyrgotidae. Most Bombyliidae are parasitic, and parasitic species occur in many of the families that principally have other habits. However, the tachinids constitute the major parasitic family, both in numbers of species and in economic importance, and have been the only group utilized extensively in biological control importation projects.

The tachinids. This is a very large family (Tachinidae), containing a number of subfamilies having species important in natural and applied

Fig.2.18. (*a*) A small platygasterid parasite, *Amitus spiniferus*, that is an effective enemy of the woolly whitefly. (*b*) Mummified woolly whitefly pupal cases showing exit holes through which *Amitus* adults have emerged after developing as an internal parasite.

(*a*)

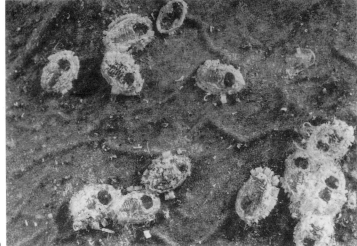

(*b*)

biological control. There are no hyperparasitic species. As would be expected in a large group, host preferences cover a wide range. An appreciable number of species are highly specific in their host preferences, whereas others are very polyphagous. Some have the greatest host range known in any parasitic group. For instance, *Compsilura concinnata*, a parasite imported against the gypsy moth, has been recorded from some 100 species of hosts just in the United States. All in all, it appears that tachinids are less host-specific than are the hymenopterous parasites. Fig. 2.14 shows a tachinid fly. Another species is shown in Fig. 2.20, and another one in Fig. 5.10 (p. 177).

The subfamily Exoristinae is the most important, both numerically and economically. It includes many well-known genera that attack important crop pests. The hosts are predominantly lepidopterous larvae, and adult beetles in the scarab, chrysomelid and carabid families. The Tachininae principally include species that parasitize boring or otherwise concealed lepidopterous larvae. The Gymnosomatinae largely parasitize adult hosts, but sometimes nymphs, of the pentatomid, pyrrhocorid and coreid hemipterous families. Scarab grubs are commonly attacked by species of Dexiinae, which also include in their host preferences adult scarabs and boring larvae of beetles and moths.

Fig.2.19. An adult female of the scelionid *Trissolcus basalis*, in the act of chemically marking an egg of the green vegetable bug to inform other females that the host egg has already been parasitized. (Photo courtesy Division of Entomology, CSIRO, Canberra, Australia.)

Predatory arthropods

Arthropods serve as prey to an enormous array of predatory animals, ranging from numerous other arthropods to various vertebrate species. Some of the latter have been occasionally used in the biological control of arthropod pests: certain fish and lizards against mosquitoes, giant toads against scarabaeids, mynah birds against a locust, ducks against planthoppers, etc. Fig. 2.21 shows the mosquitofish *Gambusia*, which has been used extensively in many parts of the world for the biological control of mosquito larvae. However, of greatest importance in applied biological control have undoubtedly been various insect and acarine predators, especially coccinellid and carabid beetles, lacewings and hemipterans, as well as phytoseiid mites.

Occasional predatory or cannibalistic behavior is quite common among phytophagous arthropods, whereas obligatory predators are present in

Fig.2.20. Adult of a parasitic tachinid fly. (From van den Bosch & Hagen, 1966.)

almost all the main orders of insects and mites. Accordingly, some predators use biting or chewing mouthparts to devour their prey – e.g. praying mantids, dragonflies and beetles – whereas others, such as hemipterans, neuropteran larvae, flies and certain mites, use piercing and sucking mouthparts to feed upon the body fluids of their prey.

Many predators are agile, ferocious hunters, actively seeking their prey on the ground or on vegetation, as do beetles, lacewing larvae and mites, or catching it in flight, as do dragonflies and robber flies. Certain hunters have specially adapted seizing organs, such as the barbed forelegs of mantids, shown in Fig. 2.22, or the labial 'mask' of aquatic dragonfly nymphs (see Fig. 2.32, p. 74).

Many species are predaceous in both the larval and adult stages, although not necessarily on the same kinds of prey. Others are predaceous only as larvae, whereas the adults may feed on nectar, honeydew etc., and among these it is often the non-predaceous adult female that seeks the prey for her larvae by depositing eggs among the prey, because their larvae are sometimes incapable of finding it on their own. Such division of labor is common among species that attack sessile, colonial prey such as aphids or coccids. Syrphid flies and cecidomyiid midges, for instance, oviposit in the vicinity of aphid colonies, which then serve as an ample source of food to their eyeless, legless larvae. An extreme case is that of various solitary wasps, in which the adult female actively hunts, overpowers and paralyzes the prey on which her progeny eventually feed.

Fig. 2.21. The mosquitofish, *Gambusia affinis*, about to catch a mosquito larva. This species has been used extensively in biological control in many parts of the world. (Photo courtesy Paul H. Rosenfeld.)

Other predators may use various traps to catch their prey, of which the webs of spiders and the sand pitfalls of ant lions are perhaps the best known. Still others may lure their prey in various elaborate ways. However, in general, predators lack the highly specialized adaptations which are correlated with the parasitic mode of life. Moreover, as a general rule it is those predators having relatively simple, straightforward lifestyles that have proved most useful in applied biological control projects.

Such a wide variety of taxonomic groups of diverse habits and prey preferences occur among arthropod predators, that only a few of the larger and more ecologically important ones can be touched upon.

COLEOPTERA. The beetles contain many families with a large number of predaceous species and, all told, probably more than half of all insect predators are in this order. The principal predaceous families are the Coccinellidae (the well-known ladybeetles, also called ladybirds), Silphidae, Staphylinidae, Histeridae, Lampyridae, Cleridae, Cantharidae, Meloidae, Cicindelidae, Carabidae, Dysticidae, Hydrophilidae and Gyrinidae.

Fig.2.22. Chinese mantid, *Tenodera aridifolia sinensis*, feeding on a long-horned grasshopper. Note the elongated, barbed forelegs it uses to capture and hold its prey. (Photo courtesy James R. Carey.)

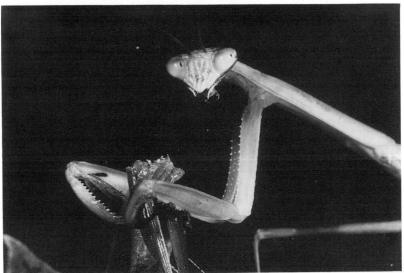

The coccinellids and the carabids are of most importance in biological control of agricultural pests. The ladybeetles especially have produced outstanding results in biological control importation projects in a number of cases, and the vedalia ladybeetle, *Rodolia cardinalis*, was responsible for the first outstanding success in biological control. Fig. 2.23 shows an

Fig.2.23. Adult (*a*) and larva (*b*) of the vedalia ladybeetle, *Rodolia cardinalis*, feeding on the cottony-cushion scale. This beetle produced the first outstanding case of biological control.

(*a*)

(*b*)

adult and larva of the vedalia beetle feeding on the cottony-cushion scale (see also Fig. 5.2, p. 143). This common, worldwide family specializes mainly in feeding upon scale insects, mealybugs, aphids and whiteflies. Fig. 2.24 shows an adult and larva of the common ladybeetle, *Hippodamia*, feeding on aphids. Comprehensive reviews of the biology of coccinellids were made by Hodek (1967, 1973).

Other families having rather specialized feeding habits include the lampyrids, which commonly feed both as larvae and adults on snails and earthworms; the meloids that prey upon the eggs of locusts in the soil, or develop in the cells of bees; the generally nocturnal carabid ground beetles that commonly feed on lepidopterous larvae and pupae, and the checkered clerid beetles, which predominantly feed both as adults and

Fig.2.24. A common ladybeetle, *Hippodamia*, feeding on aphids. (*a*) Adult. (From van den Bosch & Hagen, 1966.) (*b*) larva. (Photo by Jack K. Clark.)

(*a*)

(*b*)

larvae on scolytid and other wood-boring beetles, and are an important factor in the control of these forest pests. Other families tend to be restricted to particular habitats but may be rather generalized feeders. Thus the dytiscid diving beetles are aquatic and feed on nearly everything of the right size, including various aquatic insects, snails, tadpoles and earthworms, although most feed on the immature stages of dragonflies, mayflies or aquatic bugs. The larvae of the aquatic Hydrophilidae also feed upon a similar range of prey. Fig. 2.25 shows a hydrophilid 'water tiger' larva preying upon a mosquito larva. The agile cicindelid tiger beetles are predominantly terrestrial and favor a wide variety of prey in exposed habitats such as open sandy areas, roadways or paths. A large number of the staphylinid rove beetles prey upon dipterous larvae in manure and other refuse, in decomposing animal carcasses and in the soil.

NEUROPTERA. Most of the species of this order are predatory. It contains such familiar forms as the common green lacewings and the ant lions whose pits are seen so often in sandy or dusty areas. The larvae of nearly

Fig.2.25. An aquatic predator, 'water tiger' larva (the beetle, *Tropisternus lateralis*, family Hydrophilidae), seizing a mosquito larva under water. (Photo courtesy Paul H. Rosenfeld.)

all neuropterans possess long curved mandibles, which act like forceps to grasp and pierce the prey and suck out the body fluids.

The most important families in biological control have been the Chrysopidae (green lacewings) and Hemerobiidae (brown lacewings), which attack many agricultural pests including scale insects, mealybugs, aphids, whiteflies, mites and a variety of others. The eggs of the green lacewings are characteristically borne on long slender stalks and commonly in groups. The larvae when magnified appear ferocious, which indeed they are, and nearly dragon-like because of their large scimitar-shaped mandibles. They feed mainly on aphids and mealybugs, but on a variety of other insects and mites as well. The beautifully bright green adults frequently have golden eyes. They are often abundant in orchards, fields and gardens. The adults tend to feed on the same kinds of prey as the larvae. Fig. 2.26 shows the stalked eggs, larva and adult of a green lacewing.

The brown lacewings eat aphids, mealybugs, whiteflies, kermesids, scale insects and a variety of others. The adults appear similar to the green lacewings in form, perhaps smaller, but are brown and drab in appearance as compared to the former. Population studies have shown them to be quite effective in natural biological control of mealybugs.

HYMENOPTERA. About one-quarter of the families of this order are strictly predaceous. Most of the predatory forms tend to be social and live in colonies. The ants (Formicoidea) probably comprise the most important predatory group and are important in natural control, even though many are classed as pests. They are fostered in Europe in the biological control of forest insects, and in the Orient for biological control of certain citrus pests. Ants are particularly effective against soil-inhabiting larvae, pupae or adults of a diverse variety of forms.

The colonial vespid wasps provision their paper nests with the bodies of caterpillars and other soft-bodied insects, on which their progeny feed. The family is important in natural control, especially of exposed foliage-feeding larvae. Species of *Polistes* have been used in several biological control projects.

The sphecoid wasps likewise provision their nests with other insects, but tend to be much less social than the vespids. The nests range from one cell to a group. They are commonly built in the soil, or else in stems of plants. Some construct nests of mud or sand. Host preferences are very diverse and include prey in all the common insect orders, especially the

Fig.2.26. (*a*) The stalked eggs, (*b*) a larva feeding on an aphid and (*c*) an adult of a green lacewing, *Chrysopa* (now known as *Chrysoperla*). (From van den Bosch & Hagen, 1966.)

Orthoptera, Hemiptera, Homoptera, Hymenoptera, Lepidoptera and Diptera. When large hosts are used, only a single one may be placed in a cell, hence the developing larva is parasitic in this case; usually, however, more than one prey individual is used to provision a cell, so they qualify as predators. They generally paralyze the prey for a greater or lesser duration of time. Fig. 2.27 shows a sphecoid wasp dragging a paralyzed grasshopper into its burrow.

DIPTERA. Quite a few families are entirely predaceous, and many others contain predatory species. Either or both larvae and adults may be predators, generally on noxious insects. They are of decided importance in natural control, and some have been utilized in applied biological control. The more common or important economic families include the Asilidae, Syrphidae, Cecidomyiidae, Bombyliidae, Anthomyiidae, Calliphoridae, and Sarcophagidae. Of the latter four, many are predaceous in the egg masses of grasshoppers. The asilid robber flies prey as adults on many flying insects, often including beneficial ones, but the larvae feed in the soil or in decaying wood, largely on pest insects. Fig. 2.28 shows an adult asilid fly feeding on an aphid. The syrphid larvae are common and valuable predators of aphids especially, although some attack scale insects and related forms. Fig. 2.29 shows an adult and larva of a syrphid fly. The

Fig.2.27. A sphecoid wasp dragging a paralyzed grasshopper into its burrow. This is the great golden digger, *Sphex ichneumoneus*. (Photo courtesy James R. Carey.)

tiny cecidomyiid gnats or midges include a substantial number of species whose larvae prey upon aphids, scale insects, whiteflies, thrips and mites, especially the eggs or the younger stages. Fig. 2.30 shows a cecidomyiid larva feeding on an aphid.

HEMIPTERA. Although this order is predominantly plant-feeding, a substantial number of species in various families have evolved to become predaceous. Many attack important economic crop insects, and are considered to be of great importance in natural control. A predatory hemipteran, *Geocoris*, is shown sucking the body fluids from a leafhopper through its proboscis in Fig. 2.31(*a*). Some have been used in biological control importation projects, of which *Tytthus mundulus*, which controlled the sugar-cane leafhopper in Hawaii (see Chapter 5), is an outstanding example. It feeds exclusively on the leafhopper's eggs. A considerable number, if not most, of the species of the same family, the Miridae, are predaceous. Among the anthocorids, the best known predator is *Orius insidiosus* which sucks the body fluid of a variety of prey including thrips, Homoptera, Hemiptera, Lepidoptera and mites. It is a fairly effective predator of corn earworm eggs and young larvae, as well as of various thrips and mites. The reduviid assassin bugs prey on a wide

Fig.2.28. Robber fly (family Asilidae) feeding on an aphid. (Photo courtesy James R. Carey.)

range of plant-feeding insects, including caterpillars, leafhoppers, and aphids. There are several families primarily aquatic in habit, which prey upon aquatic insects and other organisms. These are included in the superfamily Gerroidea. Fig. 2.31(*b*) shows a backswimmer, *Notonecta*, feeding on a mosquito larva.

ODONATA. The well-known large dragonflies and damselflies number several thousand species, all of which are predaceous both as nymphs and as adults. The nymphs are virtually all aquatic and feed on a wide range of aquatic insects and other organisms. They have specialized mouthparts, equipped with a labial 'mask' that they use for capturing prey. Fig. 2.32 shows a dragonfly nymph seizing a mosquito larva. The adults capture

Fig.2.29. Syrphid flies. (*a*) Adult, non-predaceous. (*b*) Larva, feeding on an aphid. (Photos by Jack K. Clark.)

their prey on the wing and tend to prefer mosquitoes and other flies, Lepidoptera and Hymenoptera. Active predation on swarming termites has been recorded. The biology of Odonata has been reviewed by Corbet (1980).

ACARI. Mites of numerous families are predaceous, feeding on a wide range of small insects and mites. Comprehensive discussions of mites as biological control agents are found in Hoy, Cunningham & Knutson (1983). The most important families are Phytoseiidae, Hemisarcoptidae, Cheyletidae and Pyemotidae. Phytoseiids prey upon red spider mites, eriophyid rust and gall mites, etc., but many of them can use small insects such as scale-insect crawlers, and even pollen, as alternative food. Several species of *Amblyseius*, *Typhlodromus*, *Phytoseiulus* and *Metaseiulus* are highly effective in the natural control of red spider mites in various agro-ecosystems, and some have been used in applied biological control programs. Fig. 2.33 shows the phytoseiid *Metaseiulus occidentalis* feeding on the two-spotted spider mite. For recent reviews of phytoseiid biology see Helle & Sabelis (1985). Hemisarcoptids may sometimes complete their development on a single armored scale insect, and in such cases would fit the definition of an ectoparasite rather than a predator. These mites are frequently transported underneath the elytra of coccinellid beetles. *Hemisarcoptes malus* was used in the biological control of the oystershell scale (*Lepidosaphes ulmi*) in Canada. (See Gerson, O Connor & Houck, 1990.) Other predaceous mites have been tried against such

Fig.2.30. Larva of a cecidomyiid midge, *Aphidoletes meridionalis*, feeding on an aphid. (From Davis, 1916.)

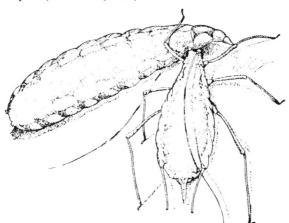

(a)

(b)

Fig.2.31. Predatory hemipterans. (*a*) A big-eyed bug (*Geocoris*, family Lygaeidae) sucking the body fluids from a leafhopper through its proboscis. (From van den Bosch & Hagen, 1966.) (*b*) A backswimmer (*Notonecta*, family Notonectidae) feeding on a mosquito larva under water. (Photo courtesy Paul H. Rosenfeld.)

(a)

(b)

Fig.2.32. A dragonfly nymph seizing a mosquito larva under water with its labial 'mask'. (a) Laterial view; (b) Ventral view. (Photos courtesy Paul H. Rosenfeld.)

diverse pests as the lucerne flea (*Sminthurus viridis*), the imported fire ant (*Solenopsis invicta*) and filth-breeding flies.

SPIDERS. This ubiquitous group of arthropods comprises more than 30 000 known species, all of which are predators, feeding almost exclusively on insects. They have been rather neglected in biological control, perhaps mainly because they are usually very general predators, attacking various prey species in direct proportion to their relative abundance. However, numerous studies have indicated that spiders may be important predators in various agro-ecosystems, especially if undisturbed by chemical pesticides. Although individual species may be incapable of controlling a given pest, spider assemblages may have a pronounced stabilizing effect on pest populations. Fig. 2.34 shows a spider feeding on a moth. The role of spiders as biological control agents is reviewed by Riechert & Lockley (1984).

Pathogenic micro-organisms

Insects and mites are probably subject to as wide a variety of diseases as are the vertebrates, and the pathogenic micro-organisms attacking them have life cycles more or less similar to those of the pathogens developing in other groups of animals. However, with the exception of the rickettsiae, virtually none of the arthropod pathogens

Fig.2.33. A phytoseiid mite, *Metaseiulus occidentalis*, feeding on the two-spotted spider mite. (From Roush & Hoy, 1980.)

also occurs in mammals, and none has been recorded from man. Thus they are safe to use in applied biological control, even in large-scale microbial spraying operations.

In terms of numbers of species, relatively few insect pathogens – about 1500 – are known as compared to the numbers of described entomophagous species. However, numerous new ones are discovered every year, and many of them attack a wide range of host insects. Their effects are sometimes spectacular in natural control, especially when epizootics occur to devastate great outbreaks of pest insects. On the other hand, pathogens have the handicap that they are not active searchers for the host or prey as are entomophagous insects, hence generally do not limit host population densities at low levels over a period of time.

In classical biological control importation projects, very little has been done regarding introduction of new exotic insect or mite pathogens. Of those purposely imported, we know of only one case where a pathogen alone has produced satisfactory biological control at low levels. This

Fig.2.34. A wolf spider (*Lycosa pseudoannulata*, family Lycosidae) feeding on the striped rice stem borer (*Chilo suppressalis*). (Photo courtesy B. Merle Shepard, IRRI, Los Baños, The Philippines.)

project, the control of the rhinoceros beetle in the South Pacific by an imported virus, is discussed in Chapter 6. Certain other viruses have also played a prominent role in importation projects, but the greatest emphasis with pathogens in recent years has been with their mass production and application in the field as microbial insecticides. This offers much promise in integrated control programs, because microbial insecticides appear to have none of the toxic residue drawbacks of chemical pesticides. However, the possibility exists that resistance may develop. Several microbial preparations are now available commercially for the control of various important pests, and quite a few others are being developed.

Insect pathology is a whole field in itself and probably has nearly as many research specialists as are engaged in any other major area of biological control. Important references include: *Microbial Control of Insects and Mites*, ed. Burges & Hussey (1971), *Microbial Control of Pests and Plant Diseases 1970–1980*, ed. Burges (1981), *Insect Diseases*, ed. Cantwell (1974), *Microbial Control of Plant Pests and Diseases* by Deacon (1983), *Biological Control of Insect Pests and Weeds*, ed. DeBach (1964a, Chapters 18–21), *Epizootiology of Insect Diseases*, ed. Fuxa & Tanada (1987), *Microbial and Viral Pesticides*, ed. Kurstak (1982), *Proceedings of the Summer Institute on Biological Control of Plant Insects and Diseases*, ed. Maxwell & Harris (1974), *Laboratory Guide to Insect Pathogens and Parasites* by Poinar & Thomas (1984) and, of course, the classics: *Insect Microbiology* by Steinhaus (1946), *Principles of Insect Pathology* by Steinhaus (1949) and *Insect Pathology, an Advanced Treatise*, ed. Steinhaus (1963).

The most common diseases of insects are caused by viruses, bacteria, fungi, protozoa and nematodes. Except for the fungi and some nematodes, disease organisms gain entry and infect the host via the mouth and the digestive tract. That is, the insect host must eat plant or other food contaminated with the pathogen. Because of their small size, usually only the symptoms will be seen by the non-specialist.

VIRUSES. Several hundred viruses, belonging to several different groups, are known to infect arthropods. Most of them differ from the viruses of other animals and plants in that the virions (viral particles) are embedded in a proteinaceous envelope, or *inclusion body*, which may be seen under an ordinary light microscope. All are intra-cellular parasites, replicating in, and eventually destroying, live cells. Most important among them are the *baculoviruses*, characterized by rod-shaped virions. They include the *nuclear polyhedrosis viruses* (NPV), with polyhedral inclusion bodies containing numerous virions that develop in the nuclei of infected cells, and the *granulosis viruses* (GV), with small, oval inclusion bodies containing a single virion each, that develop in the nuclei or in the cytoplasm immediately surrounding them. Another group of great potential importance are the *cytoplasmic polyhedrosis viruses* (CPV), characterized by spherical virions and polyhedral inclusion bodies, that develop in the cytoplasm of insect gut cells. Fig. 2.35 shows examples of these three groups, as well as a diseased caterpillar dying of a granulosis.

Only immature stages of insects are noticeably susceptible to viral disease; adults may carry the pathogen and transmit it to their progeny via the eggs or spread it through their faeces, but they seldom show any severe symptoms. Typically, when the inclusion bodies are ingested by the host larva they dissolve in its gut, and the virions then invade the cells of the gut and various other tissues and replicate in them. The infected cells disintegrate, the host dies, and eventually its integument ruptures and the liquefied body contents, containing numerous inclusion bodies, spread over the plants.

For various reasons, governmental agencies have been reluctant to approve registration of insect viruses for commercial use. This has been unfortunate, because viruses are highly specific and therefore offer much hope as selective microbial pesticides. However, it currently appears that this problem is being solved, and four NPVs are already registered in the United States for use against various lepidopterous pests of cotton and forest trees (see Table 8.1, p. 308). Recent reviews of insect viruses are found in Entwistle & Evans (1985) and Maramorosch & Sherman (1985).

BACTERIA. Not all the bacteria that infect insects are obligate pathogens. Of the latter, spore-forming bacteria of the genus *Bacillus* are the most important because their spores, like the inclusion bodies of most insect viruses, are resistant forms that enable them to survive for long periods outside the host. Some of them produce, along with the spore, a protein crystal ('parasporal body') that acts as a powerful stomach poison when ingested by an insect host. These crystalliferous spore-forming bacteria, of which *Bacillus thuringiensis* is the best known, are currently the most widely used agents of microbial control.

Bacteria have been used successfully in pest control for decades. Since the 1940s, infestations of the Japanese beetle (*Popillia japonica*) have been controlled effectively in the eastern United States by applying dusts containing the spores of milky disease bacteria, *Bacillus popilliae* and *B. lentimorbus*, against the soil-inhabiting grubs. More recently, since the early 1960s, various commercial preparations of *Bacillus thuringiensis* (Bt) have been available for control of numerous lepidopterous and other pests. Some insects are more susceptible to the poison crystals, which

Fig.2.35. Viral pathogens. (*a*) Nuclear polyhedrosis virus: a polyhedral inclusion body containing bundles of rod-shaped virions (some are seen in cross section). (*b*) Granulosis virus of the codling moth: oval inclusion bodies, each containing a single rod-shaped virion. (*c*) Cytoplasmic polyhedrosis virus of the Zebra caterpillar: a polyhedral inclusion body containing spherical virions. (Photos courtesy Jean R. Adams, USDA.) (*d*) Caterpillar of the large white butterfly (*Pieris brassicae*) dying of a granulosis. The hanging position is typical of nuclear polyhedroses and granuloses. (Photo courtesy Margaret K. Arnold, NERC Institute of Virology, Oxford.)

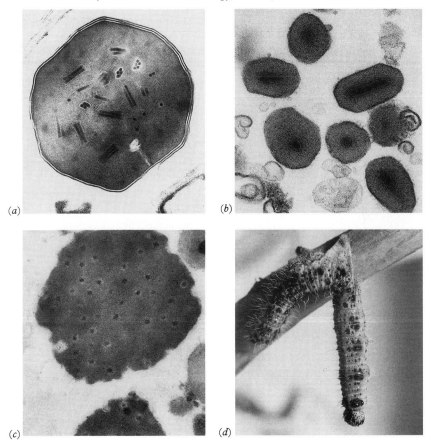

(*a*) (*b*) (*c*) (*d*)

cause gut paralysis, others to the germinating spores, which cause a bacterial septicemia, but non-target organisms are usually not affected by these selective pesticides. Different strains of '*Bt*' are effective against different insects. *Bacillus thuringiensis* var. *israelensis* (*Bti*) is currently the most promising agent for microbial control of mosquito larvae.

Fig. 2.36 shows the spores and parasporal bodies of *Bacillus popilliae* and *B. thuringiensis*. A review of bacteria as microbial control agents was recently made by Lüthy (1986).

RICKETTSIAE. Unfortunately, some of these bacteria-like, obligate intracellular parasites that cause disease in insects are also pathogenic to

Fig.2.36. Bacterial pathogens. (*a*) Spores and attached parasporal bodies of *Bacillus popilliae*. This bacterium was used extensively in the biological control of the Japanese beetle (*Popillia japonica*) in the United States. (Photo courtesy Insect Pathology Laboratory, USDA.) (*b*) Section through a sporulating rod of *Bacillus thuringiensis*, showing an oval spore (left) and a crystalline parasporal body. (Photo courtesy J. R. Norris.)

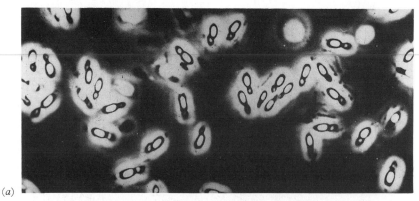

(*a*)

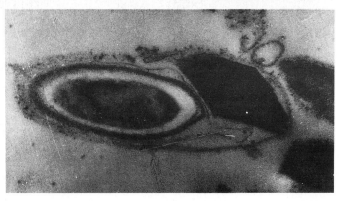

(*b*)

vertebrates, including man. For this reason, no use has been made so far of any of them in microbial control.

SPIROPLASMAS. This interesting family of micro-organisms of the myco-plasma class, characterized by their helical form, has been recognized only recently. They cause numerous plant diseases, some of them fatal. What makes them so interesting is that some of them may be detrimental to certain insect vectors. Fig. 2.37 shows the corn stunt spiroplasma, a pathogen of corn plants that is also pathogenic to *Dalbulus elimatus*, one of its leafhopper vectors, whereas another vector, *D. maidis*, is not affected. Other spiroplasmas cause diseases in honey bees and in *Drosophila* vinegar flies. A review of this little-known group was recently made by Maramorosch (1981).

FUNGI. Although fungi were perhaps the first entomopathogenic orga-nisms to be recognized and utilized (see Chapter 4), their actual use in

Fig.2.37. The corn stunt spiroplasma, a pathogen of corn plants that is also pathogenic to *Dalbulus elimatus*, one of its leafhopper vectors. (Photo courtesy Karl Maramorosch.)

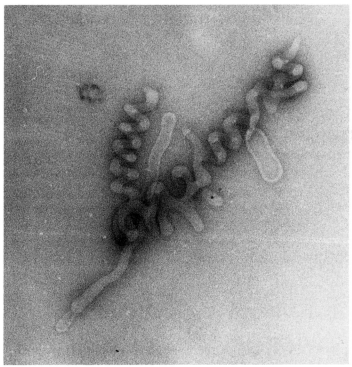

microbial control has been rather modest. Fungi usually gain entrance into the insect host's body through the integument, and free water or high humidity is generally required. Thus they tend to be restricted to moist environments. However, fungi do have the advantage of attacking also sucking insects which, because of the nature of their feeding on sap, tend to be fairly free of diseases caused by other micro-organisms because they rarely ingest them. They are considerably less specific than viruses or even bacteria, and some of them, such as *Entomophthora sphaerosperma*, *Beauveria bassiana* ('white muscardine') or *Metarrhizium anisopliae* ('green muscardine'), attack a large range of host species.

Several entomopathogenic fungi have been successfully imported and established in new habitats, and many successful trials have been made in recent years using fungi as microbial pesticides. Boverin, a preparation containing spores of *Beauveria bassiana*, has been used commercially for some time in the Soviet Union against various pests. Recently, a pesticide containing *Hirsutella thompsoni* has been registered in the United States for use against the citrus rust mite (*Phyllocoptruta oleivora*). Four examples of entomopathogenic fungi are shown in Fig. 2.38. Comprehensive reviews of fungi as insect pathogens have been made by Ferron (1978, 1985).

PROTOZOA. Although numerous protozoans infect insects, they have not figured prominently in applied biological control because their virulence is usually low and the diseases caused by them tend to be chronic. Greatest attention has been given to spore-forming species of the orders Microsporidia and Neogregarina. Like viruses and rickettsiae they are obligate intra-cellular parasites and can only be propagated in live hosts. The fact that some of them attack beneficial insects such as silkworms or honey bees may also limit their application. Nevertheless, some field trials – especially with *Nosema locustae*, which has been registered in the United States as a microbial insecticide against grasshoppers – have been encouraging and have indicated that certain protozoa may be integrated with other control agents in long-term IPM programs. For recent reviews see Canning (1981) and Henry (1981).

NEMATODES. Although many nematodes are rather macroscopic worms, not micro-organisms, they are usually included among the pathogens because they are used in a similar manner. Numerous nematodes are parasitic in insects. Some are ingested by the prospective hosts, others penetrate it actively through the spiracles or other natural openings, or directly through the integument. Some, like the Mermithidae, are obligate parasites, developing in the living host and killing it upon emergence. Others, like the Steinernematidae and Heterorhabditidae, are facultative parasites. Upon penetrating the host, they introduce into it a symbiotic bacterium (*Xenorhabdus*) that multiplies and causes septicemia that kills it within 48 hours. The nematodes then continue to develop for several generations in the host's dead body, feeding upon the bacteria. Both types have been used in field trials against a variety of pest insects,

Fig.2.38. Fungal pathogens. (*a*) *Paecilomyces tenuipes* sprouting from a lepidopteran pupa buried in the soil. Similar pathogens were misinterpreted by early naturalists to represent strange organisms, combining a plant and an animal. (Photo courtesy Michiel C. Rombach, IRRI.) (*b*) White muscardine, *Beauveria bassiana*, on the cocoa mirid, a pest of cocoa in Papua New Guinea. (Photo courtesy Chris Prior, CIBC.) (*c*) *Hirsutella citriformis* on the rice brown planthopper, a major pest of rice in Asia. (Photo courtesy Michiel C. Rombach, IRRI.) (*d*) Green muscardine, *Metarrhizium anisopliae*, on a scarabaeid beetle grub. (Photo courtesy Insect Pathology Laboratory, USDA.)

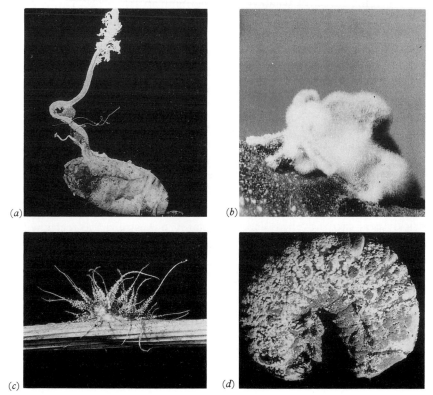

(*a*) (*b*)

(*c*) (*d*)

and the results have been quite encouraging, especially in moist or protected habitats, because of the high moisture requirements of the nematodes. The infective stages (i.e. larvae that penetrate, or are ingested by, the host) are quite resistant, may be stored for several months, and may be applied with conventional spraying apparatus. Several commercial preparations are available for use against various pests (see Table 8.2, p. 309).

Fig. 2.39 shows a mermithid nematode (*Romanomermis*) parasitic in a mosquito larva, and a steinernematid nematode (*Steinernema*) developing in the dead body of a caterpillar host. For extensive reviews of the insect-parasitic nematodes see Nickle (1984) and Poinar (1979).

Natural enemies of weeds

The insect natural enemies of weeds do not fall into specialized taxonomic groups, as is true of the entomophagous insects. Instead, they are just as diverse as the entire range of phytophagous insects, and any weed-feeding insect of whatever feeding habit in any order, family or genus can be considered potentially important in biological control of weeds. Thus we cannot categorize any particular order or family as being of outstanding importance. The critical aspect in biological weed control is the choice, by careful screening, of weed-feeders which are so highly adapted to the weed species that they are unable to develop on any other plants – at least any economic ones. This often means the choice of an insect with specialized habits, but it can just as well mean one with very high host (weed) specificity. Thus insects in such diverse groups as the moths, thrips, mealybugs, scale insects, wasps, chrysomelid and other beetles, sucking bugs as well as dipterous leaf miners, gall midges and others have been more or less successfully used in biological control of weeds; some outstandingly so, such as the *Cactoblastis* moth which killed prickly pear cactus over millions of hectares in Australia and reclaimed the land for agricultural use. Fig. 2.40 shows larvae of *Cactoblastis* feeding in a pad of *Opuntia* cactus; see Chapter 5 for further discussion of this project. Another spectacular success, the control of Klamath weed in California by *Chrysolina* leaf beetles, is described in Chapter 6.

Diseases of weeds offer great potential, but until recently there was reluctance to try them in fear that they might also attack desirable plants. However, there should be no more hazard here than in the case of weed-feeding insects, provided proper screening tests are carried out to determine near absolute specificity of the pathogen for the weed in question. The effects of accidentally imported plant pathogens, albeit on certain desirable plants, demonstrate not only that the method is feasible but that native species of weeds may be controlled. This is illustrated by the 'control' of chestnut trees in the eastern United States forests by the accidentally introduced fungus, *Endothia parasitica*, from Asia. Not only

Fig.2.39. Nematodes. (*a, b*) A mermithid nematode developing in a live mosquito larva. This is *Romanomermis culicivorax*, an obligatory parasite, shown here (*a*) coiled inside the host's thorax, and (*b*) emerging, killing the host as it does so. (Photos courtesy Paul H. Rosenfeld.) (*c*) A steinernematid nematode, *Steinernema feltiae* (previously known as *Neoaplectana carpocapsae*), a facultative parasite, developing in the dead body of the greater wax moth caterpillar. (Photo courtesy Sam R. Dutky, USDA.)

(*a*)   (*b*)

(*c*)

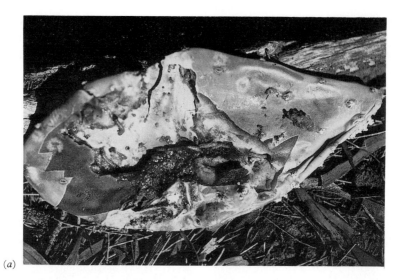

(a)

(b)

Fig.2.40. *Cactoblastis cactorum*, an important natural enemy of cactus weeds.
(a) An *Opuntia* cactus pad destroyed by *Cactoblastis* larvae. (b) Close-up of the
feeding larvae. (Photos by John Green, courtesy CSIRO Division of
Entomology, Canberra, Australia.)

was a native plant controlled, but the pathogen was highly specific to it. This important field has been developing rapidly in recent years and several fungi have been used successfully for biological control of weeds, both in classical importation projects and as microbial herbicides (see Charudattan, 1985).

Other organisms have been little utilized but may offer considerable potential in some cases. Carp have been used for control of aquatic weeds, and snails may offer promise in this regard. Even the manatee or sea cow has been considered for this purpose. Ducks, geese and goats have also been used. Mites and nematodes have shown possibilities, and parasitic plants such as dodders and mistletoes have been considered, along with so-called allelopathic crop plants, which may release chemicals that are toxic to certain weeds.

For a comprehensive review see Rosenthal, Maddox & Brunetti (1985).

3

Biological control ecology

Biological control is a part of the broader overall phenomenon of natural control. Natural control may be defined as the regulation of populations within certain more or less regular upper and lower limits over periods of time by any one or any combination of natural factors. Such factors have sometimes been classed into two groups, biotic (living) and abiotic (non-living). However, rarely, if ever, do they act alone, although any one may be the key regulatory factor largely responsible for a particular density in a given situation. In the following discussion natural and biological control of insects are stressed, but the principles involved apply broadly to all organisms.

Natural control

Practically all living organisms produce more progeny than would be required to maintain steady population densities. In other words, in any given species there is on average in each generation, initially, a surplus of progeny over parents, often a great surplus. Their populations therefore have an inherent *biotic potential* for unlimited, exponential growth. Yet, quite obviously, the populations of living organisms do not go on increasing indefinitely. To the contrary, they are usually rather stable and maintain characteristic densities which, although fluctuating between certain upper and lower limits, do not change much in a given ecosystem over long periods. This state of balance, or dynamic equilibrium, is produced by the interaction of two opposing natural forces: the biotic potential of the species on the one hand, and *environmental resistance* on the other. The latter term refers to the factors of natural control, that is, all the environmental factors that may act to limit population growth, whether through reducing natality or through increasing mortality or dispersal.

The most important factors in natural control are: (1) natural enemies (parasites, predators and pathogens), (2) weather and other physical factors, (3) food (quantity and quality), (4) interspecific competition (other than natural enemies), (5) intraspecific competition, and (6) spatial or territorial requirements. Population movement – emigration and immigration – is a complicating factor applying to some organisms which we shall not deal with.

Over the years there has been considerable disagreement and misunderstanding between two so-called schools of thought, wherein certain ecologists have stressed population regulation by weather and others have stressed regulation by natural enemies. Perhaps a third group that stresses competition should be distinguished. Actually, in the broad sense, competition between individuals of the same or different species for some common requisite or other is generally involved in all regulatory processes, whether they entail weather, natural enemies, food or spatial needs. All of the factors mentioned are important, and all are capable of regulating the population density of an organism. However, their mode of action and interaction is not the same, and the degree of natural control of a given species attributable to any one factor may differ markedly between different habitats or even microhabitats.

The main distinction here is that some environmental factors, such as physical factors, are *density-independent*, i.e. the severity of their effect is not directly related to the density of the population upon which they operate; whereas others, such as natural enemies, are *density-dependent*, increasing in intensity and destroying a larger proportion of the population upon which they operate as the density of that population increases, and decreasing in intensity as the density decreases. A severe and widespread hailstorm, for instance, may act in a density-independent manner, destroying a given percentage of individuals in a population regardless of their number. Thus, if 50 percent of the individuals in a low-density population are killed by severe hail, then approximately 50 percent will be killed by the same severity of hail in a high-density population, assuming other conditions are equal. The effect of food quantity, on the other hand, may act in a density-dependent manner: a given amount of available food in a given habitat may be sufficient for a small population, but as the density of that population increases, the amount of food available to each individual is reduced, until food shortage becomes a factor limiting further population increase. In other words, the quantity of available food may become a limiting factor only when the population exceeds a certain density threshold, and its severity increases as the density continues to increase beyond that level. As we shall see, effective natural enemies operate in a similar density-dependent manner.

Both density-independent and density-dependent processes are important factors in environmental resistance. However, their role in *reducing* population densities should not be confused with their potential role in their *regulation*, i.e. in the maintenance of a dynamic equilibrium. Only a directly density-dependent factor, capable of responding to changes in population density by increasing its own intensity as the density increases and relieving the pressure as it declines, will be able to effect population regulation, or natural control, *on its own*, as natural enemies do.

Certain other factors, although not directly density-dependent, may act in a density-dependent manner. Weather may produce natural control in the sense defined. This has been demonstrated experimentally with several insects, including both pests and natural enemies (DeBach, 1958*b*). Proof can be obtained by making paired plot comparisons in which other factors, such as natural enemies, are excluded or can be shown to have absolutely no effect in population regulation. For example, starting with a low population density so that competition is not a factor, natural enemies can be eliminated in one plot by an insecticide which is non-toxic to the pest, but not decimated in an adjacent plot. If no appreciable population increase of the pest occurs in either plot over a representative period, one can conclude that natural enemies were inconsequential and what weather therefore was responsible for the observed low population density. Of course, weather must be severe on occasion at least, to have any effect. Otherwise it will be continuously permissive, i.e. more or less ideal or optimum, and the organism will increase until checked by other factors.

The way weather acts to regulate insect population densities is by interacting with the other physical and biotic aspects of a habitat. Severe weather (in the relative sense) operates to restrict the number, size and quality of inhabitable spots in a given habitat, and may produce its results by direct effects on food or shelter or by increasing competition for such requisites. For instance, the higher the temperature and lower the humidity, the fewer are the places on a plant that a phytophagous insect may be able to survive. Therefore, like a density-dependent factor, given weather conditions will kill a higher proportion of a dense population than a sparse one, because in the dense population a greater percentage will be unable to find suitably sheltered microhabitats because the number of sheltered sites tends to be fixed. But weather in itself is not density-dependent. What we have here is a density-independent factor – weather conditions – combining with a density-dependent factor – competition for shelters – to act in a regulatory manner.

Of course, weather or other physical factors always act ultimately to determine the limits of distribution of an organism. There are certain tolerable limits for each species. If weather conditions in a given area are generally unfavorable to a particular species, it may not be able to maintain a continuous existence there. Within an organism's range of distribution, weather determines in any given habitat the *ultimate* basic limits within which certain 'average' but fluctuating population densities may be obtained. Such natural control may occur at very high or at very low densities, because natural control does not involve any characteristic density that is inherent for a species.

If weather is permissive or optimal and natural enemies absent or ineffective, than ultimately an insect population will be regulated by the quantity of food or other requisites interacting with interspecific and/or intraspecific competition. However, the last three factors usually do not control insect pest populations at sufficiently low levels to be economically satisfactory, hence are not generally important factors in applied insect ecology. In other words, if a plant-feeding insect becomes sufficiently abundant to reduce or destroy its food supply, thus limiting its further increase, this would be a case of natural control of interest to the basic ecologist but not to the farmer or applied entomologist, except that they would seek additional, more effective means of regulation.

Thus, it seems obvious that in agricultural crops, in forests and other habitats of economic significance to man, if weather is generally favorable for a pest, *natural control of an economically satisfactory nature must generally derive from the actions of natural enemies.*

Biological control and population phenomena

Biological control in an ecological sense can be defined as the regulation *by natural enemies* of another organism's population density at a lower average than would otherwise occur. Fluctuations about the mean are expected, and may either be marked or essentially unnoticeable. This falls within the definition of natural control and, in that case, does not attempt to prescribe the population densities involved – only that they be lower than would be true if the responsible factors were withdrawn or made ineffective. Thus, this definition does not cover the degree of biological control in the economic sense, which must be stated separately in relation to any particular pest organism. Nor does the definition necessarily imply any activity by man, although man may import and establish new natural enemies and thereby achieve biological control. However, a vast amount of biological control is natural, as will be shown in the next section. To go a step further, applied biological control can be achieved in differing degrees of economic importance, which have been distinguished as partial, substantial or complete. These qualifications can also be applied to natural biological control of potential pests. For instance, if the natural enemies of a potential pest under complete biological control are killed off by insecticides, a great upset will occur; if the potential pest is under partial biological control, only a mild upset may occur.

Thus far we have talked about the fact of biological control and what it is, but not how the remarkable interactions which bring it about occur. Scientists engaged in research on biological control have contributed a great amount to an understanding of how it occurs, as well as to other theories and principles of population ecology.

The mechanisms of regulation by insect natural enemies, and of climate by comparison, were visualized partially by Woodworth (1908), by Howard & Fiske (1911) in connection with the work on the gypsy moth parasites, by F. Muir working with parasites and predators in Hawaii within the next decade or so, then, and most importantly, by A. J. Nicholson in Australia and H. S. Smith in California mainly in the early 1930s. Professor Smith developed the important concept of density-dependence to explain how predator–prey or host–parasite systems interact to produce biological control. Carl B. Huffaker in recent years has done much to clarify this concept and bring together opposing views. This subject is covered in some detail by Huffaker, Luck & Messenger (1977), Huffaker & Messenger (1964a, 1964b), Huffaker, Messenger & DeBach (1971), and Huffaker & Rabb (1984).

Effective natural enemies are able to regulate the prey or host population* because they act in a density-dependent manner. Their mortality-causing actions intensify as the host or prey population increases and are relaxed as the host population density falls, so that an increase in host population beyond a characteristic high is prevented and

---

* In this discussion predator, parasite and natural enemy are often used interchangeably, as are prey and host.

a decrease to extinction likewise is prevented (except perhaps locally on occasion). A negative feedback process between population density and rate of increase is involved. This reciprocal interaction also results in regulation of the enemy's own population, because as it reduces the host population during one phase of the cycle it necessarily reduces its own population. Then, as its repressive pressure reaches a minimum and the host population again increases, the enemy population correspondingly is enabled to increase until it again overtakes the host and the cycle is repeated.

These corresponding changes in the natural enemy's population density are referred to in the biological control literature as 'numerical response'. Such response results from the effect of prey availability on the reproduction of natural enemies, and may be enhanced by the tendency of many natural enemies to aggregate in the sites where their prey is most abundant. In addition, part of the density-dependent action of natural enemies derives from their so-called 'functional response': as prey density increases, each individual natural enemy spends less time searching for it, thereby attacking more prey during its lifetime. The opposite occurs when prey becomes scarce.

Such reciprocal interaction results in the achievement of a typical average population density or 'balance' in a given habitat or area. Although the concept of density-dependence explains how natural enemies can regulate host population, it does not explain at what density they will do so. The balance may occur at high, low or intermediate population densities. The determination of this rests primarily with the inherent characteristics of the natural enemy, and secondarily with how these innate capabilities may be restricted or reduced by adverse environmental conditions. Characteristics of the host may also be modifying.

All natural enemies having an appreciable degree of prey specificity will exhibit density-dependence with respect to that prey. A natural enemy that is highly density-dependent can regulate the prey (host) population density at very low levels, whereas one that is weakly density-dependent will only do so at high levels. Very closely related species may have quite different capabilities in this regard. This is another way of describing an effective enemy as compared to an ineffective one. The difference is that the effective enemy responds very rapidly to any tendency of the prey (host) population to increase, so that its relative reproductive rate quickly increases with respect to the prey and it overtakes and reverses the trend before any important increase has occurred. An ineffective enemy responds slowly and shows much greater lag effects, thus permitting the prey to achieve much higher maxima and mean population densities.

The degree of density-dependence or efficiency of any enemy is linked to its searching ability, more so than anything else. Only an enemy that has a high searching ability can find prey when they are scarce, and this is absolutely necessary if it is to be able to regulate the prey population at low levels. It also is able to respond rapidly by finding proportionately more prey as the prey population tends to increase. Poor searchers can never do so.

In addition to searching ability, the qualities most important to the effectiveness of a natural enemy in prey population regulation are: (1) a high degree of prey (host) specificity, (2) a high reproductive capacity with respect to the prey, and (3) good adaptation to, and tolerance of, as broad a range of environmental conditions as the prey. Item (1), high prey specificity, is strongly correlated with nearly all natural enemies that have produced outstanding results in biological control. Items (2) and (3) must be favorable to the natural enemy in order for it to be highly effective, but they can also be possessed by a poor searcher and they alone will not make an efficient enemy of it. A review of these and other attributes of natural enemies was made by Rosen & Huffaker (1983).

Relatively little is known about the actual details of searching and prey-selection behavior in natural enemies. Whereas searching by a population of natural enemies is random, that of an individual enemy may be rather directional, at least at close range. It evidently responds to various physical and chemical cues (the latter are known as semiochemicals, or behavior-modifying chemicals) which may emanate from the prey itself, or from the host plant or substrate, or from their interaction (e.g. from signs of feeding by the prey). The process of prey selection comprises four steps: (1) habitat selection – seeking a certain environment, a given host plant, etc., (2) prey finding – locating prey individuals in that habitat, (3) prey acceptance – close examination leading to attack upon, or rejection of, a prey individual, and (4) prey suitability – a passive step as far as the natural enemy is concerned, which may nevertheless determine the final outcome of the interaction. Detailed reviews of these subjects are found in Nordlund, Jones & Lewis (1981) and Vinson (1985).

Thus far we do not know how to measure with any accuracy the searching ability of a natural enemy, or its potential effectiveness, except by the effect it has in prey population regulation. In other words, it must be tested or evaluated in the field. Laboratory studies have done no more than give an idea of some of the processes involved in searching, and none have yet explained why, of two closely related or even sibling species of natural enemies, one is efficient in prey population regulation and the other poor. More often than not, detailed laboratory studies of the bio-ecological characteristics of newly imported natural enemies have led to predictions of future field results that were the opposite of those obtained. We can get a good idea of the degree of prey specificity from field and laboratory studies and of potential reproductive capacity in the laboratory, as well as a fair idea – but often misleading – of adaptability to environmental conditions. How to put everything together to choose a potentially effective enemy *beforehand* with certainty is still beyond our grasp.

Additionally, even if a natural enemy possesses all the inherent qualities to be effective in prey population regulation, adverse environmental conditions can so reduce its potential as to transform it into an ineffective enemy. Weather extremes can do this and, as we have seen in Chapter 1, so can insecticides. Also shortage of requisites such as food (honeydew, pollen, nectar, etc.) or free water for adult parasites or predators, or of alternate prey or host plants, can produce the same effects. What happens is that the enemy's density-dependent response is weakened, so that it has a lessened ability to cause an increasing rate of mortality of the prey population as prey population density tends to increase. This results in the prey population achieving 'equilibrium' or 'balance' at a higher average density than otherwise would be the case.

We can perhaps envision this better from a simple numerical example. Assume that a given prey species in a particular, climatically ideal habitat, requires 90 percent mortality of the immature stages, on average, in each generation to remain in equilibrium. This would apply if the species possessed a 50:50 sex ratio and the females consistently produced 20 eggs each. Thus, starting at any point in time at an equilibrium density of 200 prey adults (100 females), 2000 eggs would be laid (100 females × 20 eggs each) which, subjected to a total of 90 percent mortality from parasites and other factors through the egg, larval and pupal stages would result in the survival of 200 prey adults (100 females and 100 males) in the next generation. The same would be repeated in each subsequent generation to maintain equilibrium.

However, if instead we arbitrarily started with an equilibrium density of 2000 adults and assume no other changes, the same 90 percent mortality would maintain equilibrium. This makes it clear that a certain total percentage mortality is necessary, on average, to maintain the equilibrium position of a given species in a particular habitat, and that the characteristics of that species, as finally expressed by its actual birth rate, determine the level of mortality needed to maintain balance, i.e. mortality must equal natality. It is also evident from the example that percentage mortality does not determine the *equilibrium position* or average density, but only that equilibrium will occur at *some* density if an average mortality that equals natality occurs generation after generation.

What then does determine the average population density? Broadly, and as already pointed out, it is the degree of density response of the environmental resistance factors. If, for example, the mortality factors involved in the preceding example are exclusively parasites, and the parasites were adversely affected by a spray, beginning in a given host generation, so that they then caused only 80 percent host mortality instead of 90, the adult host population would double in the next generation to 4000. If, then, as the spray dissipated, the parasites recovered quickly and increased their progeny production in the next generation in response to the greater host density so that they caused 95 percent parasitization, the adult host population would again return to 2000 at the beginning of the next generation. Thereafter, the parasites would produce 90 percent mortality and the system would remain in balance, given that no more spraying occurred. The numbers would be as follows: original generation, 2000 host adults → 20 000 progeny which are 80 percent parasitized, giving in generation 1, 4000 host adults → 40 000 progeny which are 95 percent parasitized, giving in generation 2, 2000 host adults. The fluctuation in host adults would be 2000 → 4000 → 2000. This would be an example of high density-dependence of a parasite, as expressed by a high relative rate of increase of host mortality with respect to increasing host density. In another case, consider that the parasite was affected by the spray for a longer period, so that its response to increasing host density was slowed after its parasitization was first reduced to 80 percent and the next host generation became doubled, so that parasitization increased in the next generation only to 90 percent. The adult host population therefore would remain at 4000. Only an increase to 95 percent in the third generation would reduce the adult host population again to 2000, but meanwhile a higher average host density would have been achieved for the period in question. The numbers would be as follows: original generation, 2000 host adults → 20 000 progeny which are 80 percent parasitized, giving in generation 1, 4000 host adults → 40 000 progeny which are 90 percent parasitized, giving in generation 2, 4000 host adults → 40 000 progeny which are 95 percent parasitized, giving in generation 3, 2000 host adults. The fluctuation in host adults would be 2000 → 4000 → 4000 → 2000 for a total of 12 000 adults during the four generations, or a mean density of 3000. In the first case only 10 000 adults would be produced during the same four generations, for a mean density of 2500. Thus we see how insecticides or other adverse factors can make a poor natural enemy out of an efficient one, merely by reducing its ability to respond rapidly to increases in host density. If we visualize the two preceding cases as representing different parasite species attacking the same host in different habitats, it is clear that the relative ability of the two species to cause increasing rates of host mortality with respect to increasing host density would determine the average host population density.

The theory of *r*- and *K*- selection has attracted some attention among biological control workers. As proposed by MacArthur and Wilson (1967), this theory attempts to explain the process taking place during the colonization of oceanic islands. In the early phases of colonization, when the environment abounds with unexploited resources, so-called *r*-selected species, exhibiting the highest rates of increase and capable of rapid resource exploitation, tend to be dominant (*r* denotes the potential rate of increase in ecological equations). In later phases, as the environment becomes more and more saturated and competition for limited resources increases, the *K*-selected species become dominant as these are better, more highly specialized competitors, utilizing resources more efficiently and producing fewer progeny. (*K* stands for the carrying capacity of the environment.) Colonization of natural enemies in a new habitat is in many ways similar to colonization of an island, and similar successions may indeed have taken place in certain biological control projects. This seems to have been the case with the successful biological control of the Florida red scale, *Chrysomphalus aonidum*, on citrus in Israel (see Chapter 6).

This theory is of much ecological interest. However, some authors have suggested that the 'evolutionary strategy' of natural enemies serve as the basis for importation policy, and that only '*r*-strategists' be employed in biological control (Force, 1974). We cannot accept this suggestion. The distinction is not clear-cut and is open to misinterpretation, with most species probably exhibiting various combinations of r- and K-selected traits. Selecting an r-strategist from among a complex of natural enemies would probably be at least as difficult as selecting the best searcher. Moreover, both types of traits evidently contribute to the density-dependence of natural enemies, and both would be advantageous in different phases of a biological control program, or in different habitats.

Many population phenomena have been illustrated by biological control research, some coming as challenges to popular ideas. Thus, a numerically rare enemy species is not necessarily an ineffective one. The best natural enemies are often the rarest, because they regulate prey populations at low levels and thereby regulate their own at even lower levels. By doing so they tend to eliminate competitors which can only exist at high prey densities. By the same token, some of the commonest, most numerous, parasites or predators are the least effective in prey population regulation. With abundant prey they are enabled to remain abundant. Weather interactions may modify otherwise expected effects. If the weather is suboptimal for the prey but even more so for a potentially effective enemy, the prey may escape regulation and become abundant, while the enemy remains less common. Thus, both would be more common than they would be in a climatic zone optimal for both, where the enemy could regulate at low densities. In the latter instances we have the anomaly of scarcity being correlated with ideal weather conditions.

A few ecologists have propounded the hypothesis that species have so evolved to fit their environment that this results in a certain characteristic population density. There is, of course, usually a characteristic average population density in a given habitat – this is natural control. But there is no inherent characteristic density for a species. All of the outstanding examples of applied biological control refute this, because by modifying the habitat, i.e. by introducing and establishing a new mortality factor, natural enemies, the equilibrium population density of numerous pest species has been greatly reduced. It is easy, in fact, by appropriate experimental techniques to maintain different population densities of a phytophagous species on adjacent plants or, with small more-or-less sessile species, on different portions of the same plant, or even on different parts of the same leaf. This is done by modifying the microhabitat either to favor the prey, as for instance by excluding or inhibiting enemies, or perhaps by favoring enemies by provision of necessary adult food which otherwise might be lacking. Of course, all the cases of 'upsets' of equilibrium of non-target species by insecticidal interference in crops further demonstrate that it is the properties of the environment, both biotic and abiotic, that determine the population density of a species. Thus, there are no organisms that are inherent pests.

Additionally, some workers have emphasized genetic feedback in the determination of population densities, i.e. that selection for better adaptation may change the density at which the populations are regulated. We see this possibility as an evolutionary process but not as a regulatory process (Huffaker, Messenger & DeBach, 1971, pp. 56–61).

The important ecological principle of competitive displacement between ecological homologs (i.e. species having identical requisites, such as different parasite species requiring the same stage of the same host species) has probably been more thoroughly demonstrated or proven with cases of natural enemies seeking to utilize the same identical host, than with any other group of organisms. Studies with scale insect parasites (DeBach & Sundby, 1963) showed that ecological homologs cannot coexist, rather that one species will eliminate the other but the food (host or prey) must be *identical*, even to the point of its being the same larval stage, or the egg or pupal stage. It was further shown that species whose food differs slightly can coexist, thus egg or larval or pupal parasites of the same host species do coexist, and in effect do not compete as ecological homologs. This has an important bearing on biological control importation policy.

When individual larvae of two species of parasites occur simultaneously in the same individual host insect, this is known as multiple parasitism in the biological control literature, and such parasites are ecological homologs and cannot coexist in the same habitat unless they possess different alternative hosts or there is some other modifying factor. Concern among biological control research workers about multiple parasitism and the associated displacement of one parasite species by another began in the 1920s. They thought that an efficient parasite might be displaced by the introduction of one less effective in host population regulation. This idea has been repeated by some until recently, but with rare theoretical exceptions all data show that if one parasite species displaces another, the winner is the one more efficient in host population regulation, and better control will result. Even during the highly competitive process of displacement, no ill effects on host population regulation will occur. If a second parasite species is imported that attacks a different host stage and becomes established, it will add to the effectiveness of control, as was independently shown deductively in the late 1920s and early 1930s by H. S. Smith and A. J. Nicholson, and has been demonstrated conclusively in various biological control projects.

Many, probably most, ecologists subscribe to the idea that species diversity is highly correlated with community population stability, that is with natural control, and this includes biological control. As a consequence, it is commonly held that this principle applies to biological control of single-species populations. In fact, L. O. Howard and W. E. Fiske propounded the sequence theory more than 70 years ago which suggested that a sequence of parasite species including egg, larval and pupal parasites was probably most likely to achieve satisfactory biological control. Actually, this has proven to be true in only a few cases of outstanding applied biological control, where the prey has been reduced to consistently low population densities. The opposite is true in the great majority of highly successful cases, in which one highly specific natural enemy species acting alone, or two acting in combination, have been responsible rather than several or many. As has been previously pointed out, a diversity of several established enemy species has often been replaced by a single newly imported, highly effective one, with the result that better prey population regulation is obtained. Thus, the theory of biological control must include the precept that, in general, increasing species diversity of natural enemies per given prey species is inversely correlated with regulation of a prey population at low levels. In other words, we should seek to *reduce* the number of natural enemy species established against a given prey species by introducing a highly specific, more effective enemy, which is capable of replacing them and as a result producing better biological control. (We should like to emphasize here, however, that inasmuch as such an effective enemy cannot be recognized beforehand, the only practical way would be to import and release as many enemies as possible, in the hope that it might prove to be among them.) If we pursue this thought a little further, one might begin to wonder whether species diversity in a community, especially with insects, might be the *result* of many separate, more or less independent, cases of natural biological control, rather than being the *cause* of community stability. When many species are held under biological control in a community, this would conserve the plant food resources, thereby enabling more species to utilize it.

Plant diversity also is generally held by ecologists to contribute to stability of phytophagous insect populations by providing better environmental conditions, including alternative prey as well as better chronological continuity of prey for entomophagous forms. A good many examples have been cited in the literature, where diversification of crops, as opposed to monoculture, has resulted in achievement of better biological control or better integrated control. (See Altieri & Letourneau, 1982, for a recent review.) However, it seems that one should not generalize regarding the advantages of crop diversification in enhancement of biological control. An interesting study by Robinson, Young & Morrison (1972) has shown that, where corn, alfalfa or peanuts were planted on either side of cotton, there was greater damage by the bollworm in the cotton compared to a check having no other crop planted on either side. Only when sorghum was planted adjacent to cotton was there any indication that bollworm damage in the cotton might be reduced over the monoculture check. It is interesting that the adjacent planting of corn resulted in the greatest bollworm damage and the least seed cotton, because corn is commonly used as a 'trap crop' for the bollworm. The cotton with sorghum and the check produced the most seed cotton, and the cotton with sorghum had the most predators. Various other examples could be cited, showing increases in certain pest problems to be correlated with diversification of crops. Thus, each habitat must be viewed separately and the best procedures for each case determined by appropriate experimental and observational studies.

One very important aspect of biological control ecology concerns the means of evaluating the relative and absolute importance of factors responsible for prey population regulation. Indeed, this is a critical problem in all population ecology. We shall cover the subject of means of evaluation of effectiveness of natural enemies in some detail in Chapter 7, so will only comment briefly here.

There has been much controversy, and doubtless it will continue for some time, as to how best to measure the impact of environmental factors on populations and to determine which factors, or groups, are regulatory and which are mainly, or entirely, responsible for an observed degree of natural control or biological control. A substantial number of ecologists have attempted to do this quantitatively by correlation analysis, utilizing periodic censuses and life-table data, and thereby determine the 'key factor'. We and others (DeBach, Huffaker & MacPhee, 1976; Huffaker & Gutierrez, 1990) have tried these approaches but have concluded that, used alone, they are quite inadequate for the job described above. In practice it appears impossible to obtain the highly accurate and detailed mortality data necessary to be statistically meaningful, and in any case a positive correlation between, say, prey and predator population trends will not show which is cause and which is effect. Modeling methods and regression analysis, however, can be of considerable value in understanding other aspects of the problem. (See, for instance, Hassell & Waage, 1984.)

Experimental 'check methods' appear to be the only satisfactory means of rating the efficiency of natural enemies, or of weather and other factors for that matter. These involve paired-plot comparisons of a suitable size and number of replicates to fit the occasion. One series of unmodified plots serve as the check, whereas the other series have one parameter, such as natural enemies, excluded or inhibited *without* disturbing anything else or otherwise modifying the microhabitat. If no subsequent change in prey population density occurs in the two series, it can be concluded that natural enemies were inconsequential and non-regulatory. If, on the other hand, the prey population rises to a new average density, this is proof that the natural enemies were responsible for the original observed density. This method isolates particular parameters and shows what their total effect is, but not how they produce this effect. The method can just as well measure the effect of climate, by isolating or removing other regulatory factors which are nearly always natural enemies, if we select insects occurring at reasonably low densities so that food shortage is not a factor. If we then exclude natural enemies from the system and no change in density occurs, it is climate that is responsible for maintenance of the originally observed density.

The differences in the philosophy and ecological basis of biological control and chemical control have been touched on before in Chapter 1. From an ecological standpoint, classical biological control is aimed at permanently lowering the equilibrium position of a pest, whereas chemical control is aimed at the population peaks. As far as man is concerned, biological control has no ecological drawbacks that we recognize, whereas chemical control has many. A comparison is shown in Table 3.1.

The extent of natural biological control

This aspect of biological control cannot be overemphasized. Upon it rests our entire ability to successfully grow crops, because without it the *potential* pests would overwhelm us. It is variously estimated that from less than one percent to perhaps two percent of phytophagous insects that are potential pests ever achieve the status of

Table 3.1. *A Comparison of biological and chemical control: the disadvantages.*

Category	Biological control	Chemical control
Environmental pollution; danger to man, wildlife, other non-target organisms, soil, etc.	None	Considerable
Upsets in natural balance and other ecological disruptions	None	Common
Permanency of control	Permanent	Temporary – must repeat one to many times annually
Development of resistance to the mortality factor	Extremely rare, if ever	Common, rapidly increasing
General applicability to broad-spectrum pest control	Theoretically unlimited but not expected to apply to all pests. Still underdeveloped. Initial control may take 1–2 years but then pest remains reduced	Applies empirically to nearly all insects but not satisfactory with some. Can rapidly reduce outbreaks but they rebound. Psychologically satisfying to the user at first

even minor pests. By and large, this is due to natural enemies. However, only field ecologists or naturalists who have made detailed studies can fully appreciate the fact. Most of the work of natural enemies goes unnoticed, but especially so where a non-economic insect is concerned. Enough cases of highly effective natural biological control have been studied, however, to indicate that the figure of 99 percent or more potential pests being under natural control is close to the truth. At a Conference of Experts in Biological Control in Canada as early as 1936, it was stated that only about one percent of insects became pests, and the 1968 *Bulletin of the Entomological Society of America* carried figures showing that only 1425 out of 85 000, or 1.7 percent, of known United States and Canadian insects were important enough to have common names, i.e. in general to have achieved some sort of pest status, although some beneficial insects also are included in the list. Dr J. D. Torthill, famous for his achievements in biological control in Fiji, considered it to be a well-established fact that virtually all insects in their native homes are controlled by natural enemies, and L. O. Howard and W. F. Fiske, the former, the Chief of the United States Department of Agriculture, Bureau of Entomology, published in 1911 that 99 percent of phytophagous insects were under natural biological control. The famous French entomologist, Professor Paul Marchal, wrote that if phytophagous insects could multiply without hindrance in proportion to their natural reproductive powers, in a short time they could eliminate all species of terrestrial vegetation. However, he stated that the increase of the destructive insects is fortunately kept within limits by parasitic and predatory insects.

We concur that these estimates are substantially accurate. We have been amazed during many years of field research at the vast amount of natural biological control that occurs. However, one virtually has to live with the insects to see or to measure it, unless it is brought into brilliant focus by chemically induced upsets, which are so common now as to further emphasize the extent of natural biological control. Yet another form of proof of the natural occurrence of biological control comes from the mounting number of successful cases of applied biological control resulting from importation of enemies, because these represent cases of natural control in the original home.

The ratio of actual pests to potential pests under natural biological control can be compared to an iceberg. The tip of the iceberg (like our pests) may appear very large, but it represents only a small fraction of the great bulk (represented by natural control) which is hidden from view and hardly believable even to the more than casual observer. Viewed ecologically, this is a tremendous asset which obviously must be preserved. In our efforts to control the approximately one percent of insects that are pests, we must make every effort to conserve the natural enemies of the estimated 99 percent that are under natural biological control.

Even a cross-section list of some selected, documented cases of natural control would be unwieldy. A fairly detailed coverage of some cases in North America will be found in articles by Hagen, van den Bosch & Dahlsten (1971), Rabb (1971) and MacPhee & MacLellan (1971). A worldwide list, wherein over 25 species of predatory mites are considered to be important predators of many phytophagous mites on a variety of crops, is given by McMurtry, Huffaker & van den Vrie (1970); additional records are listed by McMurtry (1982). Also, a worldwide record of natural biological control of phytophagous mites is given by Huffaker, van den Vrie & McMurtry (1970). A multitude of other cases is scattered throughout the literature, which we shall not attempt to review. Although most of the cases documenting natural control by enemies are based on careful quantitative studies generally backed by expert experience, relatively few have been subjected to proof by the use of experimental check methods. It may come as a surprise that the majority of natural control cases studied and demonstrated have been with 'so-called' pests. This is because there has been more motivation and research money available. However, either they were not actual pests when studied under undisturbed conditions of no chemical applications or they were only occasional pests normally under biological control for lengthy periods of time, or else they were pests in one area but under biological control in others.

The San Jose scale (*Quadraspidiotus perniciosus*) is a good example. A major pest of deciduous fruit trees in North America, Europe and elsewhere, it is usually rare and of little economic importance in southern California. Careful studies, including check-method experiments, have demonstrated that this is due to natural biological control, mainly by the ectoparasite *Aphytis aonidiae*. (See Gulmahamad & DeBach, 1978.)

Some idea of the generality of natural biological control can be had by mentally extrapolating from the number of cases of naturally occurring biological control personally documented by the authors. This is based on field studies mainly involving homopterous insects on citrus in California, Israel and abroad. In California, most of these cases have been proved by use of experimental check methods (often replicated in different localities and by the use of different check methods). Abroad, we have relied on experience, observations of chemical upsets of non-target organisms, the observed quantitative actions of natural enemies with respect to host density in the field, and the use of naturally-occurring experimental check comparisons, such as one tree with honeydew-seeking ants (which interfere with enemies) compared to another without ants, or to natural microexclusion 'cages' or other protected situations, as spider webbing on certain leaves, which will exclude parasites from scales or mealybugs. Table 3.2 does not include cases which might have been judged to represent natural biological control based upon experience but the observation or information was insufficiently thorough.

One might ask the question: if natural biological control is so

Table 3.2. *Cases of natural biological control documented by the authors.**

Species	Common name	Host plant	Country or area	Type of natural enemy
Acaudaleyrodes citri	Citrus black aleyrodid	Citrus	Israel	Parasites
Aleurocanthus woglumi	Citrus blackfly	Citrus	India and Orient	Parasites
Aleurothrixus floccosus	Woolly whitefly	Citrus	Mexico, Brazil	Parasites
Aonidiella aurantii	California red scale	Citrus	Japan to India	Parasites
Aonidiella citrina	Yellow scale	Citrus	Japan to Pakistan	Parasites
Aspidiotus nerii	Oleander scale	Olive, lemons, ornamentals	Cosmopolitan	Parasites
Chrysomphalus aonidum	Florida red scale	Citrus	Hong Kong and Orient	Parasites
Coccus hesperidum	Brown soft scale	Citrus, ornamentals	Cosmopolitan	Parasites
Coccus viridis	Green scale	Citrus	Orient	Parasites
Dialeurodes citri	Citrus whitefly	Citrus	Japan to Pakistan	Parasites
Diaspis echinocacti	Cactus scale	Cacti	California, Mexico	Parasites
Diaspis simmondsiae	Jojoba scale	Jojoba	Mexico	Parasites
Ferrisia virgata	Striped mealybug	Citrus, avocado, ornamentals	Mexico	Parasites
Hemiberlesia lataniae	Latania scale	Avocado ornamentals	California, Israel, Greece	Parasites
Hemiberlesia rapax	Greedy scale	Citrus, ornamentals	California, Israel, Greece	Parasites
Lepidosaphes beckii	Purple scale	Citrus	Hong Kong to India	Parasites
Panonychus citri	Citrus red mite	Citrus	California	Predators
Parlatoria oleae	Olive scale	Olive	Pakistan, Mid-East	Parasites
Planococcus citri	Citrus mealybug	Citrus	Cosmopolitan	Parasites and predators
Pseudococcus longispinus	Long-tailed mealybug	Citrus, avocado, ornamentals	Cosmopolitan	Parasites and predators

Table 3.2. (*cont.*)

Species	Common name	Host plant	Country or area	Type of natural enemy
Quadraspidiotus perniciosus	San Jose scale	Deciduous fruit trees, ornamentals	California	Parasites
Toxoptera aurantii	Black citrus aphid	Citrus	California, Israel	Parasites and predators
Unaspis citri	Snow scale	Citrus	Hong Kong	Parasites

*Certain of these have also been documented or confirmed by others.

commonplace, why don't we hear more about it? Why is it so overlooked, even by many professional entomologists and ecologists? Perhaps the main reason is that in applied research, most scientists are paid to work on pests and they rarely have had the opportunity or motivation to study natural control. Happily this is changing, but the great population studies conducted on forest insects in Canada, Europe and elsewhere in recent years furnish a case in point. The stress has been on the study of periodic major outbreaks or epizootics and the factors involved in their diminution, but little emphasis has been placed on their population dynamics during the long non-outbreak periods of endemicity, or on other related non-outbreak species. Whatever meager information is available about natural control in forests has derived mainly from occasional studies of disturbed situations, such as pesticide-induced upsets (Turnock *et al.*, 1976). In crops, the stress has naturally been on the major pests, i.e. ones that often lack effective established enemies. As a result, such studies have emphasized the population dynamics of the few insects and mites that lack effective natural enemies or whose enemies are periodically depressed by unusual conditions, but these very studies have been most used to furnish generalizations regarding population theory and the presumed paucity of effective biological control. Attention was called earlier to some of the weaknesses in evaluating the role of natural enemies, especially in crops, but this applies to a considerable extent to forests and other natural habitats. The general lack of adequate experimental evaluation techniques, and the reliance upon quantitative methods of correlation analysis, do not permit an accurate interpretation of the role of enemies.

Professor Alvah Peterson, late of Ohio State University, covered these points very succinctly as long ago as 1936, as follows:

'I sometimes wonder if most entomologists, including myself, have not approached the question of biological control from a too narrow and restricted point of view. For the most part, all of our biological control studies today have been directed toward and with insects of economic importance. All of us know the reasons for this situation. Has this approach distorted our view or concept of the entire biological control picture? Are we not restricting ourselves too closely when we work with so-called economic species only? By so doing, are we not attempting to develop biological control principles and theories from insect hosts that are more successful in their environment than many closely related species which are held in check by natural enemies or environmental factors?

Briefly stated, it may be said that many economic insects are misfits so far as the average balance of nature is concerned. Economic pests frequently develop in numbers and extent far beyond what usually takes place in nature. In other words, as investigators we are studying for the most part unsuccessful cases of biological control.

If we consider the numerical relationship of economic species to all other insects in almost any geographical area, a very conservative estimate will show less than one insect pest to one hundred non-destructive species. As economic entomologists, we devote all our time to one percent or less of all insects.

Supposing in our biological control studies we reversed the picture and devoted some or all of our time to some of the 99 percent of insects that are apparently unimportant. There is very little doubt in my mind that if we did this we would find many outstanding examples of biological control in nature, and from such studies of naturally successful cases we would learn fundamental principles, and probably learn them more rapidly than from economic pests that are out of control. In simple language, would it not be a wise policy to devote some time studying successful examples of biological control in nature? In other lines of study and in the business field, investigators frequently devote much time analyzing successful ventures, in order to learn principles and gain ideas that will improve less profitable lines.'

Unfortunately, times have not changed to a great extent since Professor Peterson spoke these words.

Fortuitous biological control

Strictly speaking, natural biological control may be limited, by definition, to cases where an indigenous natural enemy controls an indigenous insect. There is another category, however, which has been termed fortuitous biological control, that does not involve any manipulation by man, hence in this sense is also natural biological control. This includes all cases where biological control has occurred as a result of the accidental immigration and establishment (ecesis) of an exotic natural enemy, or conversely of ecesis of an exotic pest which is then attacked and controlled by indigenous enemies. Fortuitous biological control is probably more overlooked than natural biological control *per se*, although the number of such cases must be very large. Thus, out of 271 beneficial immigrant arthrophods established in the United States (including parasites, predators and pollinators), 118 species have been accidentally introduced. Similarly, it has been estimated that the number of accidentally introduced insect parasites of insects established in Canada equals that of colonized species. (See Beirne, 1985; Sailer, 1978.) The results of classical biological control from purposeful importation of enemies are often self-evident. However, equally successful results can occur accidentally and go unheralded, even though their importance in pest control is just as great. We know of no general review of such cases, so again perhaps we can extrapolate and generalize from data on fortuitous biological control acquired during our studies on biological control of scale insects by species of the parasite genus *Aphytis* (DeBach, 1971; Rosen & DeBach, 1979).

The purple scale, *Lepidosaphes beckii*, has been a notorious citrus pest in many countries in past years. Its dominantly important parasite, *Aphytis lepidosaphes*, was discovered and imported in 1948–9 into California from Hong Kong and Taiwan, where it is native. It was not then known to occur elsewhere outside its range in the Orient, but it may possibly have already spread undetected to one or more other countries.

Subsequent to the parasite's successful establishment in California we have sent it to 13 other countries, with establishment being recorded from nine thus far. However, we have found or recorded the parasite to be accidentally established in 14 other countries to which it was not sent or purposely imported, and in nearly all of which it is producing excellent fortuitous biological control. These include such diverse and distant places as Hawaii, Florida, Turkey, Israel, Puerto Rico, Guadaloupe, Jamaica, Australia, New Caledonia, El Salvador and Argentina. In nearly all of these places the presence of the parasite and its good work was unknown until specifically sought out. In Israel, for instance, thorough studies carried out in the 1930s revealed no *Aphytis* parasites attacking the purple scale, yet in the early 1960s *A. lepidosaphes* was widespread and purple scale populations were lower than ever before. These fortuitous cases are just as valuable as if they had been accomplished purposely, and are currently saving millions of dollars a year just with the purple scale alone.

There is so much movement of plant products internationally today, that it is reasonable that sometimes natural enemies accompany their prey or subsequently reach it accidentally. If they move with their prey originally, fortuitous biological control may occur from the beginning and the presence of the *potential* pest may remain undetected. Regarding the accidental movement of parasites, we regularly find live parasitized scale insects on citrus fruit from Texas and Florida in California food markets. We have even started parasite cultures from such. Obviously, establishment of these parasites in the field could easily occur accidentally.

Another parasite studied by us, *Aphytis chrysomphali*, has become nearly cosmopolitan by accidental spread. Its presumed native home area is the Mediterranean basin, but it is now definitely known to occur also in Japan, Hong Kong, Taiwan, the Philippines, Hawaii, Tahiti, New Caledonia, Australia, South Africa, Iran, Argentina, Chile, Brazil, Peru, Panama, Costa Rica, El Salvador, Mexico, the Caribbean, Florida, Texas and California. It attacks the California red scale and others, which have been moved worldwide with their host plants. It has been responsible for appreciable fortuitous biological control in various places. Yet another parasite of the California red scale, *Aphytis lingnanensis*, was dispersed accidentally from its home in the Orient to Mexico and then to southern Texas; originally perhaps to Mexico many years ago via the Manila galleons. It was unknown when we imported it to California from southern China in 1948, and it turned out to be much more effective in biological control of the California red scale than *Aphytis chrysomphali*, yet for unknown years it had occurred at Hermosillo, Sonora, Mexico, only about 500 airline kilometers from the citrus areas of San Diego County of California. A survey throughout all Mexico and the citrus area of Texas, made shortly after the importation into California from China, showed the parasite to be thoroughly established everywhere in those areas and made it clear that the parasite had been in Mexico for many years. It was responsible for fortuitous biological control in most Mexican citrus areas, but its presence was unknown there or in Texas. This species was also reared (but misidentifed) in Queensland, Australia, as early as 1930, and had become widespread there by the early 1970s.

Aphytis chilensis primarily attacks the oleander scale, *Aspidiotus nerii*, usually a minor or non-economic insect, and was also considered to be native to the Mediterranean area. However, a very closely related species occurs only in Australia, so it is possible that *chilensis* also originated there. Be that as it may, it has spread accidentally and is thoroughly established in Europe, the Middle East, North and South Africa, North America, South America, Australia and New Zealand.

Aphytis mytilaspidis, an important parasite of the oystershell scale (*Lepidosapes ulmi*) and the San Jose scale (*Quadraspidiotus perniciosus*) of fruit trees over much of North America, is an accidental import from Europe, even though it was first described from Illinois by William Le Baron in 1870. This parasite probably involves the first written record of a case of fortuitous biological control, although Le Baron (1870) doubtless thought it to be a native parasite at the time. He stated:

> 'The Oyster-shell Bark-louse of the apple tree has, for a number of years past, been gradually disappearing, so that it no longer occupies the rank which it has heretofore so preëminently held, of a first-class noxious insect . . . Already the smoother bark, the greener foliage, and the fairer fruit, proclaim to the orchadist that this deadly insect is loosening its hold upon the apple tree; and many, no doubt, have prided themselves upon the successful application of some infallible wash, or patent nostrum; but underneath all this goodly show, busily intent upon the accomplishment of her own curious economy, and heedless of the momentous results she is effecting in human interest, works unseen our infinitesimal friend, the Apple-tree Bark-louse parasite (*Chalcis* [*Aphelinus*] *mytilaspidis*).'

These only represent part of the records we have for ecesis and fortuitous biological control by species of only one genus. Were the same information available for all parasite and predator genera, it is obvious that a fantastic amount of fortuitous biological control has occurred and is going on unnoticed. These results suggest some interesting conclusions: (1) since fortuitous biological control has occurred accidentally time and again, man could have brought about the result much sooner in each case, thereby providing substantial financial savings in reduced chemical pest control costs in addition to the reduction in environmental pollution; (2) sometimes exotic natural enemies may be more easily and cheaply found and imported from a country where they have been accidentally established than from their native home. For example, the purple scale parasite was in Hawaii and the California red scale parasite was in nearby Mexico, yet California went to much more trouble and expense to seek them out and obtain them from their native home in the Orient.

The ecological basis of classical biological control

Classical biological control seems to be an apt term to cover the applied phase involving the discovery, importation and establishment of exotic natural enemies. Most applied biological control successes have resulted from this approach; in fact, it is probably thought of by many as constituting essentially all there is to biological control. This supposition and other aspects will be dealt with in Chapters 5, 6 and 7.

The broad ecological aspects of the process are relatively simple, and are shown in Fig. 3.1. The basic theory is that the average population density of an organism may be lowered by the introduction of additional density-dependent mortality factors into its environment. In most cases, the target organisms are exotic insects that have invaded a new habitat (or country) and become pests. When such is the case, the assumption is usually correct that essential natural enemies have been left behind. Their discovery abroad, importation and establishment by man constitutes an

Fig.3.1. The introduction of an exotic natural enemy, or any manipulation increasing the efficacy of an existing enemy, may result in complete biological control of a pest by permanently lowering its equilibrium position and population fluctuations below the economic threshold. (From Smith & van den Bosch, 1967; courtesy of Academic Press, Inc.)

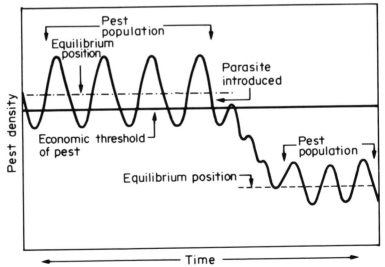

attempt to reduce the pest to a non-economic status, and achievement of success constitutes a case of classical biological control.

The ecology involved applies as well to indigenous (native) pests. Natural enemies obtained from related pests abroad can be imported and established to constitute a new regulatory factor. This was the case with the famous coconut moth in Fiji, which is discussed in Chapter 5. There are a number of other well-documented cases in the literature. Obviously one should not assume, as some do, that native pests are not amenable to classical biological control.

Management of natural enemy populations

The discussion thus far in this chapter of the ecology of natural enemies has related principally to their operation in a more or less optimal undisturbed environment. Often this does not occur in nature or in agro-ecosystems. Also, the only manipulation of enemies by man was that discussed in the last section – importation and establishment of new exotic enemies.

However, the ecology of *established* enemies also can be manipulated to man's advantage in pest control. This can be likened to wildlife management in many respects, and is assuming an important role in biological control. As discussed earlier in this chapter, and as shown by numerous examples of pesticide interference in Chapter 1, inherently effective enemies can be rendered ineffective in prey population regulation by a variety of adverse environmental phenomena. In addition to the effect of insecticides, which is by far the most adverse factor in agro-ecosystems, these include, among others: weather extremes, lack of food or other requisites for adults, lack of alternative prey which are necessary to provide year-around propagation because of periodic shortages of the primary prey, asynchronization of natural enemy and prey life-cycles, disturbing effects of ants on natural enemies, and killing effect of airborne dust. Precise determination of which factors are adverse or which requisites are insufficient, coupled with the formulation of methods of alleviating them, constitute natural enemy management. A most important premise to bear in mind is that the enemy we choose to manipulate should be an *inherently* effective one. Management requires considerable research and expenditure of money, and in general should not be wasted on an enemy which is intrinsically incapable of satisfactory biological control under favorable conditions.

The determination of the impact of adverse environmental conditions on natural enemy efficiency depends on basic bio-ecological studies involving periodic field censuses, life-table construction and analysis, and experimental check methods conducted in adequate-sized plots that have not, or will not, receive chemical pesticidal treatment (if at all possible), preferably for several years. Sufficient time is essential to determine basic undisturbed predator–prey population interactions, as well as the long-term effect of other parameters. Biological, behavioral and even physiological studies in the laboratory will be desirable and necessary to supplement the field information. Only after the adverse factors are pinpointed can the natural enemies themselves, or the environment, be knowledgeably manipulated to favor natural enemies. It is to be expected, however, that in some instances the information obtained will indicate (1) that the natural enemy(ies) is inherently ineffective and manipulation should not be tried, or (2) that the type of manipulation needed is impossible or impracticable to attain. In these cases, major emphasis should be placed on the discovery and importation of new exotic enemies.

The latter recourse turned out to be the case with our project on biological control of California red scale on citrus. A tiny parasite (1 mm in length), *Aphytis lingnanensis*, had been imported from south China and was found to be highly effective in scale population regulation in the mild coastal areas of southern California, but it proved to be much less effective in intermediate areas and generally ineffective in the more climatically extreme interior areas of Riverside and San Bernardino. Basic bio-ecological studies showed that weather extremes, especially low winter temperatures, but also to an extent high summer temperatures combined with low humidities, caused such great periodic mortalities of the parasite that it was rendered ineffective in controlling the host, which was much less affected by the same conditions.

We found many interesting and unsuspected facets of the effect of weather. Prolonged periods of cool winter weather (days below 18 °C (65 °F) maximum for 1 or 2 weeks) proved to be as severe in causing heavy mortality to all stages as several nights of rather heavy frost. It was found that the sperm stored in the spermatheca of the female would be killed by subjection to − 1 °C (30 °F) for eight hours (not uncommon during the winter in Riverside), hence they were sterilized to become 'factitious virgins' and would produce only male progeny. Furthermore, they would not mate again. The effect on males was as bad or worse; all sperm were killed and no regeneration occurred, so that they were permanently sterilized. It should be noted that to all intents and purposes these adults appeared to be normal insects after they were sterilized. Even more surprisingly, it was found that short exposures (24 hours) to temperatures in the 15 °C (60 °F) range altered sex ratios unfavorably by increasing the proportion of males produced and also severely reduced total progeny production. The *effective* progeny production (number of female progeny per female parent) was cut from a normal of 21.4 to only 4.5 by such an exposure – a drastic reduction of about 78 percent. The ability or inability to control the host population because of differences in local weather were found to occur within distances of just a few miles in some instances.

As a result of the knowledge obtained that *Aphytis lingnanensis* was prevented from controlling its host in certain areas by adverse effects of weather, it was reasoned that insectary mass production and periodic colonization of the parasite in the field following unfavorable periods might solve the problem. This was tried and found to be satisfactory and economically feasible in intermediate climatic areas, but results were not consistently reliable in interior areas. Hence, additional natural enemies were sought and discovered in the Orient and one, *Aphytis melinus* (Fig. 2.4, p. 42–43), has very nearly solved the problem because it is inherently better adapted to interior area conditions than was *A. lingnanensis.*

Management of enemies can be divided into two major headings: Augmentation and Conservation. Research and application toward these ends will be covered more fully in Chapters 7 and 10. The periodic colonization technique just mentioned is a type of augmentation, in that the natural enemy itself is manipulated. Another type of augmentation would involve selective breeding for genetic improvement, or the use of chemical cues (kairomones) to enhance natural enemy activity. Conservation involves manipulation of the environment to favor the natural enemy. This includes elimination or mitigation of adverse factors, or provision of necessary requisites that are lacking. Aside from the importation of new natural enemies in classical biological control, conservation is by far the biggest, most important, aspect of biological control. Successful conservation permits natural enemies to operate to their full effectiveness; its lack more often than not denies the possibility of obtaining satisfactory natural or applied biological control.

4

Early naturalists and experiments

The concept and development of biological control necessarily occurred within the framework of the gradual accumulation of biological and ecological knowledge as man's civilization progressed. The history of biological control is the history of early naturalists, biologists and experimental scientists. The information in this chapter has been derived from various sources, but the following especially have been drawn upon: Bodenheimer (1931), Doutt (1964*b*), Essig (1931), Howard & Fiske (1911), Riley (1893), Silvestri (1909) and Steinhaus (1956).

The first use of predators

There were some apparently empirical developments that preceded even the earliest intellectuals and skilled professionals. Probably primitive agriculturists, who usually were surprisingly good naturalists, observed the more obvious insect predators feeding on phytophagous insects. This would be straightforward and understandable, as it would were they to observe a lion preying on a gazelle. It took much more biological sophistication to perceive and understand insect parasitism, as we shall see.

The first record of the use of predatory insects by man is lost in history, but it is known that the ancient Chinese fostered the ant *Oecophylla smaragdina* in their citrus trees to control caterpillars and large boring beetles. This ant builds great paper nests in trees, containing thousands of individuals. These colonies not only could be purchased or else moved from wild trees, but the movement of ants between cultivated trees was engendered by placing bamboo runways from one tree to another. According to the Ministry of Agriculture of the People's Republic of China, the earliest record of this practice was in 324 BC (Coulson *et al.*, 1982). One of the authors (PD) observed it still being continued in the

Shan States of North Burma in the 1950s, and it reportedly continues in China. A similar development was recorded among the Yemenite date growers of Arabia by P. Forskål in 1775. Each year they brought colonies of a predatory ant down from the mountains and put them among their date palms to control pest insects. The use of spiders for pest control in China also dates back 2000 years (Sparks, Ables & Jones, 1982).

One of the first to suggest the use of natural enemies in modern times was no other than Carl Linnaeus, who apparently had given the matter some thought. In a lecture presented in 1752, he made the following explicit statement (see Hörstadius, 1974):

> 'Since people noted the damage done by insects, thought has been given to ways of getting rid of them, but so far nobody has thought of getting rid of insects with insects. Every insect has its predator which follows and destroys it. Such predatory insects should be caught and used for disinfesting crop-plants.'

The first known successful introduction of a natural enemy from one country to another occurred in the eighteenth century, and indicates that the idea of using predators may have been more than a rarity even by then. The red locust, *Nomadacris septemfasciata*, was the most serious pest of agriculture in Mauritius in those days. In order to possibly solve the problem, the mynah bird was introduced from India by the Count de Maudave in 1762. By 1770 it was credited with the successful control of the locust. The predaceous pentatomid bug *Picromerus bidens* of Europe not only is an important predator of various Lepidoptera and other foliage feeders, but has a preference for bedbugs. Its use for biological control of bedbugs was apparently tried as early as 1776. A few specimens placed in an infested room reportedly cleaned out the bedbugs in a matter of weeks.

Observations during the late Renaissance

Meanwhile, the necessary biological basis for the scientific development of modern biological control was accumulating in Europe. It took most of the seventeenth century for learned men to pass from the first vague observations of insect parasitism to an understanding of the process. Silvestri (1909) and Bodenheimer (1931) pointed out that U. Aldrovandi, in 1602, was the first person to publish an observation of insect parasitism. He recorded the exit of the parasitic larvae of *Apanteles glomeratus* from the common cabbage butterfly larva in order to spin their external cocoons, but he mistakenly supposed these to be eggs laid by the caterpillar, and wrote:

> 'Twice have I also observed the cabbage caterpillar laying yellow eggs covered with delicate wool, and afterwards transforming itself into a yellowish pupa, marked with green and black. What appeared peculiar to me was that from these eggs emerged small winged animalcules, so small that they could barely be seen . . .'

Later, F. Redi published the same observation in 1668, as well as another involving insects of different species being born from the same pupa – again a misunderstood observation of parasitic insects. Another parasite of butterflies was noted in print in 1662 by Johannes Goedaert in Volume 1 (Plate 77, p. 175) of *Metamorphosis et Historia Naturalis Insectorum*. It is not clear that Goedaert fully understood what he observed. In one instance he wrote:

> 'Out of one caterpillar, which had pupated on June 12, emerged on the 30th the butterfly. But out of another caterpillar, which had pupated on July 13, emerged after pupation 82 small flies. Thus we have here, again, two entirely different metamorphoses, to a beautiful butterfly and to 82 small flies.'

Goedaert's figure (see Fig. 4.1) clearly shows small gregarious adult parasites emerging from a butterfly pupa. Inasmuch as there is no known parasitic tachinid or other parasitic fly of this size having such gregarious habits, these 'flies' are undoubtedly parasitic hymenopterans, probably pteromalids. This is probably the first published illustration of a hymenopterous parasite.

The first to offer a correct interpretation of the phenomenon of insect parasitism apparently was the British physician Martin Lister. In letters published in 1670–1 in the *Philosophical Transactions* of the Royal Society of London, he had already suggested that some 'ichneumones' lay their eggs in the bodies of living caterpillars, and in 1685, in annotations to an edition of Goedaert's book, he stated:

> 'The 82 flies that emerged from the pupa are the progeny of an ichneumon fly, which had gotten into the caterpillar in a manner that is still not entirely clear to me. In all likelihood they were laid right there by the mother fly . . .'

Antoni van Leeuwenhoek, famous for his pioneering microscopic work, in 1700 illustrated a female parasite (the aphidiid, *Aphidius*; see Fig. 4.2) in some detail. In an accompanying letter to the *Philosophical Transactions* he described rearing these 'flies' from dead aphids, and recorded the manner in which they laid their eggs in live aphids. A few years later, in 1706, Antonio Vallisnieri of Padua also correctly interpreted the phenomenon and wrote about the form and biology of many parasites he discovered. A colleague of Vallisnieri's, one Cestoni, also corresponded with Vallisnieri, discussing parasites of the cabbage aphid, the cabbage butterfly and the cabbage whitefly.

Knowledge of parasites rapidly increased thereafter. Among the leaders, R. A. F. de Réaumur (who also invented the Réaumur thermometer) published principally from 1734 to 1742 (*Mémoires pour Servir à l'Histoire des Insectes,* in six volumes) and Baron Carl DeGeer, a pupil of Linnaues, from 1752 to 1778. Both carefully worked out the biologies of a number of parasitic as well as predatory insects. Réaumur's

Fig.4.1. Parasites emerging from a butterfly pupa (center left). Published by Johannes Goedaert, 1662, this is probably the first published illustration of a hymenopterous parasite.

beautifully detailed and accurate illustrations of 1748 (see Figs 4.3 and 4.4) show how significantly the knowledge of entomophagous insects increased between 1700 and 1750.

The recognition and understanding of insect diseases was developing during this same general period of the late Renaissance (see Steinhaus, 1956). This came about largely from man's study of the diseases of the silkworm, mainly in Europe, but also in Japan and China. However,

Fig.4.2. A parasite of aphids (*Aphidius* sp., family Aphidiidae), from a letter by Antoni van Leeuwenhoek to the *Philosophical Transactions* of the Royal Society of London, Vol. 22, 1700. Note that the abdomen is drawn in the position held while ovipositing in an aphid host (Compare with Fig. 2.7, p. 45). 'S T V represents a little Silver Instrument, to which the Fly, being dead, was fastened . . .'

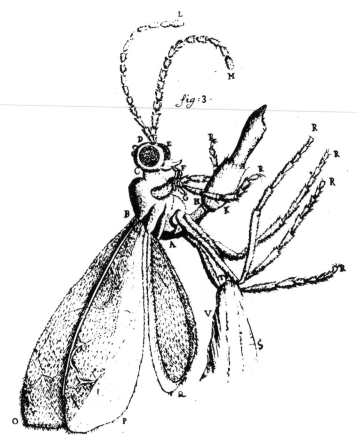

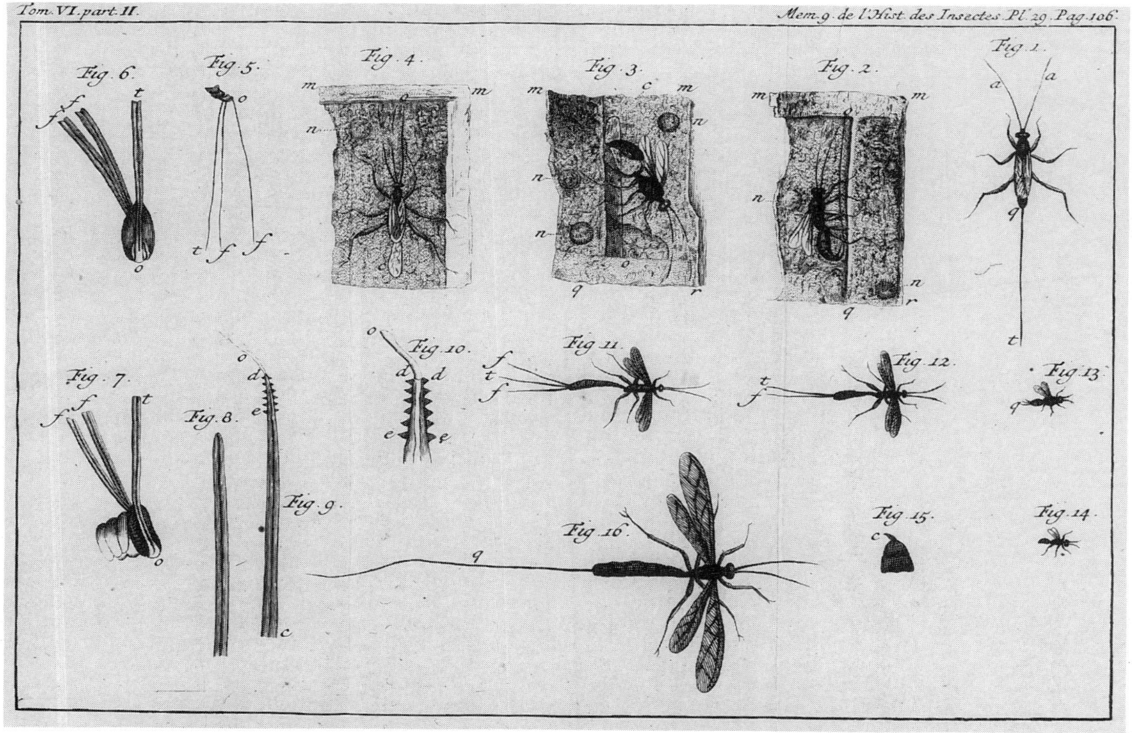

Fig.4.3. Details of an ichneumonid parasite ovipositing (top center), details of the ovipositor (left), and other parasites (bottom right). (From de Réaumur's *Mém. de l'Hist. des Insectes*, vol. 6, Part 2, 1748.)

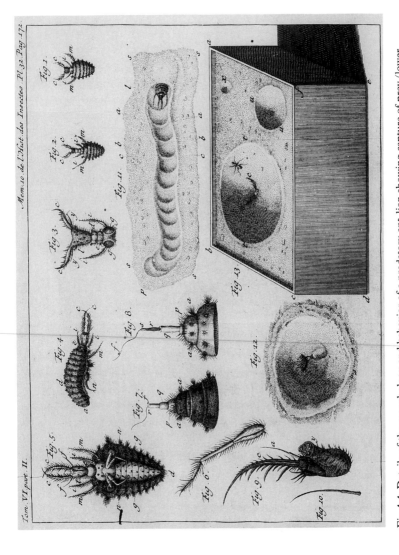

Fig. 4.4. Details of the morphology and behavior of a predatory ant-lion showing capture of prey (lower left) and a laboratory observation unit (lower right). (From de Réaumur's *Mém. de l'Hist. des Insectes*, vol. 6, Part 2, 1748.)

diseases of bees were known to the Greeks and were commented on by Aristotle. Pliny also wrote about bee maladies in AD 77.

The Italian poet Vida published a poem in 1527 on the silkworm, which contained descriptions and remedies for diseases. During the sixteenth, seventeenth and eighteenth centuries, literally hundreds of works were written on silkworm diseases in Europe. Silkworm culture, of course, was a major industry. Vallisnieri, who correctly observed insect parasitism in 1706, also was the first to mention the disease now known as muscardine. However, the first actual publication depicting an observed disease *organism* was by de Réaumur, who in 1726 included drawings of insect larvae killed by a fungus which we now know to belong to the genus *Cordyceps*. DeGeer also comes into the picture again, with the publication in 1776 of the first description of what is now known to be an *Empusa* infection in flies. The microbial nature of these diseases was not actually realized, however. Several early naturalists, observing fungal pathogens sprouting from dead insects (see Fig. 2.38(*a*), p. 83), misinterpreted them as strange plant-insect combinations and referred to them as 'plant worms' or 'vegetable wasps'.

Biological knowledge matures during the 1800s

In the nineteenth century biology, including biological control, came of age. The philosophical and biological foundations of modern insect pathology were laid, beginning principally with William Kirby's chapter on 'Diseases of Insects' in Volume 4 (1826) of *An Introduction to Entomology* by Kirby & Spence. This was the first at all comprehensive account of diseases of insects in general, and by this time Kirby and others recognized that true fungi actually grew in the bodies of some insects, some clearly as saprophytes but others, they believed, possibly as parasites.

Agostino Bassi of Lodi, Italy, is considered the father of insect pathology and by some the founder of the doctrine of pathogenic microbes. In this respect his studies helped overthrow the theory of spontaneous generation. He was the first to demonstrate experimentally the *parasitic* nature of white muscardine diseases of the silkworm, an account of which was published in 1835. Thus Bassi was the first to show experimentally that a micro-organism (subsequently named *Beauveria bassiana*, see Fig. 2.38(*b*), p. 83) was the cause of an infectious disease of an animal.

This brief account of basic contributions must be terminated with the great work of Louis Pasteur, who began his career by saving the silk industry of France. Like Bassi, his later contributions to higher animal diseases and pathology began with research on silkworm diseases, which were ravaging the great French industry. He spent about five years (1865–70) on these studies and published, among others, the famous two-volume treatise *Etudes sur la Maladie des Vers à Soie*. His was apparently the first laboratory to use a microsope for the diagnosis of infectious diseases. He finally differentiated two main diseases, pebrine and flacherie. With pebrine he observed that the pathogen (a protozoan) could be transmitted through the silkworm egg, as well as by contact and ingestion of contaminated food. By microscopic examination of the moth that laid a given batch of eggs, he could determine if disease corpuscles were present. If so, the eggs and the moth were burned; if not, he could obtain eggs yielding pebrine-free silkworms. Distribution of such eggs to producers, along with sanitary precautions, gradually enabled the industry to control the disease. With Pasteur's work, following Bassi's, the real scientific development of insect pathology was initiated.

Meanwhile, beginning about the mid-1800s, the idea of using micro-organisms to control insects began to appear. This concept grew out of the increasing knowledge that insect diseases were infectious and contagious and could be transmitted from diseased to normal individuals. Again Bassi appears first on the scene. In 1836 he published a suggestion to utilize putrified liquids to spray the leaves of plants to kill pest larvae. Not until 1873, apparently, was such a suggestion again made. In August of that year, at the Annual Meeting of the American Association for the Advancement of Science, the American entomologist J. L. LeConte read a paper in which he proposed the production of diseases, especially the muscardine fungus of the silkworm, and its artificial transmission in the field to other lepidopterous larvae. Then, in 1874, Pasteur suggested the possible use of pebrine against the notorious grape phylloxera in France, and again in 1882 he noted that the grape phylloxera must have its own diseases, which should be sought out and utilized artificially by man. About the same time, suggestions and some laboratory experiments were made in America with commercial yeast as a possible insect pathogen, but, as would be expected today, they were fruitless.

In the meantime, probably unbeknownst to the Americans and others, Elie Metchnikoff began conducting experiments in 1878–9 with three diseases of the wheat cockchafer (*Anisoplia austriaca*), a serious pest of cereal crops in his area of Odessa, Russia. In 1879 he published an important paper on the green muscardine fungus, *Metarrhizium anisopliae* (see Fig. 2.38(*d*), p. 83). He envisioned and recommended the practical application by man of disease organisms in insect control and successfully tested this possibility experimentally in the laboratory. His observations on natural epizootics led him to believe that pathogens could not be depended upon alone but that with man's application, effective control might be obtained. Toward this end he discovered a method of producing spores artificially on sterilized beer mash. Thus the main ingredients were at hand for the application of insect pathology as it is mainly practiced today.

Metchnikoff's suggestions and experiments led Isaak Krassilstschick to develop a small *Metarrhizium* spore production plant in Smela, the Ukraine, in 1884. Over 120 lb of spores were produced, and subsequent field tests caused mortality of sugar-beet curculio larvae (*Cleonus punctiventris*) of from 55 to 80 percent from green muscardine. From this time on, interest quickened in the use of fungi to control insects, and ultimately to today's actual or potential use of a wide variety of pathogens.

Ideas and efforts directed toward the biological control of weeds apparently did not arise until after 1850. The American entomologist Asa Fitch was probably the first to suggest the biological control of weeds in about 1855, when he observed that the toad-flax, a European weed very destructive in New York pastures, had no American insects feeding on it and speculated that the importation of insects feeding on it in Europe might solve the problem.

The first practical attempt at biological control of weeds dates from 1863, when stem segments of the prickly pear cactus *Opuntia vulgaris*, infested with an imported cochineal insect, *Dactylopius ceylonicus*, were distributed for cactus control in southern India after they unexpectedly had been observed to decimate cultivated plantings of this same cactus in northern India (Goeden, 1978). In 1865, the first successful international importation for biological weed control took place, when this same insect was transferrerd from India to Sri Lanka, where in a few years time widespread feral populations of this same cactus were effectively controlled. (An earlier attempt, to import another species of *Dactylopius* from South Africa into India in 1836, failed.) It was earlier supposed that the first use of insects to control a weed occurred in Hawaii in 1902 against *Lantana camara*, an introduced ornamental plant that escaped and ran wild (see Holloway, 1964).

Developments with insect parasites and predators moved considerably more rapidly than with pathogens during the 1800s, as might be expected if only because of the technical difficulties of studying minute pathogenic micro-organisms in those days. Work with weeds was probably restricted because of the fear that the introduced insects might transfer to cultivated plants. In view of the limited knowledge of that period, it is as well that more wasn't attempted. Today's sophisticated studies of weed-feeding insects exclude the importation of any that might be able to feed on economically valuable plants.

Basic studies of the taxonomy and biology of insect parasites and predators, and later of their ecology, were a necessary prelude to the scientifically sound, practical application of biological control. Among an increasing number of workers who published good taxonomic studies of parasites, should be mentioned M. M. Spinola who published his *Insectorum Liguriae* in Genoa in 1806, J. W. Dalman of Sweden who published on *Encyrtus* in 1820, J. L. C. Gravenhorst who described 1300 European Ichneumonidae about the same period, J. O. Westwood, the great English entomologist who published voluminously from 1827 on through most of the 1800s, Francis Walker, who specialized in the Chalcidoidea from 1833 to 1861 and erected the family Encyrtidae in 1837, the Italian C. Rondani (publ. 1840–60) who also tabulated host-parasite relations, the German A. Förster (publ. 1841 to the 1880s), and J. T. C. Ratzeburg, another German who published between 1837 and 1852. His *Die Ichneumonen der Forstinsecten* (1844) was the best European contribution devoted to the biology of hymenopterous parasites and remained a standard work for many years. Other leaders in the taxonomy of parasites, some of whom continued well into the 1900s, can only be mentioned. They include Holmgren and Thomson (Sweden), Mayr (Austria), Motschulsky (Russia), Hartig and Schmiedeknecht (Germany), Wesmael (Belgium), Haliday, Marshall and Cameron (England), Brulle, Giraud and Decaux (France), Provancher (Canada), Cresson (the pioneer American hymenopterist who described many parasites and published a serial list of the Ichneumonidae of North America), Riley, Howard and Ashmead (USA).

Practical ideas and tests gradually accelerated as the nineteenth century advanced. Dr. Erasmus Darwin, the grandfather of Charles Darwin, started the century off by stressing the controlling effect of certain parasites and suggesting in his book on agriculture and gardening, *Phytologia*, published in London in 1800, that aphids in hothouses might be controlled by the artificial use of predaceous syrphid fly larvae. Kirby

& Spence (1815) devoted considerable space in their *Introduction to Entomology* to strongly stressing the good work of parasites and predators, as well as to a discussion of the major groups and their habits. They state:

> 'I observed in a former letter, that the devastations of insects are not the same in every season, their power of mischief being evident only at certain times, when . . . an unusual increase of their numbers [may occur] . . . The great agents preventing this increase, and keeping the noxious species within proper limits, are other insects; and to these I shall now call your attention.
>
> Numerous are the tribes upon which this important task devolves, and incalculable are the benefits which they are the means of bestowing upon us; for to them we are indebted . . . that our crops and grain, our cattle, our fruit and forest-trees, our pulse and flowers, and even the verdant covering of the earth, are not totally destroyed.'

Noting the value of the common English ladybeetle (or ladybird) in destroying the hop aphis, they suggested the desirability of discovering means of artificially increasing the ladybeetles for use in aphid control on hops and in glasshouses, and also suggested that certain hemipterous predators of bedbugs ought to be encouraged because six or eight of the predators shut up in a room swarming with bedbugs would completely 'extirpate' the latter in several weeks. Obviously, this referred to the small test of 1776 with *Picromerus bidens*, mentioned earlier in this chapter. They noted that gardeners and florists in England had used ladybeetles for *very many* years by transferring them from one place to another, and that the authors also had done this to good effect. Thus, this empirical use of predators possibly may antedate that of the date growers in Yemen or of the mynah bird in Mauritius.

Kirby & Spence used the term parasite as we use it today, but insect predators they referred to, curiously, as imparasites. In their discussion of the types, habits, and effectiveness of parasites, they covered the ground surprisingly well, including, it appears, all the major families or higher groups we know today (i.e. their Ichneumonidae equals today's Ichneumonoidea). Apropos of habits and effectiveness, they say:

> 'The habits of the whole of this tribe, which properly includes several families (*Ichneumonidae, Chalcididae*, &c.) and a great number of distinct genera, are similar. They all oviposit in living insects, chiefly while in the larva state, sometimes while pupae . . .; at others while in the egg state . . . The eggs thus deposited soon hatch into grubs, which immediately attack their victim, and in the end insure its destruction.

The number of eggs committed to each individual varies according to its size, and that of the grubs which are to spring from them; being in most cases one only, but in others amounting to some hundreds.

From the observations hitherto made by entomologists, the great body of the Ichneumon tribe is principally employed in keeping within their proper limits the infinite host of *lepidopterous* larvae, destroying, however, many insects of other orders; and, perhaps, if the larvae of these last fell equally under our observation with those of the former, we might discover that few exist uninfested by their appropriate parasite. Such is the activity and address of the Ichneumonidans, and their *minute* allies . . ., that scarcely any concealment, except, perhaps, the waters, can secure their prey from them; and neither bulk, nor ferocity avail to terrify them from effecting their purpose. They attack the ruthless spider in his toils; they discover the retreat of the little bee, that for safety bores deep into timber; and though its enemy Ichneumon cannot enter its cell, by means of her long ovipositor she reaches the helpless grub, which its parent vainly thought secured from every foe, and deposits in it an egg, which produces a larva that destroys it. In vain does the destructive *Cecidomyia* of the wheat conceal its larvae within the glumes that so closely cover the grain; three species of these minute benefactors of our race, sent in mercy by Heaven, know how to introduce their eggs into them, thus preventing the mischief they would otherwise occasion, and saving mankind from the horrors of famine . . . Even the clover-weevil is not secure within the legumen of that plant; nor the wire-worm in the earth, from their ichneumonidan foes . . . The ichneumonidan parasites are either external or internal . . . But the great majority of these animals oviposit within the body of the insect to which they are assigned, from whence, after having consumed the interior and become pupae, they emerge in their perfect state. An idea of the services rendered to us by those Ichneumons which prey upon noxious larvae may be formed from the fact, that out of thirty individuals of the common cabbage caterpillar . . . which Réaumur put into a glass to feed, twenty-five were fatally pierced by an Ichneumon . . . The parasites are not wholly confined to the order Hymenoptera: a considerable number are also found amongst the tribe of flies, many of the species of the Dipterous genera *Tachina* Meig.; . . . depositing their eggs in caterpillars and other larvae, often in such great numbers, that from a larva of *Sphinx atropos* . . . which had sufficient strength to assume the pupa state, not fewer than eighty flies of *Senometopia atropivora* came out of it . . . Generally speaking, parasitic larvae do not attack insects in their perfect state, but to this rule there are several exceptions.'

Theodor Hartig of Germany, when only 22 years of age in 1827, perhaps was the first to suggest the construction of rearing cages for parasitized caterpillars in order to recover the parasites in large numbers for release where needed. Vincent Kollär, also of Germany, clearly understood the value of entomophagous organisms, especially insect parasites and predators, in natural control. He stated that we often owe the preservation of our agricultural products to natural enemies. The Austrian Emperor Francis I sponsored the publication of his work in 1837. Kollär described in rather accurate detail the biology and habits of various predators and parasites, including egg parasites, and stressed the need for ample knowledge of natural enemies in order to protect man from injurious insects. For example, he states:

> 'The manner in which the Ichneumonidae accomplish their work of destruction is highly curious and interesting. All the species are furnished at the end of the body with an ovipositor, composed of several bristles attached together, with which they pierce the larvae of other insects, and introduce their eggs into the flesh of the wounded animals. In some this sting is longer than the whole body, sometimes more than an inch long, namely, in those species which seek the object of their persecution in the interior of trees or wood that has been much and deeply perforated by the insects which reside therein. They perceive, either by their sense of smelling or by their antennae, that their prey is at hand, and introduce their eggs, not without difficulty, into the bodies of the larvae living in the wood. Some attack caterpillars feeding openly on plants, others perforate the various excrescences, or gall-nuts, which also contain larvae; there are even many species, scarcely visible to the naked eye, which lay their eggs in the eggs of other insects, such as butterflies, and thus anticipate their destruction.
>
> The eggs are hatched within the body of the living insect, and the young parasites, in the most literal sense, fatten on the entrails of their prey. At last the wounded caterpillar sinks, the enemies escape through the skin, and become pupae; or the caterpillar, notwithstanding its internal parasites, enters the pupa state, but instead of a butterfly, one or more Ichneumonidae appear. To these wonderful animals we often owe the preservation of our orchards, woods and grain.'

Kollär concluded that:

> 'We can only protect ourselves from the injurious influence of insects by an ample knowledge of the reciprocal relation in which one stands to another, and in order to obtain this, it is essentially necessary to acquire a

knowledge of those kinds which are directly or indirectly injurious to man, their different stages of life, their nourishment, propagation, duration, and finally their natural enemies.'

Another student of natural enemies, the Italian Rondani, proposed the use of parasitic insects for the control of pests in the 1850s.

Actual tests using insect predators, collected and translocated in the field, were reported in Europe in the 1840s. Professor Boisgiraud of Poitiers, France, used the predatory carabid *Calosoma sycophanta* in 1840 in an attempt to control gypsy moth larvae on willows. He also tried predatory staphylinids against earwigs in his garden, and claimed success in both cases. This led an Italian Society for the Promotion of Arts and Crafts, in 1843, to offer a gold medal for presentation in 1845 to anyone who meanwhile had conducted successful tests with artificial breeding of predatory insects for control of agricultural pests. Antonio Villa, just preceding the deadline, presented a written account of such tests to the Society on December 26, 1844. He used climbing carabid predators for phytophagous larvae, staphylinids for flower insects and ground carabids for soil-inhabiting insects. He claimed that most of the predators used increased and gave good control. As a result he was awarded the gold medal.

The second *international* transfer of a predatory bird, or apparently of any natural enemy for that matter, took place in 1849. The Spanish governor of the Philippines, Juan Martinez, introduced the Chinese starling or crested mynah (*Aethopsar cristatellus*) from China or India for the control of migratory locusts. It became established in Luzon, with unreported results (Baltazar, 1981).

The scene now switches largely to the United States, where by the 1850s the ravages of accidentally imported insect pests were beginning to bestir ideas of *importing new* natural enemies from their homeland. Some American entomologists were quite cognizant of the fact that many pests of European origin were considered to be held under control by natural enemies in Europe.

Vigorous suggestions and actual attempts to import the European parasites of the disastrous pest, the wheat midge *Sitodiplosis mosellana*, were made by Asa Fitch, famous State Entomologist of New York, in 1855. He had documented the spread of the midge, studied it carefully in the field, correctly identified it as the European species, observed a lack of effective parasites in New York and knew that it was held under natural control in Europe. In his own words:

'There must be a cause for this remarkable difference. What can that cause be? I can impute it to only one thing; we here are destitute of nature's appointed means for repressing and subduing this insect. Those other insects which have been created for the purpose of quelling this species and keeping it restrained within its appropriate sphere have never yet reached our shores. We have received the evil without the remedy. And thus the midge is able to multiply and flourish, to revel and riot, year after year, without let or hindrance. This certainly would seem to be the principal if not the sole cause why the career of this insect here is so very different from what it is in the Old World.'

As a consequence, in May 1855 he wrote to Mr John Curtis, President of the London Entomological Society, informing him of the damage in the United States and the lack of effective enemies, and asking co-operation in obtaining shipments of parasites. He even suggested how to collect and ship parasitized larvae. The request was received with favor by the London society, but no shipments ever resulted. Although the Rev. C. J. S. Bethune, a Canadian entomologist, was credited by C. V. Riley as probably being the first to suggest the importation of the European parasites of the wheat midge, his suggestion apparently was not made until 1864. Again nothing came of the idea.

Benjamin D. Walsh, as State Entomologist of Illinois, subsequently actively continued the written campaign started by Fitch to obtain wheat midge parasites, and apparently tried to obtain them through some English friends. In several publications he strongly criticized the inactivity of the government and State Entomological Societies for failing to import parasites. He pointed out that the damage in New York State alone amounted to fifteen million US dollars in one year and that the midge frequently took half of the crop, whereas in England the largest amount ever destroyed was five percent. He stressed that there were no parasites in this country but at least three species in England, and this was the reason for the difference in damage. His efforts are summed up in a nutshell by the statements published in the September 29, 1866 issue of the *Practical Entomologist*:

'The plain common sense remedy . . . is, by artificial means to import the European parasites, that in their own country prey upon the Wheat Midge, the Hessian Fly and other imported insects that afflict the North American farmer. Accident has furnished us with the bane; science must furnish us with the remedy. [He went on to say:] Vaccination, Gas, the Steam-engine, the Steam-boat, the Rail-road, the Electric Telegraph, have all been successively the laughing-stock of the vulgar, and have all by slow degrees fought their way into general adoption. So will it be with the artificial importation of parasitic insects. Our grandchildren will perhaps be the first to reap the benefit of a plan, which we ourselves might, just as well as not, adopt at the present day. The simplicity and comparative cheapness of the remedy, but more than anything else the ridicule which attaches, in the popular mind, to the very names of 'Bugs' and 'Bug-hunters', are the principal obstacles to its adoption. Let a man profess to have discovered some new Patent powder pimperlimpimp, a single pinch of which being thrown into each corner of a field will kill every bug throughout its whole extent, and people will listen to him with attention and respect. But tell them of any simple commonsense plan, based upon correct scientific principles, to check and keep within reasonable bounds the insect foes of the Farmer, and they will laugh you to scorn. [and finally:] But we should not stop here. [with introduction of parasites of the wheat midge.] The principle is of general application; and wherever a Noxious European Insect becomes accidentally domiciled among us, we should at once import the parasites and Cannibals that prey upon it at home.'

Obviously, Walsh was vigorously proposing classical biological control, but unfortunately again nothing practical came of his efforts.

The well-established and older Walsh appears to have transmitted his enthusiasm for biological control to young C. V. Riley who, in 1868, when only 25 years of age, was appointed State Entomologist of Missouri during the height of Walsh's campaign in nearby Illinois to import parasites. Riley began his career reporting on economic entomology for the *Prairie Farmer* in Chicago in 1864, and probably got to know Walsh at once. In fact, they published 478 papers together, and jointly started the *American Entomologist* in 1868.

Riley appears to have been the first to *transport insect parasites* from one locality to another. In studying the parasites of the plum curculio (*Conotrachelus nenuphar*) in 1868–70 he determined that they could be easily collected and distributed from Kirkwood, Missouri, and sent two species of parasites to several correspondents in other parts of the State. Then, in 1871, he proposed *conservation* of parasites of the rascal leaf-crumpler of apple and other fruit trees (*Acrobasis indigenella*) by collecting larvae in their cases in mid-winter and then removing them a short distance from the trees, so that the larvae could not return but the adult parasites emerging from the parasitized ones the next spring easily could.

In writing of this procedure toward the close of his career in 1893, Riley considered that conservation of natural enemies was one of

> 'but two methods by which these insect friends of the farmer can be effectually utilized or encouraged, as, for the most part, they perform their work unseen and unheeded by him, and are practically beyond his control. These methods consist in the intelligent protection of those species which already exist in a given locality, and in the introduction of desirable species which do not already exist there.'

In this same 1893 paper Riley was certainly among, if not *the* first, to recognize the adverse ecological effects deriving from the use of insecticides, and to visualize the interactions of natural enemy-prey populations to produce natural biological control. He states:

> 'Year in and year out, with the conditions of life unchanged by man's actions, the relations between the plant-feeder and the predaceous and parasitic species of its own class remain substantially the same, whatever the fluctuations between them for any given year. This is a necessary result in the economy of nature; for the ascendancy of one or the other of the opposing forces involves a corresponding fluctuation on the decreasing side, and there is a necessary relation between the plant feeder and its enemies, which, normally, must be to the slight advantage of the former and only exceptionally to the great advantage of the latter.
>
> This law is recognized by all close students of nature, and has often been illustrated and insisted upon by entomologists in particular, as the most graphic exemplifications of it occur in insect life, in which fecundity is such that the balance is regained with marvelous rapidity, even after approximate annhilation of any particular species. But it is doubtful whether another equally logical deduction from the prevalence of this law has been sufficiently recognized by us, and this is, that our artificial insecticide methods have little or no effect upon the multiplication of an injurious species, except for the particular occasion which calls them forth, and that occasions often arise when it were wiser to refrain from the use of such insecticides and to leave the field to the parasitic and predaceous forms.
>
> It is generally when a particular injurious insect has reached the zenith of its increase and has accomplished its greatest harm that the farmer is led to bestir himself to suppress it, and yet it is equally true that it is just at this time that nature is about to relieve him in striking the balance by checks which are violent and effective in proportion to the exceptional increase of and consequent exceptional injury done by the injurious species. Now the insecticide method of routing this last, under such

circumstances, too often involves, also, the destruction of the parasitic and predaceous species, and does more harm than good. This is particularly true of those of our Coccidae and Aphididae and those of our Lepidopterous larvae which have numerous natural enemies of their own class; and it not only emphasizes the importance of preventive measures [cultural control], which we are all agreed to urge for other cogent reasons, and which do not to the same extent destroy the parasites, but it affords another explanation of the reason why the fight with insecticides must be kept up year after year, and has little cumulative value.'

Riley's remarks concerning insecticides are even more true today.

One cannot help but reflect that during this early period the great American economic entomologists and national leaders in the profession were primarily strong advocates of biological control, and in several cases were accomplished expert taxonomists in the parasitic Hymenoptera as well. Both Riley and his successor as Chief of the United States Bureau of Entomology, L. O. Howard, outstandingly filled both categories. Then recall also the great early State Entomologists, Asa Fitch of New York and Benjamin Walsh of Illinois (as well as Riley in Missouri), plus University advocates such as J. H. Comstock of Cornell. Relatively speaking, we are afraid that the very deep appreciation, and actual study, of natural biological control by many of the top men of those early days were gradually replaced by entomologists having more and more one-sided interest in insecticidal control. Hopefully, today this trend is being reversed toward a state of broader ecological understanding and appreciation of the need to rebalance the overuse of chemicals, with adequate stress on non-chemical approaches to pest control.

Just after Riley's 1871 proposal for conservation of natural enemies, F. Decaux, a French entomologist, made essentially the same suggestion in 1872. He was impressed by the large number of parasites that emerged from apple buds attacked by the *Anthonomus* weevil and advised, instead of collecting and burning these buds, as was generally done, placing them in boxes covered with gauze and then opening the boxes from time to time to permit the parasites to escape. He later tested the idea personally in 1880, by collecting more than one million *Anthonomus*-infested buds, from which he obtained and liberated about 250 000 parasites which during the following year aided in the control of the weevil. He repeated the experiment again that year, and reportedly this stopped all serious damage in the treated orchards for ten years. Decaux also made some tests in 1872 in transporting parasites from one locality to another.

During the same period of the early 1870s, Dr William Le Baron, State Entomologist of Illinois, was conducting intensive studies on the oyster-shell scale of apple (*Lepidosaphes ulmi*) and its parasite, *Aphytis mytilaspidis*. In the winter of 1871–2 he moved twigs bearing parasitized scales from one locality, Galena, to another, Geneva, where the parasite did not exist. At the end of 1872, the parasite was established in the new location.

Although these tests by Riley, Le Baron and Decaux were progressive for their time and a sign of things to come, they were not well-designed experiments in the modern sense. In the transfer tests, the possibility existed that the parasites actually were already present in low numbers in the new locality, and in the conservation of parasites by collecting parasitized material and permitting the parasites but not the host to escape, there were no adequate checks.

Riley comes back into the picture in 1873 by making the first successful international transfer of an *arthropod predator* by sending the predaceous mite, *Tyroglyphus phylloxerae*, of the grape phylloxera (*Daktulosphaira vitifolii*) to Planchon in France. The mite, although established, did not prove to be of importance in control. However, a puzzling post-script to this record was provided by systematic acarologists, who some time later suggested the synonymy of *T. phylloxerae* with *Rhyzoglyphus echinopus*, a phytophagous mite already present in Europe and known as a pest of various bulbs and roots (Michael, 1903). If this is correct and the two mites are indeed the same, then this early project may serve as an outstanding illustration of the great importance of systematic and biological knowledge to applied biological control. (See Chapter 7 for further discussion of these subjects.)

International shipments of aphid predators, and possibly also para-sites, were made from England to New Zealand in 1874 but without any startling results, although the ladybeetle *Coccinella undecimpunctata* was reported to have become established. This would represent the first overseas establishment of a predaceous *insect*.

Another international movement of a vertebrate predator of insects was made in 1875. The Surinam toad *Bufo marinus* was introduced into Bermuda from British Guyana by Captain Nathaniel Vesey, especially for control of roaches. It became established, and is now abundant. It is a very general predator, which has been distributed into various Caribbean islands.

The first authentic *inter-country* transfer of a *parasitic* insect appears to have been the importation of *Trichogramma* from the United States to Canada by W. Saunders, President of the Entomological Society of Ontario, in 1882 for the control of the gooseberry sawfly (*Nematus ribesii*). Results, if any, are unknown, and it is quite possible that the parasite had been present there beforehand.

It will be recalled that the first parasite species recorded by man was by Aldrovandi in 1602, although he did not recognize it as such. However, it is evident from his description that the parasite was *Apanteles glomeratus*, a common and fairly effective parasite of the cabbage butterfly. Two hundred and eighty-one years later, in 1883, this became the first *parasite species* to be *shipped intercontinentally* and to become successfully *established*. Riley, with the co-operation of G. C. Bignell of Plymouth, England, was responsible for this importation to the United States. The species was colonized in the District of Columbia, Iowa, Nebraska and Missouri and rapidly spread throughout the country. It remains today one of the more effective enemies of the cabbage butterfly.

The stage is now set for the first great success in classical biological control – that of the cottony-cushion scale in California, which firmly established biological control as a major pest control method. The story is related in the next chapter. Riley again was the mastermind; in fact, he had begun gathering data on the cottony-cushion scale and its enemies in California as early as 1872, and over the years into the mid-1880s continued to seek information on the native home of the scale in order to obtain effective natural enemies. In his writings one can almost feel the surge of his great drive, enthusiasm and confidence in the yet unproven biological control method. As will be related, he was more successful than he could ever have dreamed.

5

The first foreign explorers

Foreign exploration for new exotic natural enemies is the classical approach to biological control, and still remains the major one. It furnishes the big pay-off. Its evolution was sketched in the preceding chapter. Its beginnings often were fraught with adventure, danger and hardships for the explorers, as will be related in some of the individual stories that follow.

These early projects cover the period from the first great success, beginning in 1888–9 to about the beginning of World War II, i.e. 1938 or 1939. During this period, nearly all foreign exploration and importation of natural enemies entailed lengthy trips, primarily involving steamship and railway travel, supplemented by local and even long-distance travel by foot, ox-cart, camel, horse and, if lucky, sometimes by automobile. In those days, the explorer often had to accompany his shipments home on lengthy sea voyages in order to maintain the live cultures *en route*. Some illustrative views from various trips made by Harold Compere for the University of California at Riverside during the 1920s and 30s are shown in Fig. 5.1. The commercial development of international and overseas air transport marked the end of this era. The first long-distance international air shipment of natural enemies we have documented was from Japan to Israel in 1939, although some shorter air shipments occurred in the mid-1930s.

We have chosen the following 15 projects from among many. Most, but not all, represent great successes, and between them illustrate the main principles and practices of classical biological control. Projects involving classical biological control, it will be recalled, should be triggered when a new exotic insect invades and becomes a pest. In order to establish a new natural balance below the pest-status level, the origin of the pest is sought

(a)

(b)

(c)

(d)

(e)

(f)

(g)

Fig.5.1. Aspects of foreign exploration for natural enemies by Harold Compere *circa* 1930. (*a*) field collecting in Eritrea, (*b*) cages containing parasitized black scale on host plants being transported to ship on camels, Eritrea, (*c*) cages being loaded on board S.S *President Polk*, Red Sea, (*d*) cages *en route* on deck, (*e*) protection of cages on deck from salt spray and hot sun, Red Sea, (*f*) cage repair in India, (*g*) transport in India.

out, and a search made for its natural enemies. When found, these are imported and colonized and, if they prove to be efficient, successful biological control occurs, permanently solving the problem.

Some foreign explorers and famous projects
Cottony-cushion scale in California, 1888–9

This project established the biological control method like a shot heard around the world. About 1868, the cottony-cushion scale, *Icerya purchasi* Maskell, was discovered on *Acacia* in Menlo Park in northern California and spread rapidly. By 1886 its effect on the new and growing citrus industry in southern California was devastating to the point of destruction. Growers tried washes and HCN fumigation, but these were not sufficiently effective. Damage was so extensive that many growers pulled out or burned their trees, and land values plummeted.

The alarmed California horticultural officials started inquiries, and early in the game enlisted the aid of Charles Valentine Riley, Chief of the United States Department of Agriculture Division of Entomology. Riley emerged as the dominant and guiding figure in this odyssey, but others made more important active contributions. Riley's earlier interest and experiments in biological control have already been mentioned in Chapter 4.

Albert Koebele was the foreign explorer involved. He became famous and followed the same career for many years. D. W. Coquillett, the third man of the team so to speak, played the important role of receiving and colonizing the imported natural enemies.

Setting the stage began in 1885, when Riley sent his assistant, Koebele, to Alameda, California, at a salary of $100.00 per month to conduct an 'investigation of the history and habits of insects of California'. About the same time he appointed Coquillett, a skilled amateur entomologist residing in Anaheim, California, as field agent to work on the control of the cottony-cushion scale, *Icerya purchasi*, around Los Angeles. In February 1886, both men were assigned to work together at Los Angeles on this latter problem.

Meanwhile, Riley apparently had been endeavoring by correspondence to determine the native home of the *Icerya*, in order to discover and import its natural enemies, and probably – at least by 1887 – California officials were doing likewise. Possibly stimulated by such correspondence (here the story is cloudy), Frazer Crawford, an entomologist of Adelaide, Australia, discovered in 1886 the parasitic fly *Cryptochetum iceryae*

(Williston), an effective natural enemy of the cottony-cushion scale. Early in 1887, Crawford wrote to Riley that *Icerya* in Adelaide was destroyed by a dipterous parasite, and sent drawings of the fly and also specimens which the United States Department of Agriculture (USDA) received in February 1887. At first Riley was dubious of its parasitic status, because no true dipterous fly parasites of scale insects were then known. Later he became so convinced of its importance, that this assumed the main objective in Koebele's future trip.

In his Annual Report of 1886, Riley recommended that the natural enemies of the cottony-cushion scale be investigated in Australia and introduced into California. The same year the California Fruit Growers' Convention petitioned Congress to appropriate funds for the USDA to do this work. Politics and pressures then were much like now. However, Congress not only refused, but maintained the regulation that USDA funds could not be spent in foreign travel.

In April 1887, as invited speaker at the California Fruit Growers' Convention, Riley was asked to provide a remedy for the cottony-cushion scale scourge. He stated his belief that the scale came from Australia, where it was harmless, and probably not from New Zealand, where it was recorded as a serious pest. He assumed that parasites kept the scale in check in Australia, and again recommended that those be sought out and imported into California. He offered to send an entomologist to do this, but said that the US Congress would laugh at the idea and asked California, or even Los Angeles County, to appropriate a couple of thousand dollars to import the parasites. Although the Convention again adopted a resolution in favor of sending someone to Australia for natural enemies, no money was forthcoming from California.

About this same time, W. G. Klee, the California State Inspector of Fruit Pests, corresponded with W. M. Maskell in Auckland, New Zealand (Maskell described the cottony-cushion scale as a new species from Auckland in 1878) and with Frazer Crawford in Adelaide, Australia. Maskell told Klee positively that Australia was the native home (letter published in *Pacific Rural Press*, May 7, 1887). Subsequently Riley, who meanwhile had had second thoughts as to the country of origin and was suggesting Mauritius (letter in *Pacific Rural Press*, June 4, 1887), agreed that Australia was probably the native home (letter in *Pacific Rural Press*, March 4, 1888).

Late in 1887, as a result of Klee's correspondence, Frazer Crawford, with considerable effort, collected and sent some live *Cryptochetum* to

Klee, who liberated the flies on cottony-cushion scale in San Mateo County near San Francisco in early 1888, *before* Koebele sailed for Australia – ostensibly to get the same flies. There is good reason to believe that this resulted in establishment, because it eventually became common in California and there is doubt whether Koebele's later shipments to Los Angeles survived after release.

The financing for Koebele's trip to Australia came about through some adroit political manoeuvering. In 1888, an International Exposition was to be held in Melbourne, Australia, and a US exhibit was planned through the US State Department. Through the efforts of Riley, N. J. Coleman, the California Commissioner of Agriculture and others, the US Secretary of State was persuaded to set aside $2000.00 to pay the travel expenses of an entomologist who, ostensibly, was to represent the US State Department at the Exposition. Riley selected Albert Koebele, who sailed from San Francisco, August 25, 1888.

In Australia, Koebele had none of the problems that beset some of the later foreign explorers. As an official representative of the US State Department and the USDA, he received utmost co-operation and was nearly everywhere accompanied by knowledgeable local entomologists or growers, who often led him to known pockets of the otherwise rare cottony-cushion scale. Additionally, the State railways throughout Australia furnished him with free passes.

Arriving in Sydney, September 20, 1888, Koebele searched for four days and found only a few *Icerya* and no natural enemies. The local orange growers had no knowledge of the scale. Proceeding to Melbourne by train, he searched for some six days but found no *Icerya*. Next he went to Adelaide with a letter of introduction to Frazer Crawford (the discoverer of *Cryptochetum* in 1886). The next day, in gardens in Adelaide, they found *Icerya* and the very first scale examined contained nine pupae of the parasitic fly. In fact, nearly all the scales examined were parasitized. We can well imagine Koebele's excitement, especially when the scale was so rare that most local entomologists hardly knew of its existence.

On October 15, while collecting scales for shipment to California with Crawford in a North Adelaide garden, Koebele relates 'I discovered there, for the first time, feeding upon a large female *Icerya*, the Lady-bird, which will become famed in the United States – *Vedalia cardinalis*.' (Note that this report was written in July 1889, after his return and after vedalia was showing its potential in California.) Fig. 5.2 shows an adult of the vedalia ladybeetle feeding on the cottony-cushion scale. (Incidentally, the scientific name of the vedalia was subsequently changed to *Rodolia cardinalis*). Koebele wrote to Riley about this discovery, and Riley replied that *Cryptochetum* was probably the most promising, but to try others as well. As we shall see, he was both right and wrong regarding *Cryptochetum*.

Searching further afield for more scales, Koebele went to Mannum in the Murray River valley, where so much of today's oranges are grown. There he found the scale along with *Cryptochetum*, the ladybeetle (ladybird) vedalia, and a predatory green lacewing. He returned within a week to Adelaide with considerable material, which was placed in a cool cellar to await shipment to California. On October 24–25 he collected more scales in North Adelaide, along with many parasitic flies and green lacewings. The first shipment was described by Koebele thus:

> 'I finished collecting for my first shipment on the 25th and estimated that I had about 6000 Icerya, which in return would produce at an average about four parasites [Lestophonus = *Cryptochetum*] each. They were packed partly in wooden and partly in tin boxes. Small branches generally full of scales were cut so as to fit exactly lengthwise into the box. With these the boxes were filled and all loose scales placed in between, plenty of space remaining for any of the insects within to move

Fig.5.2. Adult of the vedalia ladybeetle feeding on the cottony-cushion scale. (For close-ups of the beetle and its larva, see Fig. 2.23, p. 64.)

freely without danger of being crushed by loose sticks. Salicylic acid was used in small quantities in the tin boxes to prevent mold, yet these, as I have been informed by Mr Coquillett, arrived in a more or less moldy condition, while those in wooden boxes always arrived safe. In addition, Dr Schomburgh, director of the botanical gardens at Adelaide, kindly fitted up for me a Wardian case which was filled with living plants of orange and *Pittosporum* in pots. Large numbers of Icerya were placed in this, and such larvae as were found feeding upon them . . . The object of this was to have the *Lestophonus* go on breeding within the case during the voyage. No doubt many infested scales arrived in Los Angeles.

I found [later] on examining the tree [in Los Angeles], on April 12, 1889, under which this case had been placed with a tent over it, that from several of the Iceryas the *Lestophonus* had issued. This case, as Mr Coquillett informed me in a letter of November 30, arrived in good condition, except that the putty had been knocked off in several places, leaving holes large enough for the parasites to escape. Before opening the case he found two coccinellid larvae crawling on the outside, and these when placed with the Icerya attacked it at once. He further said that there were only about half a dozen living *Chrysopa* adults. This would show that the Lestophonus was still issuing on arrival in California and all turned out more favorably than I had anticipated on seeing the box handled in such a rough manner by the steamer hands at Sydney, to which point I accompanied this as well as all the subsequent shipments. I expected little good would come out of this method of sending and therefore concluded to send only small parcels on ice thereafter, as had been partly done at first. If once the insects could be placed in good condition in the ice-house on the steamer just before leaving, where a temperature of 38° Fah. at first and about 46° Fah. on arrival in San Francisco existed, they must arrive safely. To accomplish this, the parasites with their hosts were all collected the last three days before leaving Adelaide, and on arriving home were immediately placed in a cool cellar. On the trip from Adelaide to Sydney, which takes two days by train, my insects came generally in an ice-box on the sleeping car.'

Following this he surveyed other areas of Victoria and New South Wales, but concluded that the Adelaide area was best, so returned on November 8, 1888. After collecting about 6000 scales in five days and making a trip to Melbourne for additional material, he left Adelaide for Sydney with the second shipment.

'On the 26th I left Adelaide on my way to Sydney, with what I considered even a better shipment than the first. Unfortunately this lot arrived in a bad condition at San Francisco, owing to a gale on the route when the parcels fell off the shelving in the ice-house, in which they had been placed, and most of them were crushed by cakes of ice falling on them.'

A third shipment was made about the end of December. Meanwhile Koebele traveled to Brisbane, where he found only a few specimens of *Icerya*, and slowly returned to Melbourne with similar very poor collecting along the way. At Melbourne he collected *Cryptochetum* on a related scale insect, *Monophlebus* sp. Nearing the end of his stay in Australia, he collected *Icerya* with parasites and about 200 vedalia, mostly in the Sydney Town Hall garden. Either he now knew how to search better, or was luckier than during his first trip to Sydney, when he drew a blank. Then, under instructions from Riley to study *Icerya* in New Zealand on his way home, he boarded ship on January 23, 1889, with his insects in the cold room, and arrived in Auckland, New Zealand, January 28. The scales with parasites and ladybeetles were found to be in excellent condition at Auckland and were repacked in wooden boxes with fresh *Icerya* found in Auckland, and apparently sent on to California. He found no natural enemies in Auckland; however, at Napier he found large numbers of vedalia beetles feeding on *Icerya*. This was unknown before; according to Koebele the vedalia arrived in Auckland by chance, where *Icerya* was destroying everything five years previously, and there cleaned nearly the whole district around Auckland within about two years. At the time of Koebele's visit the vedalia was dispersing into new areas, hence his big collection of about 6000 specimens of vedalia at Napier. Returning to Auckland, these were placed in the ship's cool-room at 4 °C (38 °F). He left Auckland on February 25 and arrived in San Francisco Saturday evening, March 16, 1889. The material could not be sent to Coquillett at Los Angeles until the following Monday, and he received it on March 20 – 34 days after collection and 29 days on ice – yet they arrived in better condition than any shipment previously received. The specimens were liberated under the same caged tree that had received the earlier specimens. This was on the F. W. Wolfskill property in Los Angeles.

According to Coquillett's records, live vedalia arrived in Los Angeles as follows:

November 30, 1888	28 (Shipment 1)
December 29, 1888	44 (Shipment 2)
January 24, 1889	57 (Shipment 3)
Total	129

These were placed under the caged tree at Wolfskill's.

On February 21, thirty-five vedalia (shipment 4) arrived and were colonized at the J. R. Dobbins property at San Gabriel. The final shipment of 350 live vedalia, that was brought personally by Koebele on board ship, arrived and was colonized on March 20, 1889. About one-third went to Dobbins' grove, and the remainder to the large A. Scott Chapman grove in San Gabriel valley.

All told, about 12 000 *Cryptochetum* arrived in Los Angeles from Koebele. These were all put under one caged tree, and when he examined this tree on April 12, 1889 he noted that very few *Cryptochetum* remained of the vast numbers of flies received. We have found no record as to whether this resulted in establishment, because at this time vedalia was beginning to explode and clean up the scale. Apparently, everyone forgot the fly in the excitement. If this release didn't cause establishment then the 1888 release by Klee at San Mateo surely did, because the fly is common in all coastal areas today.

By early April 1889, nearly all the *Icerya* in the caged tree at Wolfskill's were destroyed by vedalia, so one side of the cage was removed and the beetles allowed to move to adjoining trees. On April 12, Coquillett began sending colonies to other parts of the State. By June 12 – two months after the cage was opened – 10 555 vedalia had been distributed to 208 different growers, and successful colonization occurred in nearly every case. At this time, i.e. within six months of the first release of 28 beetles and with a total release of only 129, the original trees in Wolfskill's were virtually clean of *Icerya* and the beetles had spread to a distance of three quarters of a mile away. In his Annual Report for 1889, Riley states that in the original orchard (Wolfskill's) practically all the scales were killed before August 1889, and further that by the end of 1889, *Icerya* was no longer a factor to be considered in citrus growing in California. The following season saw this statement completely justified.

According to Coquillett in mid-1889, regarding the San Gabriel colonization of February and March:

> 'All of these colonies have thrived exceedingly well. During a recent visit to each of these groves I found the lady-birds on trees fully one-eighth of a mile from those on which the original colonies were placed, having thus distributed themselves of their own accord. The trees I colonized them on in the grove of Dobbins were quite large and were thickly infested with the *Iceryas*, but at the time of my recent visit scarcely a living *Icerya* could be found on these and on several adjacent trees, while the dead and dry bodies of the *Iceryas* still clinging to the trees by their beaks, indicated how thickly the trees had been infested

with these pests, and how thoroughly the industrious lady-birds had done their work.'

The citrus grower, J. R. Dobbins, stated in July 1889, only some four months after the first beetles were released:

'The vedalia has multiplied in numbers and spread so rapidly that every one of my 3200 orchard trees is literally swarming with them. All of my ornamental trees, shrubs, and vines which were infested with white scale, are practically cleansed by this wonderful parasite. About one month since I made a public statement that my orchard would be free from *Icerya* by November 1 [1889], but the work has gone on with such amazing speed and thoroughness that I am today confident that the pest will have been exterminated from my trees by the middle of August. People are coming here daily, and by placing infested branches upon the ground beneath my trees for two hours, can secure colonies of thousands of the vedalia, which are there in countless numbers seeking food. Over 50 000 have been taken away to other orchards during the past week, and there are millions still remaining, and I have distributed a total of 63 000 since June 1.'

Dobbins' orchard was so completely free of *Icerya* that on July 31 he posted a notice that he had no more beetles for distribution. The other colonized grove in San Gabriel was similarly cleaned of scale. In 1888 the owner, Mr A. Scott Chapman, had stated that he was being forced to abandon citrus growing by the scales; in October 1889 he stated that the vedalia had cleaned up the scale on 150 acres. In just one year, shipments of oranges from Los Angeles County increased dramatically from 700 to 2000 carload lots.

Yet another testimonial was published by Riley in 1893, as follows:

'Mr William F. Channing, of Pasadena, son of the eminent Unitarian divine, wrote two years later [in 1891]: "We owe to the Agricultural Department the rescue of our orange culture by the importation of the Australian lady-bird, *Vedalia cardinalis.*

"The white scales were incrusting our orange trees with a hideous leprosy. They spread with wonderful rapidity and would have made citrus growth on the whole North American continent impossible within a few years. It took the Vedalia, where introduced, only a few weeks absolutely to clean out the white scale. The deliverance was more like a miracle than anything I have ever seen. In the spring of 1889 I had abandoned my young Washington navel orange trees as irrecoverable. Those same trees bore from two to three boxes of oranges apiece at the end of the season (or winter and spring of 1890). The consequence of the deliverance is that many hundreds of thousands of orange trees (navels almost exclusively) have been set out in southern California this last spring.'''

From a total original stock of 514 beetles colonized from the end of November 1888 to late March 1889, the rapidity and extent of this control was nearly unbelievable. Coquillett in a letter to Riley, October 21, 1889, summarized it as follows:

> 'The first half of the year I devoted nearly the whole of my time to propagating and distributing the Australia Lady-bird (*Vedalia cardinalis*) recently introduced by this Division. At the present time it is very difficult to find a living Fluted Scale (*Icerya purchasi* Maskell) in the vicinity of this city [Los Angeles], so thoroughly has the Lady-bird done its work; and, indeed, the same is true of nearly the entire southern part of the state, as well as of many localities in the northern part.'

By 1890, all infestations in the State had been completely decimated. The cost, aside from Koebele's and Coquillett's salaries, was about $1500.00; and all told less than $5000.00. Benefits to the citrus industry of California have amounted to millions of dollars annually ever since, and as an aftermath similar successes have been attained over the years in more than 50 countries around the world by transfer of vedalia, and to a lesser extent of *Cryptochetum*.

As mentioned earlier, Koebele immediately became famous and continued as a foreign explorer for the USDA and later for Hawaii, although he never again achieved such spectacular success. As a fitting finale to his great achievement, the Californians raised a fund and presented Koebele with a gold watch and Mrs Koebele with a pair of diamond earrings.

The sequel to this story is that, meanwhile, *Cryptochetum* had been increasing in California and eventually it became strongly dominant in all coastal areas, including Los Angeles, where vedalia attained its first fame. This part of the account rarely is heard. From the evidence we have today, it is certain that had *Cryptochetum* alone been introduced, it would have done just as spectacular a job in control of the scale in the citrus areas of 1890 as did the vedalia beetle. Today, however, the vedalia beetle remains dominant in all interior and desert citrus areas. (See Caltagirone & Doutt, 1989; Koebele 1890; Quezada & DeBach, 1973.)

Gypsy moth and brown-tail moth in New England, 1905–11

These two moth species, *Lymantria dispar* (Linnaeus) and *Euproctis chrysorrhoea* (Linnaeus), are invaders from Europe and can, if unchecked, defoliate large tracts of forest and shade trees. The gypsy moth actually was purposefully imported by an astronomer about 1869 into Medford, Massachusetts, but only to be studied in the laboratory in connection with its silk production. Unfortunately, to the everlasting embarrassment of Professor Leopold Trouvelot, it escaped and was soon wreaking havoc over the New England countryside. Today, of course, importation of potentially dangerous organisms is strictly prohibited. This dual biological control project, like quite a few that have followed, was undertaken only after attempts at chemical eradication had failed.

This was the first of only a few 'super-projects' conducted thus far in biological control, and was broadly directed by Riley's successor, Dr L. O. Howard, Chief of the Bureau of Entomology of the USDA. We say super-project because, for the period, considerable money and a very large number of co-operating entomologists were involved. Howard had great faith in the parasite method, as it was called in those days, and was a leading taxonomist in parasitic Hymenoptera. A large number of young American entomologists, who later on became famous in their own right, were employed on this dual project. These included, among others, P. H. Timberlake, Professor Harry S. Smith, Dr W. R. Thompson and Dr J. D. Tothill, all famous names in biological control.

They worked principally at two converted laboratory-insectaries: the first at North Saugus, Massachusetts, beginning in the autumn of 1905. This proved insufficient, and by the end of the 1907 summer they moved to the famous Melrose-Highlands laboratory-insectary, remodeled from a large house. Numerous innovations, rearing techniques, use of cold storage for synchronization of host and parasites, as well as exciting biological discoveries were worked out at Melrose-Highlands and served as guides for years to come. Mr E. S. G. Titus was in charge of the laboratory work, with strict emphasis on identification, quarantine and biology, until his health broke down and he was forced to resign in May 1907 due to the irritating and poisonous effects of the barbed hairs of the brown-tail moth larvae. His physician ordered this in order to save his life because of intense lung irritation. Some assistants also resigned in spite of efforts to solve the problem by the use of goggles, masks, and protective clothing. W. F. Fiske, who succeeded Titus, pretty well solved the difficulty by confining the larvae to closed cages having two sleeved openings with attached gloves, so that the material inside could be safely handled. A large number of foreign entomologists were employed to collect material following survey trips by L. O. Howard in Europe and Professor T. Kincaid, of the University of Washington, in Japan and Russia.

In general, the two moth projects were conducted simultaneously. Some of the natural enemies were common to both but most were not. As is well known today, the gypsy moth still remains a problem even though quite a few parasites and predators were established against it. According to Clausen (1978):

> 'The gypsy moth still remains a serious pest in New England, but it is believed that the natural enemies established under the biological control program have contributed substantially to a reduction in severity of the infestations.'

On the other hand, the brown-tail moth is now considered to be at best a minor problem in New England, and is reported to be under complete biological control in Canada from some of the same natural enemies originally imported into New England.

The conclusions have been drawn by some, including certain entomologists and others in decision-making positions, that biological control of the gypsy moth has been completely tried and is hopeless. True, it was given a mighty effort, but largely in the early days when a lot remained to be learned. Even L. O. Howard concluded:

> 'It has been found impossible to secure certain of the parasites in adequate numbers for colonization under satisfactory conditions. The proportion of such is very small, it is true, but at the same time it may easily be that ultimate success or failure may depend upon the establishment, not of the most important among the parasites and other natural enemies, but of a group or sequence of species which will work together harmoniously toward the common end. Viewed in this light, the importance of parasites which otherwise might be considered as of minor interest is greatly enhanced.
>
> It is impracticable to determine certain facts in the life and habits of those parasites which have been colonized under conditions believed but not known to be satisfactory. Further detailed knowledge is necessary before we can judge whether the circumstances surrounding colonization were in truth the best that could be devised. Furthermore, so long as

original research is confined to the study of material collected by foreign agents, some of whom are technically untrained, it is practically impossible to secure the evidence necessary to refute published statements concerning the importance of certain parasites abroad which the results of first-hand investigations have not served to confirm. It is believed that these statements are largely based upon false premises, but should this belief prove ungrounded it would mean that there are important parasites abroad of which little or nothing is known first-hand.'

In view of this and of what we know of the successful conclusion of other programs taking place by the discovery of new effective natural enemies many years after the projects were considered impractical of solution, and especially in view of the enormous millions of dollars spent in repetitive chemical campaigns over the years, it seems necessary that strong additional biological control research on importation of exotic natural enemies of the gypsy moth be conducted. Indeed, extensive importation of exotic parasites of the gypsy moth was renewed by the USDA in 1963, and several new species have been obtained from Europe, North Africa and the Far East.

It is known that in many countries of Europe, Asia, and especially Japan, the gypsy moth rarely, if ever, becomes a pest, due to the action of parasites. This is not universally true, for as Howard noted, in parts of Russia parasitic control was obviously inefficient and although disease became prevalent it was not sufficient to keep the gypsy moth from increasing until defoliation of large areas resulted. However, the general degree of biological control accomplished abroad, and therefore potentially possible in the United States, is evident from Howard's statements:

'The study of the tussock moth has resulted in demonstrating another fact which is of peculiar interest in this connection, which is, that the parasites which assist in effecting its control in country districts where this control is perfect are sometimes entirely absent in the city. Something of the same sort may be true of the parasites which assist in effecting the control of the gypsy moth in many localities in Europe where it is so uncommon as to make collection of material for exportation in any quantity impossible. Some of the most interesting lots of caterpillars or pupae which have been received were from such localities, and it may well be that there are parasites abroad which have not been received at the laboratory in Massachusetts in sufficient quantity for colonization, and which can never be received there until new methods for collecting and importing them are devised, but which at the same time are actually among the important species. This fact can only be determined definitely by careful study of the gypsy moth in localities where it was not sufficiently abundant to permit of its collection in large quantities. Professor Kincaid's reports upon the effectiveness of the parasites, even when taken with more than the prescribed grain of conservatism, have been so consistently optimistic as to leave no room to doubt that the parasitism to which the moth is subjected in Japan, even in localities where it is is more than normally prevalent, is sufficient to meet and overcome the rate of increase of the gypsy moth in America . . . The species . . . which have never been received in sufficient abundance to make their colonization possible . . . may, upon investigation, prove to be of more than sufficient importance to justify an attempt to secure their introduction into America . . . It will be noted that the parasites . . . which are to be considered as of some importance in effecting the control of the moth, form, when taken together, a perfect sequence, and that every stage of the moth from newly deposited egg to the pupa is subjected to attack . . . From a technical standpoint it was exceedingly interesting and valuable since there were found to be present [in Russia] in the boxes of young gypsy-moth caterpillars the cocoons of several species of hymenopterous parasites which had either not been received from other sources or which were not known to be sufficiently abundant in any other part of Europe to make possible their collection in large quantities.'

Recent studies have indeed emphasized low-density populations of the gypsy moth.

In Russia, the reports were quite variable. According to Professor Trevor Kincaid in Bessarabia:

'Bendery, Bessarabia, Russia, June 11, 1909. The season here is in full swing, but the situation causes me considerable anxiety, as the whole business is so utterly different from my experience in Japan. The damage wrought by *dispar* in the forests and orchards of Bessarabia this season is enormous and parasite control seems to be most inefficient in checking the depredations of the caterpillars. When I think of the masterly and well-ordered attack of the Japanese parasites and the splendid fashion in which they wiped out the caterpillars in large areas before depredation took place I am surprised by what I see here. [But in Kiev,] Kief, June 26, 1909. From what I can see in the field and from what I can gather from Prof. Pospielow, *dispar* was almost exterminated in this district last year through the activity of the parasites. Only a few isolated colonies seem to have survived, the most important of these being at Mishighari, a small place on the river about two hours by steamer from Kief. In this place, which is perhaps 100 acres in extent, the trees are plastered with cocoons of *Apanteles fulvipes*. The attack of the parasite was so thorough that the first generation seems to have been sufficient to wipe out the caterpillars, as I can find no large caterpillars about the place, and a few days will doubtless witness the complete wiping out of *dispar*.'

The Russian observations seemed on the whole to indicate that in that country the gypsy moth is not controlled by its parasites to an extent which serves to remove it from the ranks of a destructive pest. Again according to Howard:

> 'Conditions similar to those prevailing in Russia emphatically do not prevail in western Europe, nor, according to all accounts, in Japan. Natural conditions in western Europe and in Japan are in many aspects more like those of our own Eastern States than are those of Kharkof Province. Conditions in Kief Province, even, are much more like those of Massachusetts than are those of Kharkof, and in Kief parasite control seemed to be an accomplished fact, although of course there is no assurance that it is continuous and perfect.'

All told, Dr Howard made five foreign exploration trips to Europe, including Russia, in 1905, 1906, 1907, 1909, and 1910. During these trips he met virtually all the important and concerned entomologists of the day and through them, or agents they suggested, he arranged for the collection and shipment of the great masses of natural enemies that ultimately were received and colonized in Massachusetts. Prominent and especially helpful entomologists he consulted included Prof. F. Silvestri and Dr G. Leonardi, Portici (Naples), Italy; Prof. A. Berlese, Florence, Italy; Miss Marie Ruhl, Zurich, Switzerland; Dr. G. Horvath, Prof. A. Mocsary and Prof. Josef Jablonowski, Budapest, Hungary; Dr Paul Marchal, Paris, France; Dr Gustav Mayr, Vienna, Austria; and Prof. Sigismond Mokshetsky, Simferopol, Russia, to name but a few. He surveyed throughout the most important part of the European range of the moths, from Transylvania and Russia on the southeast to Brittany, France, on the northwest, and from Spain and Portugal on the southwest to the Baltic shores on the northeast. He rode the famous Orient Express from Paris to Budapest, waded through newly fallen volcanic ash from Vesuvius to get to Silvestri's laboratory at Portici, and even rented an orchard in Kiev, the Ukraine, for the state of Massachusetts. Ultimately he perfected arrangements so any shipment from any part of Europe to Boston only spent at most 48 hours subjected to summer temperatures, 24 in Europe and 24 in the United States. The remainder of the time was

Table 5.1. *Total numbers of gypsy moth and brown-tail moth imported and colonized in the United States, 1905–11*

	Number	
	Species	Individuals
Hymenopterous parasites	11	1 794 640
Tachinid parasites	16+	68 343
Predatory beetles	7	18 835
Total	34+	1 881 818

spent in a special low-temperature holding room in Cherbourg, France, where all European collections were consigned, or in the ship's cold-room. In the United States, official arrangements were made for passage of shipments through customs without delay or opening. Professor Trevor Kincaid, a skilful collector and investigator on leave from the University of Washington, made two foreign trips, the first to Japan in 1908 and the second to Russia in 1909. The Japanese trip was quite successful. Kincaid knew various Japanese entomologists and received great co-operation from many of them, especially from Professor S. I. Kuwana, who was officially assigned to this work by the Japanese government. A large amount of material was sent home, and many parasites colonized in New England. The Japanese devised excellent shipping boxes, and material often arrived in better condition than that from Europe. The Japanese continued to ship material over subsequent years under Dr Kuwana's supervision. Later, under a new project during the period 1922–33, much additional material was sent in from Japan.

Kincaid surveyed the provinces bordering the Black Sea in the spring and summer of 1909. As noted earlier, the work of the natural enemies was not as impressive here as elsewhere in Europe, but he made various shipments, especially from Kiev. However, owing principally to deficient transportation facilities, the material received proved to be generally unsatisfactory.

The total numbers of natural enemies of both moth species imported and colonized from 1905 to 1911 are shown in Table 5.1.

All told, and including later shipments from Japan, as during the period 1922–33, more than 40 species of natural enemies of the gypsy moth were imported and colonized. Recent efforts in 1963–85 yielded some 80 imported species. Altogether 15 species – 12 parasites and 3 predators – have become established, and between them they cause quite appreciable mortality of the gypsy moth, but generally this affords only partial control. Many parasites were reared or collected from established colonies and redistributed, so that by 1927 about 92 million parasites had been spread throughout the infested areas. Additionally, a wilt disease, or polyhedral virus, appeared in New England in 1907, doubtless introduced unknowingly with the European material. It becomes important only when infestations are damagingly heavy. The question persists, if among the remaining numerous natural enemy species not established, or others not yet discovered, the ultimate successful solution lies hidden. (See Coulson *et al.*, 1986; Doane & McManus, 1981; Howard & Fiske, 1911; Hoy, 1976*a*.)

Sugar-cane leafhopper in Hawaii, 1904–20

In the early days, sugar was king in Hawaii and any threat to it caused consternation. The sugar cane leafhopper, *Perkinsiella saccharicida* Kirkaldy, was discovered in the islands by Dr R. C. L. Perkins in 1900, and by 1902 it was doing serious damage on Oahu. By mid-1903 it had been found on all of the Hawaiian islands, and the next year a sharp drop in the total yield of sugar occurred. Large areas of sugar-cane were actually destroyed. This emergency led the Hawaiian Sugar Planters Association to create a Division of Entomology and in 1904 to employ a staff of entomologists on a full-time basis. Dr Perkins was appointed superintendent, and his staff consisted of O. H. Swezey, G. W. Kirkaldy, F. W. Terry, Alexander Craw and Albert Koebele. These men and their successors proved to be pre-eminent in the field of biological control.

Kirkaldy, a specialist in leafhoppers, determined that closely allied species occurred in Java. Then Dr Perkins found leafhoppers on cane cuttings newly arrived from Queensland, Australia, and determined by correspondence and exchange of preserved specimens that the same species as in Hawaii occurred at Cairns, Queensland, but did no noticeable damage there. Thus, the most important clue as to where to obtain effective natural enemies occurred from this simple detective work.

Dr Perkins and Koebele left for Queensland to search for natural enemies on May 11, 1904. Upon arrival in Brisbane and at Bundaberg in South Queensland they at once found the same leafhopper as in Hawaii and observed a number of parasites. Koebele soon discovered the minute and previously unknown egg parasite, *Anagrus optabilis* (Perkins), that played a dominant role in the early reduction of the leafhopper in Hawaii. It was found to be generally distributed in Queensland and several shipments were made to Honolulu, but the slow and difficult transport by ship resulted in nearly everything arriving greatly weakened or dead. However, a few direct colonizations of parasites were made by Terry on Oahu in August, October and November 1904. Then Dr Perkins stocked a breeding cage in Australia, arriving by ship in Honolulu December 14, 1904 with a few parasites surviving. In late January 1905, half of the parasites were liberated and half kept for breeding purposes. Dr Perkins' insectary culture prospered and more field liberations were made. By the end of 1905 this parasite, *Anagrus optabilis*, was recovered in the field. It became abundant and widespread in 1906, and during 1907 the leafhopper was greatly reduced by it. Meanwhile, a closely related species, *Anagrus frequens* Perkins, was found to be established, presumably from the late 1904 releases made by Terry, but it was of lesser importance. Another parasite species, *Ootetrastichus beatus* Perkins, was discovered in Fiji by Koebele during his return voyage from Australia. He returned with stocks of this in April 1905, and it rapidly became established and spread throughout the territory, but it also was of minor importance. At this time, after more than ten years of arduous foreign collecting for Hawaii, Koebele's health was failing, so he was put on a consulting basis and Dr Frederick Muir was engaged for the foreign work.

Dr Muir left nearly immediately and soon (March 1906) sent in another new parasite species from Fiji, *Haplogonatopus vitiensis* Perkins, but which again added little to the effect of the others. Then, in China, he discovered at Mei Chow another new parasite species, *Pseudogonatopus hospes* Perkins, which after much difficulty he succeeded in shipping alive to Honolulu in December 1906 and early 1907. This became established and was additionally helpful to the other parasites. From 1907 to 1916 Muir explored extensively throughout the Malay archipelago, the Philippines, Papua New Guinea (then New Guinea), Taiwan (then Formosa) and Japan in connection with a variety of projects, but always searching for sugar-cane leafhopper enemies. In Taiwan during February 1916, he discovered another new species of egg parasite, *Ootetrastichus formosanus* Timberlake, and brought it, alive, to Honolulu. This was cultured and colonized in the field and proved to be an additional help to the already established species. Thus, from 1904 to 1916, five species of parasites were successfully established, which produced excellent biological control on most of the Hawaiian sugar plantations. However, outbreaks continued to occur from time to time in certain areas, and it seemed to be particularly where heavy rainfall normally occurred during much of the year.

By 1919 it was evident that in the wet areas such as the Hilo, Hawaii district, the leafhopper remained an especial problem. Thorough trials with chemical and other control measures proved futile, and additional evidence confirmed the opinion that heavy and frequent rains seriously interfered with the egg parasites introduced by Koebele, Perkins and Muir. It was concluded that additional natural enemies that could operate efficiently under such conditions were needed.

Thus, in May 1919 Muir went to Queensland, Australia, to carry out extended and detailed field studies. Nearly a year later, he discovered *Tytthus* (*Cyrtorhinus*) *mundulus* (Breddin) to be a highly effective predator of sugar-cane leafhopper eggs, whereas previously it had been thought to be a sugar-cane-feeding insect and in fact had been recorded as a minor pest in Java. Both Koebele and Perkins, as well as Muir and others, had previously overlooked this predator, principally because it belongs to the family Miridae, in which most of the species suck plant sap.

Muir's discovery of its habits combined careful basic science and good luck – often the main ingredients in important discoveries. The story is best told in Muir's own words.

'In 1919 I went to Australia to make further investigations of the habits of a small carabid beetle which I had noted previously preying upon Perkinsiella, but owing to the very exceptionally dry season these beetles were so scarce I could make no progress with this work, so turned my attention to other phases of the question. It soon came to my notice that a very large percentage of Perkinsiella eggs were dead and attacked by a fungus, a fact that Perkins noticed in 1903–4. I found the fungus in the form of yeast-like spores present in old egg shells from which the young had hatched, which could be recognized by the egg cap being off, and also in unhatched eggs, which in itself was intriguing. In moist cells these spores gave rise to mycelia and then to fruiting bodies and yeast-like spores similar to the original ones. Further investigation showed that these spores were present in all young and adult leafhoppers in the body cavity, where they multiplied by division; that they passed through the walls of the ovarian tubes and entered the young eggs, congregating in a small round ball at one end of the egg, and eventually becoming mostly incorporated into the embryo. As these were universal, it then became evident that the fungus could not be the cause of the dead eggs, as otherwise all could be destroyed. Upon killing the egg by pricking, the spores developed. This led to observations in the field to discover what led to the death of the egg. The fact was then revealed that *Cyrtorhinus mundulus* pierced the egg and sucked it. In some cases the egg was sucked nearly dry, in others the egg was only pierced and very little sucked, but it led to the death of the egg and to the development of the fungus. Thus the fungus is symbiotic and passes from adult to embryo and is always present. Whether the leafhopper can be reinfected by spores developed outside is not known. It is highly probable that the spores play some part in the metabolism of the insect, as similar bodies are found in all species of Delphacidae and many other Homoptera.

Thus we see that the least obvious of the death factors was the keystone of the complex and one that was overlooked by observers in different parts of the world.'

How often this statement has proven to be true in succeeding projects over the years!

Muir returned with a cage of live *T. mundulus* in June 1920, and after thorough testing to be absolutely certain that it would not live on sugar-cane leaves but required leafhopper eggs for food, releases were made in July 1920. In fact, according to Dr J. G. Myers, the West Indian entomologist, Dr Muir once told him that he had spent several sleepless nights considering the risks of the introduction, should the bug not remain faithful to its egg-eating habit. C. E. Pemberton then went immediately to Fiji, where *T. mundulus* was known to occur, and sent six well-stocked cages by ship to Honolulu, from September through to November. Abundant material was released in the field, and some was cultured for future releases. It became established at once, and by 1923 the leafhopper everywhere subsided to an insignificant level, where it has remained ever since.

H. P. Agee wrote:

> 'The value of this work in economic entomology cannot be measured in millions of dollars, regardless of how high we carry the count. It saved an industry that has been the mainstay of modern Hawaii . . .'

Huffaker & Caltagirone (1986) estimate the total savings accrued by this project to be in the order of US $600 000 000. (See Muir, 1931; Pemberton, 1948.)

Sugar-cane beetle borer in Hawaii, 1907–10

This insect, *Rhabdoscelus obscurus* (Boisduval), also known as the New Guinea sugar cane weevil, was the second major pest in Hawaii to become the object of a biological control project. It was of south Pacific origin, quite possibly native to Papua New Guinea and neighboring islands, where it probably fed on sago and other palms as well as banana. It presumably adopted sugar-cane upon the introduction of that plant, and was spread from island to island by natives carrying sections of sugar-cane along in their travels. It was a pest on cane in Hawaii as early as 1865. Albert Koebele regarded this beetle as the most injurious enemy of the sugar-cane in Hawaii in 1896 because of the damage its boring did to the cane and its difficulty of control.

It was conservatively estimated by Hawaiian entomologists that before parasites were introduced, i.e. up to 1910, the annual loss from this insect amounted to between US $750 000 and US $1 000 000. It did not threaten to wipe out the industry, but was a continuing source of appreciable loss and complaint.

Dr F. A. G. Muir of the Hawaiian Sugar Planters Experiment Station commenced the foreign exploration for natural enemies of the borer from south China in January 1907. The successful completion of his mission occurred over three long years later, and entailed a nearly unique odyssey in foreign exploration for natural enemies. The story has been well told by Pemberton (1948)[*]:

> 'At the time Muir began this adventure there was some published information concerning the borer. It received its name from specimens collected on New Ireland in 1835. In 1885 it was recorded from the Island of Larat and by 1895 literature referred to its presence in New Guinea. It was also known in Tahiti and was supposed to have reached Fiji from Hawaii. It had also reached North Queensland presumably in cane brought in from New Guinea. Muir thus had some grounds for suspecting that the native home of this borer was somewhere in New Guinea or the Malay archipelago to the west. He began by spending about two months in the Federated Malay States searching for the borer but found none. He then spent 10 weeks in West Java studying borers allied to the cane borer. Here he found no true parasites; but a number of predatory insects (Histeridae and Hydrophilidae) were observed feeding on the grubs of these related borers that he found in palms and banana stumps. He failed to find the sought after cane borer in Java.

[*] Copyright © 1948, Hawaiian Planters Record, Honolulu. Reprinted by permission.

Muir then went to Borneo in July 1907 to undertake similar borer studies. He failed to find the cane borer, but again found predatory insects similar to those found in Java which fed on other borer grubs occurring in palms and banana stumps. He remained in Borneo until October 1, 1907 and then returned to Java to ship more of these predatory insects to Honolulu. These failed. Having found no promising natural enemy of various borers which he studied in Java and Borneo, Muir decided to move into regions where the cane borer was known to occur and on October 1, 1907, he left Batavia, Java and journeyed some 1500 miles eastward to the Island of Amboina, where he had reasons to believe the borer could be found. Arriving at Amboina on October 9, he spent 6 weeks searching for the borer in sugar cane, but failed to find it. He then departed for Larat, a six-day steamer trip to the southeast, where the cane borer had been found by a wandering naturalist in 1885. Arriving at Larat on November 29, 1907, he soon found the cane borer in sugar cane and in sago-palm stalks and betel-nut palm stems. A month's study of these revealed no parasites and he returned to Amboina about January 9, 1908. Having learned at Larat that the borer occurred commonly in sago-palm leaf stalks, he renewed his search for borer parasites on Amboina and was rewarded with success as soon as he visited a sago-palm swamp. Here he immediately found the borer in the palm-leaf stalks and also discovered that from 25 to 90 percent of the borer grubs were parasitized by a tachinid fly, which was later described and named *Ceromasia sphenophori* by Dr J. Villeneuve. This is the parasite which ultimately checked the borer ravages in Hawaii. It is known as *Microceromasia sphenophori* Vill. [The current name is *Lixophaga sphenophori* – authors.]

To import this fly from Amboina to Honolulu alive in those days was a disheartening problem. Transportation was very slow and irregular and all material had to be sent to Macassar, Celebes, thence to Hong Kong and from there transshipped to Honolulu. Muir remained in Amboina eight months striving in various ways to surmount these difficulties. Several shipments of the fly were attempted but all failed. Mr Terry had been sent out to Hong Kong to receive and send on the consignments from that point to Honolulu. Muir sent 18 consignments of beneficial insects in all, many of which were histerid beetles which he found preying on borer grubs. With Terry's aid in handling this material at Hong Kong many of these beetles or their larvae reached Honolulu alive and were liberated in Hawaiian cane fields, but were never seen again. None of the flies survived the trip. Finally in September 1908, Muir personally conducted a consignment of the flies from Amboina to Hong Kong, but all died en route.

Muir then decided to return east to the Island of Ceram to devise other means, if possible, by which the parasite could be introduced into Hawaii. He had as a companion J. C. Kershaw, who had the year before published an excellent book on the butterflies of Hong Kong. They worked together in Ceram for a month without success. Knowing that the cane borer occurred in New Guinea, Muir laid plans to investigate its parasites there, if such could be found. He returned again to Macassar and obtained transportation to Port Moresby, New Guinea, on April 9, 1909. Soon after reaching Port Moresby he found the borer in sugar cane some 15 miles inland and immediately discovered the same parasite, which he first found in Amboina. This he soon attempted to bring to Honolulu alive via Thursday Island and Brisbane, Queensland, but he was forced into a hospital with typhoid fever at Brisbane and his consignment of flies, which was transshipped to Honolulu by a friend, arrived dead. After 5 weeks hospitalization in Brisbane, Muir returned to Honolulu to recuperate.

On January 8, 1910, Muir left Honolulu for Queensland to resume his labors on this problem. During the same month Mr Kershaw was engaged on the entomology staff. A relay station for breeding the parasite was established at Mossman, North Queensland, with Kershaw in charge. Muir then proceeded to Port Moresby again to obtain the parasite. A mail shipment to Kershaw failed and on April 22, 1910, Muir left Port Moresby for Mossman with a large consignment of parasite material in a living condition. After unexpected delays and other difficulties he finally reached Mossman with a good lot of living parasites. The borer being common in cane at Mossman, Kershaw had experienced no difficulty in having three large cages well stocked with borers awaiting Muir's arrival. This was on May 5. Some of these parasites were used to stock the cages for the production of a new generation. Though suffering from an attack of malaria at the time, Muir took the remainder of the parasites and departed for Fiji on the first available steamer. At Fiji, where the borer was numerous in cane, a cage containing borer-infested cane had been previously prepared by the Colonial Sugar Refining Company to await Muir's arrival. Upon reaching Fiji, Muir immediately placed the parasites in the cage and then was forced to enter a hospital for treatment and rest because of his malarial condition. Kershaw reached Fiji from Mossman, Queensland, on August 9 with the remainder of the parasites and by then Muir had sufficiently recovered to depart for Honolulu with Kershaw's cages and the one he had already prepared in Fiji. Muir arrived at Honolulu August 16 with a good stock of living parasitic material and a month later Kershaw arrived from Fiji with more which he had reared in Fiji from a portion he kept out of the original lot brought by Muir.

These flies were multiplied at the Station in large cages and distributed to the plantations for more than two years, resulting in the establishment of this beneficial insect on all parts of Hawaii where sugar cane was grown, and, during the ensuing years, in the savings of millions of dollars.'

Fig. 5.3. shows various developmental stages of the beetle and the parasite. To make the story more complete, a subsequent change in variety from soft- to hard-rind canes, which are not as susceptible to the borer, along with the constant operation of the parasites from Papua New Guinea further reduced the borer so that damage became even more negligible (see Pemberton, 1948). In the 1960s, an inadvertent change to

Fig.5.3. Various developmental stages of the sugar-cane beetle borer and its parasite. (Courtesy of Asher K. Ota, Hawaiian Sugar Planters Experiment Station.)

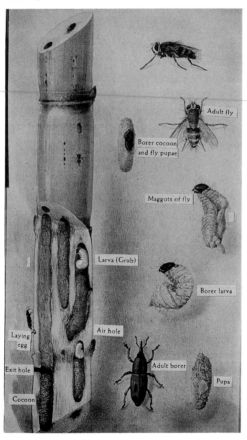

susceptible varieties on the island of Kauai again led to an increase in the borer problem, but this has been ameliorated by replanting with resistant or moderately resistant varieties. This, combined with the biological control exerted by *Lixophaga*, has kept the borer below economically damaging levels.

Citrus whitefly in Florida, 1910–11

This work was undertaken at the time the gypsy moth and brown-tail moth parasite work was at its peak and was generating considerable enthusiasm. The citrus whitefly, *Dialeurodes citri* Ashmead, then was by far the most serious pest of citrus in the gulf states, from Florida to Texas. Overall, considering the injury it caused and the difficulty in control, it could be considered the most serious pest of citrus in the entire country. In Florida it was estimated that the loss in an average infested grove was 45–50 percent of the normal value of the crop, and statewide the annual loss was over half a million dollars. This led to demands from Florida for foreign exploration to discover natural enemies of the whitefly.

Dr L. O. Howard assigned Russell S. Woglum to travel abroad and import any natural enemies he might discover. Just previously, Woglum had been perfecting the HCN fumigation method for scale insect control in California. The United States Congress, contrary to its stand when money was sought for the cottony-cushion scale project, readily made a special appropriation in 1910 for the purpose of searching the world to discover the native home of the citrus whitefly and to learn if it was held in check by natural enemies. Woglum soon decided that special emphasis should be placed on a search of tropical and subtropical oriental areas, based upon the presumption that citrus was native to the Orient and because specimens of the citrus whitefly from Japan, southern China and India were present in the collection of the Bureau of Entomology in Washington, DC. He set sail on July 31, 1910 via the Mediterranean, where he failed to locate any citrus whitefly in the citrus groves of Spain, Italy or Sicily. Proceeding to Ceylon, again no citrus whitefly was found.

His first objective in India was the great Indian Museum in Calcutta, where he soon found specimens of the citrus whitefly collected in the northwestern Himalayas about 1893, and also some from a remote locality named Kulu in northwest India. Kulu was virtually impossible to reach with the primitive transportation available, so Woglum decided to concentrate on government botanical gardens in north India, which generally had a large variety of fruit trees. The oldest botanical garden in a known citrus area was learned to be at Saharanpur. Soon after arriving, he found the citrus whitefly to be generally distributed, but rather scarce and not a pest. Nearly at once he found a tiny ladybeetle, *Cryptognatha flavescens* Motschulsky, feeding on eggs and larvae of the whitefly. After several days' collecting he carefully prepared a shipment which was forwarded to the American Consul-General at Calcutta, who placed it in charge of the captain of a cargo steamer sailing directly to the United States. Unfortunately, all specimens arrived in Florida dead. A second sending also failed.

Moving on to Lahore in the Punjab in November, Woglum found a fine botanical garden and abundant citrus. Again, the citrus whitefly was found to be generally distributed but scarce. He soon found emergence holes made in the mummified whitefly pupae by the exit of a minute parasitic wasp, but due to the cold weather no adult parasites could be found. From infested leaves sent to Dr Howard, the parasite was determined to be new to science and was named *Prospaltella lahorensis* Howard. Its current name is *Encarsia lahorensis*.

Because it was then winter and live specimens were scarce, Woglum decided to survey as much new territory as possible. Throughout India he found the whitefly wherever citrus was grown, and nearly always found evidence of the parasite as well. He proceeded on through Burma, Java, south China and the Philippines, but saw a few citrus whiteflies only at Macao, near Hong Kong. Howard had cabled Woglum to return to Lahore and prepare shipments of the natural enemies for Florida, but Woglum fell ill, probably with malaria, and had to spend a month in the hospital in Manila. He finally returned to Lahore in April 1911, and after a painstaking search found a fairly good source of whiteflies on an ornamental citrus hedge, which became his main material.

Elaborate special preparations had to be made to get the natural enemies alive from north interior India to Florida in the year of 1911. The trip from India to Florida took five to six weeks and the entire life cycle of the parasite was only about three weeks, so that the adult parasites would emerge and die *en route* unless living whitefly material was available, in which they could lay eggs and start a new generation. This meant considerable innovation in artificially infesting nursery trees with citrus whitefly in different stages and in cages, in order to exclude and eliminate other pests. These infested trees, inoculated with the natural enemies, would then be shipped in Wardian cages – a sort of portable miniature greenhouse.

With the coming of the monsoon rains in August, the fall brood of adult whiteflies emerged and laid eggs on the potted nursery trees. Good infestations developed, and by mid-October all was ready. Three Wardian cages were filled with material parasitized by *Encarsia lahorensis* and two with the predator, *Cryptognatha*. A sixth cage contained uninfested trees. Woglum's cages being started on their trip by ox cart are shown in Fig. 5.4. Leaving Lahore on October 20, Woglum proceeded by train to Bombay and thence by steamer for Europe, as there was no through vessel to New York. Careful planning enabled good connections and trans-shipments to be made at Port Said and Naples, so little delay occurred. *En route* he cared for the cages like a mother hen, finally arriving in New York exactly one month out of Bombay and five weeks from Lahore.

The cages were immediately sent by express to Orlando, Florida, arriving December 2 at the government laboratory. Examination showed 28 live healthy beetles and eight adult parasites, as well as a large number

Fig.5.4. Wardian cages on oxen used to transport infested plants bearing natural enemies of the citrus whitefly at the beginning of their long journey from Lahore, India, to Florida in 1911. (From Woglum, 1913.)

of developing immature parasites. Sufficient material was on hand for successful culture and all seemed promising.

Unfortunately, at this time the whiteflies in Florida were in the winter-dormant pupal stage which is not suitable for either the predator or parasite, which only attack younger stages; the result being that both species were lost before spring. For reasons unknown, such importations to Florida were never tried again, although Woglum outlined what he considered to be a feasible procedure.

Only today is the finale being written. The citrus whitefly was found at several places in California in 1967, and in 1968 one of the authors (PD) began importations of *Encarsia lahorensis* from India and Pakistan (then West Pakistan), only this time much more simply and rapidly, by airmail. This parasite became established in California, and was subsequently sent also to Florida, where it finally became established in 1977.

The sad part of the story is that these natural enemies could have been reimported to Florida and doubtless established many years ago, thus probably preventing considerable economic loss and reducing or elimi-nating the necessity for chemical sprays for the whitefly. (See Sailer *et al.*, 1984; Woglum, 1913.)

Spiny blackfly in Japan, 1925

This is an example of how simply some pest problems can be solved by biological control. The project was casually arranged on a co-operative basis and, as far as we can learn, cost nothing. Nonetheless, it is a great economic success, ranking among the best in the world.

The so-called spiny blackfly, *Aleurocanthus spiniferus* Quaintance, which technically is a member of the whitefly family Aleyrodidae, was discovered near Nagasaki in Japan by Professor S. I. Kuwana and Dr T. Ishii about 1922, although it probably already had been present and spreading for some time. Within a few years it became one of the serious pests of citrus trees in the southern island of Kyushu. Spraying and fumigation failed to bring relief.

No effective natural enemies occurred in Japan, and it was surmised that the pest came from south China, which is in the native home range of citrus and where the spiny blackfly was known to occur. Entomologists and horticulturalists were considering an attempt to discover natural enemies on the mainland, when Professor F. Silvestri of Portici (Naples), Italy, who was on a temporary foreign exploration mission for the University of California to find parasites of the California red scale, visited Japan. At that time he was requested by the Japanese to be on the look-out for enemies of the spiny blackfly upon his return to the Asiatic mainland.

Professor Silvestri obliged, and on his second trip to Nagasaki, Japan, May 23, 1925, brought with him a parasite, *Encarsia smithi* (Silvestri), and a ladybeetle, *Cryptognatha* sp. Professor Kuwana met him there, and on May 25 Silvestri and Ishii colonized them on heavily infested trees in the village of Ikiriki, near Nagasaki. In all there were 20 parasites and 10 beetles. The beetles apparently were never seen again, but the parasites were a different story. By the end of November, adult parasites were easily found on the leaves of the original release tree. However, because Nagasaki has a colder winter than Canton, it was feared that the parasites might not overwinter successfully.

These fears were allayed when parasite activity was noted on the release tree in June 1926, and it was found that 74 percent of the whitefly pupae had exit holes from which adult parasites had emerged. Spread occurred rapidly and was aided by distribution of leaves bearing parasitized pupae, so that within a short time the pest was almost completely eliminated. All this came about from an initial stock of only 20 parasites, hand-carried in a small vial. Control remains perfect to the present day, as we can attest from personal observations made thoughout the citrus areas of Japan in recent years. (See Kuwana, 1934.)

Prickly pear cactus in Australia, 1920–5

This was a massive, long-term project, comparable in this respect to the gypsy and brown-tail moths project, and represents the first striking success in biological control of a weed. Numerous entomologists and assistants were employed and a great deal of foreign exploration carried out in the Americas, ranging from the United States to Argentina.

Ironically, but for an unfortunate fluke, the natural enemy *Cactoblastis cactorum* (Bergroth) (see Fig. 2.40, p. 86), which caused the ultimate destruction of the prickly pear, could have done the job at least ten years earlier than it did, and with a minimum of cost. It would have been another simple, nearly miraculous success like the cottony-cushion scale–vedalia project. It seems that Mr Henry Tyron, one of a pair of members of a Queensland Prickly Pear Travelling Commission, was impressed by the potentialities of the *Cactoblastis* work he saw in Argentina and actually brought a small number of larvae back to Australia in 1914; unfortunately, he was not able to rear these through to maturity, or we might have had quite a different story. No project activity followed this until 1920, and *Cactoblastis* was not imported until 1925.

The enormity of the prickly pear problem in Australia is difficult to conceive. Evidently, various species were brought by the early settlers as ornamentals and escaped. Two species, *Opuntia inermis* and *O. stricta*, assumed major catastrophic pest proportions, but several others were pests in more localized districts. These two obtained their footing in the 1800s, when the country was being opened for grazing and population was sparse. From 1900 the cacti spread rapidly, reaching a peak about 1925, when 24 million hectares were heavily infested, 12 million of which were so dense that the land was completely useless. Actually hundreds of miles were impenetrable to man or beast. Otherwise it was grazing land, with potential for dairying and general farming. Eighty percent of the infested land was in Queensland and 20 percent in New South Wales. The cost of chemical or mechanical control was more than the land was worth, and so was out of the question. During the peak years, the rate of spread of the cactus was alarming; year by year more land becamed unoccupied and more holdings and homesteads deserted.

The project which led to ultimate and complete success began with the appointment of the Commonwealth Prickly Pear Board in 1920. Their first steps were to send entomologists to America and to establish insectaries and quarantine facilities in Australia. They early decided to concentrate on cactus-feeding insects rather than disease organisms, because disease didn't appear to be important in the field, several diseases occurred accidentally in Australia already, and there was the question of the disease perhaps transferring to other plants. The entomological work in America was centered in the United States, Mexico and Argentina, where experimental centers were maintained; however, virtually all other American countries were surveyed.

Altogether, 150 species of insects restricted to feeding on cactus of one sort or another were discovered. About 50 species were sent to Australia, totaling more than 500 000 individuals. All imported material was screened through a central quarantine and culture laboratory at Brisbane and subjected to starvation tests on a wide variety of other plants before being sent out to local breeding or colonization stations. Final tabulation showed that twelve species of prickly pear insects had become established and were exerting some measure of control, when the establishment and final success by *Cactoblastis* put an end to further trials.

Larvae of *Cactoblastis cactorum* were collected, apparently by Alan P.

Dodd, at Concordia, Argentina, in late January 1925, and taken to Buenos Aires, Argentina. It is interesting to note that they came from two different species of *Opuntia* than those they where destined to subjugate in Australia. Moths emerged in February and readily laid many eggs. Six Wardian cages were filled with *Opuntia* for food, and 3000 *Cactoblastis* eggs were placed on the prickly pear in the cages. In March, the shipment went by steamer via Cape Town to Australia. Some 250 larvae were removed for examination at Cape Town, so the original shipment then included 2750, and this gave rise to all Australian stocks. The unsupervised cages arrived in Brisbane ten weeks and 23 000 km later, in May 1925, in excellent condition, containing about half-grown larvae. This was readily possible, as cactus pads may last several months. These larvae were given food and they successfully pupated in August–September 1925, producing 1137 cocoons. One thousand and seventy moths emerged and produced 100 605 eggs, a return of nearly 36-fold. From then on success was assured.

Had this shipment not succeeded, it is problematical whether, or at least when, another would have been made. Dodd returned to North America following the shipment, because Argentina had a limited fauna of *Opuntia* insects and he had no particular reason to believe that *Cactoblastis* might be more successful than other *Opuntia*-feeding insects. According to Dodd, 'certainly, its remarkable achievements could not have been foretold'. Thus, if the one shipment had not resulted in establishment, further investigations in the Argentine may not have been undertaken for several years. Again we see the common combination for successful scientific discovery – good basic research along with a measure of serendipity.

The first colonizations of *Cactoblastis*, consisting of 2 263 150 eggs, were made in various localities in Queensland and New South Wales during February–March 1926. By March 1927, 10 196 150 eggs had been released in the field. Redistribution of eggs obtained from field material finally amounted to 389 225 520 by November 1929.

Establishment and increase occurred so rapidly, that by 1930–2 the general collapse and destruction of the original stands of prickly pear had occurred. Mile after mile of dense growth collapsed in a few months under the concentrated attack of enormous numbers of larvae. Some spectacular examples showing prickly pear infestations before and after clean-up by *Cactoblastis* are shown in Figs. 5.5 and 5.6. According to Dodd, the most optimistic scientific opinion could not have foreseen the

(a)

(b)

Fig.5.5. Biological control of prickly pear cactus by *Cactoblastis* in Queensland, Australia. (*a*) Cactus-infested range before the establishment of *Cactoblastis*, (*b*) virtual elimination of cactus following release of *Cactoblastis*. (From Dodd, 1940.)

(a)

(b)

Fig.5.6. Biological control of prickly pear cactus by *Cactoblastis* in Queensland, Australia. (*a*) Homestead abandoned because of prickly pear infestatiion, (*b*) homestead reoccupied following clean-up of cactus by *Cactoblastis*. (From Dodd, 1940.)

was known of realistic possibilities at that time, that such miscellaneous approaches were decided upon as the importation of a series of general parasites of non-related hosts from Hawaii in the hope that one or two might attack *Levuana*, trials with the predatory ant, *Oecophylla smaragdina*, which is used in the Orient to control certain citrus pests, and possible importation of coconut-inhabiting insectivorous birds from Ceylon, India, etc. These approaches either failed or were finally rejected as impractical, undesirable, or unfeasible. The part of the plan that paid off was the decision to continue the search for *Levuana* or closely related forms and their parasites, and to reattempt the introduction of the parasites of *Cathartona catoxantha*. Even the latter was a long-shot, because it meant that the parasites would have to transfer from *Cathartona* to *Levuana*, which not only is a different genus but, in fact, a different subfamily. Often host specificity prevents such transfers, but from previous studies he had made of forest moths in Canada, Tothill had learned that in certain cases the same parasites could attack distinctly different host species, hence he concluded that a similar result might be feasible with *Levuana*.

Fig.5.8. Destruction of coconut palms in Fiji by the coconut moth, *Levuana iridescens*, before the importation of natural enemies. (From Tothill, Taylor & Paine, 1930.)

Beginning in 1925, foreign exploration was conducted throughout Melanesia and Indonesia by Taylor, Tothill and Paine. Nine species related to *Levuana* ultimately were found, mostly quite rare and heavily parasitized, and Taylor even returned to Fiji once with a shipment of *Cathartona* parasites, but none survived. Because of the rarity of most species, coupled with the fact that *Cathartona catoxantha* was more readily available in Java and other parts of Indonesia and was known to have at least two promising parasites, the tachinid fly *Bessa* and the braconid wasp *Apanteles*, it was decided to concentrate on importing these two. This would be the third attempt.

H. W. Simmonds was sent to Kuala Lumpur early in 1925 to seek out a possible *Cathartona* outbreak, and meanwhile Taylor was traveling in the then Federated Malay States, when he located a small outbreak of *Cathartona* about 500 km (300 miles) from Singapore, at Batu Gajah. Both species of parasites were present. Simmonds joined him there to prepare as large a shipment as possible for Fiji. This was to be effected in 17 large Wardian-type ventilated cages, each of which would hold four or five small coconut seedlings infested with both parasitized and unparasitized larvae of *Cathartona*. The latter were to serve as hosts for egg-laying by parasites emerging *en route*. Finally, about 20 000 larvae were placed on the 85 young palms and sent the 500 km to Singapore by rail.

The difficulties facing them in getting a shipment through alive to Fiji were stupendous. It had been decided for a variety of reasons that direct shipment from Singapore to Fiji was required, but principally because of Australian quarantines. However, there were virtually no direct sailings, so most went by Australia. Also, the collections and shipments had to be made when the insects were available, but this had to coincide with the availability of a ship, which often was months apart. About the only hope was the Clan Line of cargo boats, which ran between London and Fiji, sometimes via Singapore. In June it was learned by cable that the *Clan Mackay* would sail from Singapore on July 10 for Fiji, but not direct. Meanwhile it was learned, again by cable, that the *Clan Matheson* was sailing from Java direct to Suva, Fiji, on July 10, so for an extra £250 the *Clan Mackay* was instructed to call at Surabaya, Java, to transfer the parasites to the *Clan Matheson*, which would be ready to sail. Thus several days were saved, although it still took 25 days following collection to reach Suva. Taylor accompanied this shipment to supervise and care for it *en route*, arriving in Suva on August 3, 1925, and immediately conveyed the cages to the quarantine insectary built for this purpose (see Fig. 5.9).

(a)

(b)

(c)

Fig.5.9. (*a*) Landing on Ra island of the first colony of the coconut moth parasite *Bessa remota* to be released from the Suva insectary. (*b*) The *Bessa* insectary at Suva. (*c*) *Bessa* breeding cages outside the insectary after release from quarantine. (From Tothill, Taylor & Paine, 1930.)

Examination showed about 315 live adult parasitic flies of *Bessa* (Fig. 5.10) but no *Apanteles*, which had all perished on the way. The tachinid flies were transferred to cages stocked with *Levuana* larvae and to everyone's joy proceeded to attack them at once. By August 21, the first generation of new adult flies began to emerge, and insectary culture was assured. Within six months, i.e. by January 20, 1926, over 15 000 flies had been bred and colonized over the *Levuana*-infested districts. The potential of the fly was first realized when just two months after importation, the parasites were found to be accidentally established around the insectary. From this time in October onward, the dispersal of the fly was remarkably rapid, and it was found to be established throughout all *Levuana* zones within six months of the first liberation.

The control results were just as remarkable. Within three months of liberation of the Suva colony, *Levuana* was exterminated on the original release trees. Six months after the initial introduction of *Bessa*, many of

Fig.5.10. *Bessa remota*, the parasitic tachinid fly that controlled the coconut moth in Fiji. (1) Adult, (2) cocoon of the coconut moth opened to show the skin of the larva (below) that has previously been killed by the larval tachinid parasite which has pupated (above), (3) side view of head of adult *Bessa*. (From Tothill, Taylor & Paine, 1930.)

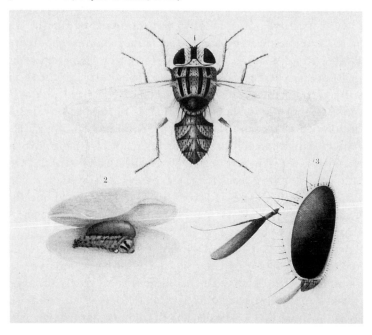

the outbreaks of *Levuana* had subsided completely. When the final report was written in 1929, there had been no new outbreak of *Levuana* for three years, and the insect had become so rare that reputedly a visiting entomologist, searching specially for it, would almost certainly fail. It seems especially significant that this single natural enemy, *Bessa remota* (Aldrich), apparently has given better control of its adopted host, *Levuana*, than it did if its native host, *Cathartona*, although such control was perfectly satisfactory. This should serve as a good indication of some of the unexplored possibilities remaining in biological control research. (See Tothill, Taylor & Paine, 1930.)

Eucalyptus snout-beetle in South Africa, 1926

In the early 1920s, species of *Eucalyptus* were highly important timber and ornamental trees in South Africa. Mining companies alone had about 4000 ha of plantations, and many times more thousands of hectares were held by municipalities and private individuals. These trees had been remarkably free of insect problems, but in 1916 the eucalyptus snout-beetle, *Gonipterus scutellatus* Gyllenhal, was discovered at Cape Town. The beetle spread rapidly, and by 1924 was found at Pietermaritz-burg, Natal, a great distance from Cape Town, as well as in the highveld of the Transvaal. By now the beetle was threatening the entire eucalyptus-growing industry of the country, and continued to rapidly invade new areas. Continued feeding damage actually killed trees, but most damage resulted from greatly reduced growth and the formation of knots and malformations in the wood. Many blocks of trees were felled as no longer worth keeping, owing to the depredations of the beetle.

C. P. Lounsbury, Chief of the Division of Entomology, believed that the solution of the problem lay in the introduction of natural enemies from the native home, Australia. Additionally, all other control measures had failed. After fruitless negotiations with Australian and New Zealand authorities to have a study made and parasites shipped, he finally got the financial backing he needed from his own government and in January 1925 appointed Dr F. G. C. Tooke to lead the project and to go to Australia in search of natural enemies in 1926.

Before leaving for Australia in June 1926, Tooke corresponded with Australian entomologists regarding the snout-beetle and its parasites, but about all he learned was that the beetle was extremely scarce, probably ranged from Queensland through New South Wales and Victoria to South Australia and Tasmania, and that nothing was known of its natural enemies. It appeared to Tooke as if he would be looking for the proverbial needle in a haystack. He arrived at Adelaide, South Australia, on July 19, 1926 and at once contacted Mr A. M. Lea, a taxonomic specialist in the genus *Gonipterus*, who had specimens of *G. scutellatus* from Tasmania, New South Wales, Victoria and South Australia. Inasmuch as *Eucalyptus viminalis* was known to be a preferred host, he searched out this species in the countryside near Penola, and 11 days after arrival, on August 1, he found some egg masses of the beetle in question. Returning to Adelaide, dissection of the eggs showed many of them to contain a small hymenopterous mymarid parasite. This proved to be new to science and was described by A. A. Girault as *Anaphoidea nitens* (see Fig. 5.11). It is now known as *Patasson nitens* (Girault). Subsequent searching around Melbourne and elsewhere failed to yield any beetles, so he returned to Penola on August 20, to collect parasitized egg masses. Enough material was collected to send off the first shipment from Adelaide to Durban, South Africa, on August 24, 1926. At this time he also found the snout-beetle larvae to be parasitized by a tachinid fly, but in low numbers.

Although Tooke scouted considerably on several occasions in Victoria and New South Wales, he never found collecting to be as good as in the *Eucalyptus* stand some ten miles from the small town of Penola, so he set up field headquarters there, where he was compelled by necessity to live and work in a small wooden shack (see Fig. 5.12). He put in the best part of a year here, which must have been a lonely, but most surely a dedicated existence. He soon learned from field and laboratory counts that the egg parasite was highly effective and lacked any hyperparasites. The latter point meant that there was no danger in making direct shipments to South Africa.

An idea of the extent of natural control of the beetle was gained from the fact that he had to search several square miles of country to collect 200 egg capsules, and this was in the Penola district, the best collecting area he discovered. In South Africa, at that time, many egg capsules could be collected from one small tree in a few minutes.

Once he had found the parasites, Tooke's big problem, like so many other foreign explorers, was how to get them home alive. The long journey by sea from Adelaide to Durban took three weeks, and at least a week was needed to collect material prior to shipment. This meant that the parasite, which had a life cycle of 17–24 days at normal temperatures, would emerge *en route*, and the adults would in all likelihood die. The scarcity of material precluded such a risk being taken. Tooke chose the alternative of using the ship's cool-room (about 2–4 °C, or 35–40 °F) to slow down the development of the egg parasite, and this proved successful using specially constructed, well-insulated teakwood boxes.

The second shipment of parasitized eggs was made on September 23, 1926 from Adelaide, and a third shipment on October 12. On February 27, 1927, Lounsbury cabled instructions to concentrate on the tachinid larval parasite, as the egg parasite *Patasson* was then definitely established in South Africa. This was just six months after the first shipment was made.

Fig.5.11. The mymarid egg parasite *Patasson nitens* that controlled the eucalyptus snout-beetle in South Africa. (*a*) Adult, (*b*) adult ovipositing in an egg mass of the snout-beetle on a eucalyptus leaf. (From Tooke, 1953.)

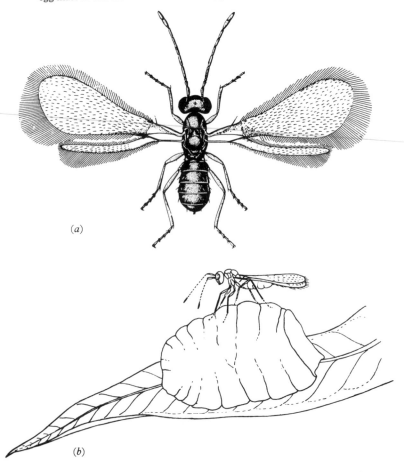

(*a*)

(*b*)

Although three shipments of the tachinid were sent to South Africa, most flies were dead on arrival and those that emerged subsequently failed to reproduce. Tooke arrived back in Cape Town, December 31, 1927 after more than a year and a half away from home.

Mass production and widespread field colonization of *Patasson* was then rapidly increased. From 1926 to 1933, some 736 395 parasites were released in hundreds of plantations throughout the Union. Its ability to spread long distances was demonstrated; in one case it dispersed more than 100 miles from the nearest site of liberation within a year.

According to Tooke, the control exercised by the parasite far exceeded expectations. Results in South Africa were spectacular in some districts, but much slower in others. This is the only known case of complete biological control where rates of control varied so greatly in different districts. This was due to differential effects of climate on the parasite's rate of increase. The period varied from three years, which is the more usual, to as much as 23 years in the most extreme climates. Perhaps a strain of the parasite slowly developed to operate successfully in the extreme climates. Complete control was achieved by 1933 in all of the southwestern and southern Cape province, which has a climate similar to that of South Australia. By 1936, complete control had occurred everywhere on all *Eucalyptus* species at elevations below 1200 m (4000 ft), as well as on 62 of the 65 species of *Eucalyptus* occurring above this altitude.

Fig.5.12. F. G. C. Tooke's living quarters and laboratory in the bush in South Australia where he spent nearly a year during 1926–7 in his quest for parasites of the eucalyptus snout-beetle. (From Tooke, 1953.)

It is significant that this is the only case of complete biological control attributed to an egg parasite acting alone. How much better and more rapid the control might have been had the tachinid parasite been successfully established remains an intriguing question. (See Tooke, 1953.)

Coconut scale in Fiji, 1928

As the biological control of the coconut moth was drawing to a successful conclusion, another serious menace to coconuts and the copra industry was rapidly increasing. The coconut scale, *Aspidiotus destructor* Signoret, first recorded as a pest of bananas in 1912, then of coconuts in 1916, was spreading at an alarming rate on coconuts by 1920. Damage was great enough to cause H. W. Simmonds, the Government Entomologist of Fiji, to seek natural enemies in Tahiti in 1920, where the scale was known to be present but much less severe. He imported and established two parasite species, *Aphytis chrysomphali* (Mercet) and *Aspidiotiphagus citrinus* Craw (now known as *Encarsia citrina*), and by 1925 had distributed them to many islands of the group. The spread of these parasites was rapid and they were considered to be doing some good, but the scale continued to spread to new islands and to cause severe outbreaks.

This insect, like all other scales, as well as whiteflies, mealybugs and aphids, sucks the plant juices. It prefers the undersides of the leaves, and in severe infestations forms a continuous overlapping crust of countless thousands per coconut leaf. This results in the leaf turning yellow, withering and even dying in severe cases.

By 1927, in spite of rigid quarantines, the scale had spread to nearly all of the Fiji island group and was the most serious pest of coconut in Fiji, since *Levuana*, the coconut moth, had been decimated by the imported parasite *Bessa remota*. The coconut scale was not quite so generally severe as the *Levuana* moth had been, but it was not unusual for trees to become completely defoliated, and many trees were killed after two or more outbreaks. Neither chemical nor mechanical control methods proved feasible for a variety of reasons.

In 1926, an attempt was made to secure additional natural enemies, this time from Java. T. H. C. Taylor was given this assignment, which was fraught with transportation difficulties because of the lengthy sea voyage and the paucity of ships sailing from Java to Fiji. However, as luck would have it, Taylor was able to make arrangements to ship his parasites and predators on a vessel carrying a cargo of other organisms to Fiji: Indian coolies. Little did these indentured laborers dream that some day they and their descendants would be the numerical and political majority in Fiji.

The coolie steamer *Ganges*, out of India, stopped in Java on January 1, 1927 and sailed on to Fiji on January 2 with Taylor bringing his infested coconut palms in Wardian cages. Three weeks later they arrived in Fiji, but a severe outbreak of smallpox on board prevented the shipment being landed at Suva but rather on the quarantine island of Nukulau, where the insects had to be tended for another five weeks with poor rearing facilities and limited stocks of scales. As a result, although some parasites and predators were colonized on Nukulau and later around Suva and some recoveries made, no permanent establishment occurred.

After mid-year 1927, it was evident that the enemies from Java either were ineffective or were not established at all. Consequently, J. D. Tothill, then Director of Agriculture for Fiji, assigned Taylor to go to Trinidad to investigate several species of ladybeetles which Urich had recorded as being important factors in the control of the coconut scale there.

Taylor started work in Trinidad in September 1927 with a survey of the natural enemies present. He found five species of ladybeetles to be more or less common and effective, even though at times they were heavily attacked by parasites of their own. These latter he subsequently took care to exclude from the shipments. By January 28, 1928, his collections of ladybeetles and his stocks of scale-infested palms were ready and were loaded on board ship. Nine large cages were used, holding from six to eight or more infested young palms. Three newly designed extra-large cages required four men to lift and move them. Additional food for the predators in the form of 80 heavily scale-infested young palms were carried along in five-gallon tins. All had to be carefully arranged to protect them from rough handling and damage from salt water during the voyage.

Although Taylor limited his initial predator stocks to 200 adult beetles, plus some larvae and pupae, per cage, the predators increased so rapidly during the voyage that many had to be removed to keep them from completely eliminating the food supply and thus starving to death before arrival in Fiji. Trans-shipment of the stock was made at Panama, February 1, 1928 to a steamer going via Tahiti to Suva, Fiji. The entire trip took just over five weeks.

All five species of ladybeetles arrived in Fiji in sufficient numbers to enable culture to be started, but more than twice as many *Cryptognatha nodiceps* Marshall (1517) as compared to all of the other four species (746) came through, which may have been an indication of the ultimate exclusively dominant role played by *Cryptognatha*.

Problems developed with insectary culture in Fiji, so with a little ingenuity Taylor devised a method of breeding in the field within cloth-sleeve-covered, heavily infested banana leaves. In each sleeve, starting with 20 beetles, he would recover about 300 in one month's time, a return of 15:1 with very little expense or labor. By September 1928, it was evident that *C. nodiceps* had by far the greatest potential, so breeding of the others was stopped and the small stocks of those remaining were liberated. Some of these became temporarily established, but evidently eventually disappeared after *Cryptognatha nodiceps* became completely dominant.

Cryptognatha nodiceps was a remarkable and spectacular success. Only nine months after the introduction of this beetle into Fiji, the coconut scale was reduced to non-economic levels in all important islands of the group. After 18 months (i.e. in September 1929), the scale was controlled completely on every single island and was so rare in many localities, where formerly it had been heavy, that it was difficult to find individual specimens. During the peak of the clean-up, adult beetles and larvae swarmed on every tree and the adults could be readily seen flying around in the air. The trees soon changed from yellow to bright green. Such rapid and thorough results exceeded all expectations.

When Taylor published his report of this success in 1935, not a single new outbreak had occurred. The ladybeetle constituted a permanent and effective check. The same results occurred on all host plants, many of which had been as badly infested as coconuts. One of the writers (PD) visited Fiji in 1969 and carefully surveyed for coconut scale. The scales – even single individuals – were virtually impossible to find on coconut, even though an occasional *Cryptognatha* adult was seen rapidly searching the clean leaves. It evidently is an excellent searcher that can survive at very low host densities, like the vedalia beetle. Additionally, it probably feeds on alternative host scales on other host plants. It is interesting to note, however, that the *Aphytis* mentioned earlier is still rather common on coconut scale in Fiji. Only one light infestation of the scale was found during that visit in 1969, and that only on avocado, but parasitization by *Aphytis* was considerable. Perhaps, after all, *Cryptognatha* should not be given all the credit, just as the vedalia beetle likewise should not in California. (See Taylor, 1935.)

Citrophilus mealybug in California, 1928

The highly successful biological control of the citrophilus mealybug, *Pseudococcus calceolariae* (Maskell) (= *gahani, fragilis*), ranks second only to that of the cottony-cushion scale in California. Today the insect is so extremely rare that few field entomologists or citrus growers have ever seen it. Yet in the 1920s it was a spreading scourge on citrus and many other plants. The discovery of its place of origin and of its parasites constitutes one of the best deductive pieces of scientific detective work in the annals of biological control.

The citrophilus mealybug was first found on citrus in southern California in 1913. How it arrived no one knew. It spread rapidly, and in a few years became a major, highly destructive, pest of citrus, other fruit trees and ornamentals in the State. Fumigation and spraying were ineffective in controlling it, so mass production of a mealybug ladybird predator, *Cryptolaemus montrouzieri* Mulsant, introduced by Koebele from Australia years before, was tried. *Cryptolaemus* did not overwinter well in California and was ineffective when unaided. The mass-release program developed under Professor Harry Smith's guidance provided a fair degree of control, but the mealybug continued to spread and each year more heavy infestations developed. In desperation, some growers even turned to high-pressure water washing in 1927, but this method was very expensive, necessitating the installation of water-pipe systems.

As the mealybug situation became more alarming in 1927, Professor Harry Smith of the University of California at Riverside (then UC Citrus Experiment Station) decided to send a collector abroad to attempt to secure effective natural enemies. Harold Compere was chosen to make the trip.

The citrophilus mealybug was obviously an invader from abroad in California, but its country of origin was unknown to entomologists. It was recorded from England and South Africa, but all evidence indicated that these also represented accidental importations. Curtis P. Clausen, then of the USDA, Bureau of Entomology, had searched unsuccessfully for it in China, Japan, the Philippines and Taiwan. Professor Filippo Silvestri, on temporary assignment for the University of California in 1924–5, had covered the same general area as Clausen, plus Indo-China, and failed to locate this species. Also Professor S. I. Kuwana, the Chief Entomologist of the Japanese Empire, and specialist on Coccoidea (including the mealybugs) of the Orient, had not seen this mealybug.

The Mediterranean citrus areas were excluded deductively, because it was reasoned that the mealybug would have been discovered if it occurred there, because of the presence of numerous competent citrus entomologists in that part of the world.

This narrowed the range down considerably. Then, the assumption was adopted that it was likely the citrophilus mealybug came from a climate similar to southern California, and from a locality closely linked by steamer to England, South Africa and California – the three places the mealybug was known to occur. The most likely common denominator appeared to be Sydney, Australia; thus this port became Compere's target.

He arrived in Sydney in September, and on the 27th, the first day spent searching, he found seven specimens of the citrophilus mealybug in the Sydney Botanical Garden. Several more weeks elapsed before a few more were found, so generally rare was the species. Drawing on past experience, Compere recalled that isolated plants sometimes become heavily infested, because the pest may find and build-up on a plant before their enemies discover it. Therefore, he deliberately made a search of the Sydney industrial district, where plants were widely scattered – only an occasional one being seen there in a garden. On the second day's search, January 18, 1928, an old mulberry tree was found completely infested in the front garden of a private house. At that time, the mealybugs appeared to be healthy and devoid of parasitism.

Meanwhile, beginning in September 1927, Compere was endeavoring to develop a small insectary, using sprouted potatoes to grow the mealybugs, but over a period of several weeks of searching around Sydney he found only 69 citrophilus mealybugs. Cultures were started from these, but some of them contained undetected, small, immature parasite larvae or eggs, and parasites got started in all the cultures. By January 1928, these parasites, mostly *Coccophagus gurneyi* Compere, later destined for fame in California, had increased to the point of overwhelming the cultures. They could not be held down sufficiently, even with constant removal of individuals. Moreover, the potato stock had deteriorated badly, so that the short-term culture situation was very precarious.

Because of these problems, Compere decided that the best plan was to personally transport the material on the first steamer leaving for California, using the many parasites and reduced numbers of mealybugs on hand, supplemented by fresh mealybug material to be collected from the heavily infested mulberry tree found on January 18. Professor Harry

Smith approved this plan by cable, and arrangements were at once made to depart on the steamship *Tahiti* on February 23, 1928.

On February 19, Compere visited the infested mulbery tree for the second time, in order to obtain the mealybug stocks to be used as hosts for the insectary-reared parasites during the voyage to California. To his surprise, close inspection showed that virtually all the mealybugs had been parasitized and contained parasite pupae. Countless thousands of parasites were present, principally *Tetracnemoidea brevicornis* (Girault), previously known as *Tetracnemus pretiosus*, which also was to become famous in California. Fortuitously, the timing was perfect for collection and shipment, inasmuch as the parasites were mainly in the pupal stage and, if cooled down, would remain so for some time before emergence. A large part of the tree was removed, and the material was packed in insect-tight boxes, which were placed in the cool-room of the steamer *Tahiti* at a temperature of about 3 °C (38 °F). However, no unparasitized mealybug stock was obtained, and this remained a problem. Meanwhile, because Australian potatoes, or citrus fruit, etc., could not be brought into California because of quarantines, Compere obtained American-grown potatoes from an American ship in Sydney Harbor just before departure, and used these to continue shipboard propagation of mealybugs. The mealybug supply problem was alleviated considerably when during a stop-over in Wellington, New Zealand, he found citrophilus mealybug in abundance on grapes and fruit trees, and obtained a box of heavily infested grapes to use as mealybug breeding stock.

During the three-week voyage to San Francisco, Compere set up a temporary insectary in a vacant hospital room on the *Tahiti*, and tended the cultures daily. New host mealybug stock was available in abundance from the material collected in Wellington. From San Francisco the material (except for the quarantined grapes) was sent on at once (about March 15) to the Riverside quarantine insectary.

Both *Coccophagus gurneyi* and *Tetracnemoidea brevicornis* were readily propagated at Riverside, their life cycles being only about three weeks. The first field colonies of *C. gurneyi* were liberated in June 1928, and the first field recovery was made on July 24, a very promising indication. Within a year, the species was thoroughly established throughout the infested area of southern California, as well as in the San Francisco Bay area, and insectary propagation was stopped. *T. brevicornis* cultures were supplied to the insectaries of Los Angeles and Orange Counties on April 23–4, and at the same time colonies were released in the field. The first recovery was made on August 15, although it was doubtless established before this, and it was soon found everywhere.

By the spring of 1929, within one year of release, there was a very appreciable reduction in the number of mealybugs in the districts where the parasites had been well established. Local insectaries which formerly had mass-produced *Cryptolaemus* continued to produce and distribute the parasites during 1929, so that by the spring of 1930 virtually the entire citrus acreage of southern California had been colonized, and both parasites thoroughly established.

In the spring of 1930, two years after the first releases, the usual 'peak hatch' of mealybugs failed to occur, and complete economic control was achieved. No infestations of any economic importance have occurred since that time. From our own experience we can safely say that today it might take a specialist anywhere from a day to a week to find even his first specimens. At the time of the successful control by parasites, it was shown by surveys that savings in Orange County alone amounted to from US $500 000 to $1 000 000 annually. This county represented about one-half of the then-infested citrus acreage (the mealybug was still spreading), so that overall savings to the citrus industry alone amounted to from $1 million to $2 million annually. The entire cost to the State of California was $1700, exclusive of Compere's salary of $150.00 per month, or all told less than $2600. (See Compere & Smith, 1932.)

Citrus blackfly in Cuba, 1930

This species, *Aleurocanthus woglumi* Ashby, like the spiny blackfly discussed earlier, is not a true fly but a member of the whitefly family Aleyrodidae. Its name derives from the shiny black color of the nymphs and pupae. Like the related scale insects, mealybugs and aphids it sucks the plant sap. Badly infested leaves wither and drop off and complete defoliation may occur in unchecked infestations, resulting in almost complete loss of crop. Before biological control was attained in Cuba, eradication was futilely attempted, and then expensive and repetitive spraying programs were conducted, but in spite of them the citrus blackfly continued increasing from year to year. This blackfly was first found in Jamaica in 1913, then in Cuba and the Bahamas in 1916, Panama in 1917, Costa Rica in 1919, and in more and more other countries and islands of the area over the years.

Because of its presence in Cuba, so close to Florida, there was real danger of its accidental introduction into the great Florida citrus-growing areas. It was reasoned by those concerned in the United States that if effective biological control could be attained in Cuba and other infested Caribbean islands, the danger of entry into Florida or other gulf states would be greatly lessened. As a consequence, the United States Department of Agriculture (USDA) entered into an agreement with the Cuban Department of Agriculture, Commerce and Labor to attack the problem co-operatively.

It was known from published records and museum specimens that the citrus blackfly was of Oriental origin, but even more importantly, Professor Filippo Silvestri had published in 1926 and 1928 that important parasites of the citrus blackfly occurred in the Orient. It will be recalled that he did this work while on foreign exploration assignment in 1924–5 for the University of California at Riverside, and that information gained on the same trip led to complete biological control of the related spiny blackfly in Japan. It is of interest that the original stimulus leading to the US–Cuba co-operative project apparently originated with Professor J. F. Tristan of San Jose, Costa Rica (where the blackfly was a pest) who, having read Silvestri's papers, set in motion a chain of correspondence in 1927 that finally culminated in the importation of parasites into Cuba. It was this background information that led C. P. Clausen, then of the USDA, to discover and import natural enemies from the Orient.

Clausen began foreign exploration in May 1929 and finished in August 1931. Two separate trips to tropical Asia were involved. P. A. Berry of the USDA was in charge of the receiving, culture and colonization work in Cuba. Clausen made studies and observations in the Philippines, Thailand, Malaysia, Java, Sumatra, Burma and Ceylon, but eventually settled on Singapore as the center for collection and shipment of parasites. Clausen found five species of parasites and nine of predators to attack the citrus blackfly. Of these, he rated three parasites, *Eretmocerus serius* Silvestri, *Encarsia divergens* (Silvestri) and *E. smithi* (Silvestri), to be important and effective in the order named, and the predators to be of little consequence.

In addition to the usual problems of maintaining living cultures of hosts and parasites during lengthy sea voyages, this work was complicated by the desire to use non-citrus host plants of the blackfly, because of the danger of introducing citrus diseases into Cuba. It was with difficulty that an infestation of citrus blackfly was obtained on mango seedlings, adequate to justify their being sent to Cuba. This first shipment was sent in Wardian cages from Singapore in March 1930, and arrived in Havana late in April. From this material, 42 female and 19 male *Eretmocerus serius* emerged, along with 34 female *Encarsia divergens*. The latter was not successfully cultured, but the former ultimately provided the control sought. Two additional shipments, emphasizing the *Encarsia* spp. and one predator, were made. Both involved more elaborate planning and preparation than the first. The second shipment was sent on citrus to New York (where citrus could be imported), with the idea of rearing the adult parasites out there and trans-shipping them on to Havana. Forty-three female *Encarsia divergens* and 51 female *E. smithi* were forwarded to Havana, but none arrived alive. However, 42 adults of the predator *Catana clauseni* Chapin were received alive in Havana and successfully cultured. They were colonized beginning December 1, 1930 and although they increased rapidly in some cases and reduced heavy infestations of the blackfly, they were unable to maintain the blackfly at low levels and were eventually supplanted by *Eretmocerus*.

The third shipment, made direct to Cuba on citrus plants originally sent from Cuba to Singapore and kept under quarantine conditions there, arrived in Havana June 3, 1931 (see Fig. 5.13). It consisted almost entirely of *Encarsia divergens*, and although 750 adults emerged in Havana, no

Fig.5.13. Twelve Wardian cages containing citrus plants infested with citrus blackfly on shipboard *en route* to Singapore to be inoculated with parasites for return to Cuba. (From Clausen & Berry, 1932.)

reproduction occurred, either in the labortory or in the field. A few *E. smithi* were also received, but also failed to reproduce.

Thus, the first small shipment was the only one of any consequence. A few of the *E. serius* from this shipment were directly released in the field but most were used for laboratory culture, which occurred readily. Soon large numbers were available for field liberation, which began in late June and early July, 1930. Shortly thereafter, the parasites became so abundant in the release sites that insectary culture no longer was necessary, it being much simpler to transfer field-reared parasites from grove to grove.

Eretmocerus increased and spread so rapidly from the first small release, that complete control occurred prior to August 1, 1931 in the first groves colonized, or in about one year. Frequently citrus groves in which *Eretmocerus* had been released six months previously would show enormous peak populations of the parasite, and three months later the groves were usually completely cleaned of the pest. It was estimated that one initial release of from 100 to 500 female parasites in the center of a two-hectare grove would provide commercial control in from eight to twelve months. Roughly, this meant the reduction during this period of the pest population, by a mere initial handful of parasites, from at least 100 million blackflies per 2 ha of heavily infested grove to just a few individuals per tree. By 1932–3 biological control in Cuba was essentially complete.

Amazingly, all this resulted from the initial stock of only 42 female *Eretmocerus* which survived the first shipment in 1930. The total cost of the entire project, as far as we can ascertain, was about US $40 000, about half being contributed by Cuba and half by the United States. This included expenses for travel and research in the Orient, collection and shipment of material, establishment and distribution of natural enemies in Cuba and some other countries.

From Cuba, *Eretmocerus* was soon sent to other islands. In April 1932, J. G. Myers introduced one large shipment of *Eretmocerus* into Jamaica from Cuba. Recoveries were made in July, and within a year it was nearly impossible to find a specimen of the citrus blackfly on the original release trees. In fact, Mr H. W. Edwards, who had received and colonized the parasites, made a standing offer of one shilling for each blackfly found on these trees. Similar results occurred in other islands. *Eretmocerus* was introduced into the Bahamas in October 1931, recovered ten months later, and was doing effective work by May 1932. It was introduced into Haiti by Dr H. L. Dozier in August 1931. Subsequently over the years it has been introduced into several other Caribbean islands or Central American countries, with generally outstanding results. The major exception has been Mexico, and that story is illustrative of how so-called biological control failures may only be indicative of lack of research emphasis or premature termination of a project.

Eretmocerus serius was a 'failure' in Mexico, at least as compared to its work in most Caribbean islands. The citrus blackfly was found in the state of Sinaloa, on the west coast of Mexico, in 1935 and, in spite of quarantines, rapidly spread during the next few years throughout the major citrus areas of all Mexico, not only causing enormous damage there, but creating consternation in the bordering citrus-growing states of California, Arizona and Texas. Again, for the same reasons as with Cuba, and after small-scale importations of *Eretmocerus* in 1938 failed to result in establishment, the USDA developed a formal co-operative project with the Mexican Department of Agriculture in 1943 to import *Eretmocerus serius*. Colonies were shipped in from the USDA Panama laboratory and released in the states of Colima, Nayarit and Sinaloa. The parasite became well established at several sites, and subsequently was generally distributed as new areas became invaded by the blackfly. Even though the parasite became abundant, it did not effect the control previously experienced in other countries, except in a few small areas of high humidity. It was concluded that climate was reducing the effectiveness of this parasite, and that additional ones should be sought abroad in areas having climatic conditions more comparable to the Mexican areas.

Therefore, in 1948–50 the search was directed to semi-arid areas of India and Pakistan, where a series of parasites was discovered and successfully imported, that gave complete biological control throughout Mexico. This truly outstanding achievement will be related in the next chapter. (See Clausen & Berry, 1932.)

Sugar-cane borer in the Caribbean area, 1932–3

This is to be but a small part of the sugar-cane borer story. Biological control has been going on for many years now with this borer, *Diatraea saccharalis* (Fabricius), and related species. It has been quite successful in some islands or districts: less so in others. Much of this variability has been due to the degree of research emphasis, or lack of it, and to differing climates and philosophies of control. Quite a number of parasites are known, and many are or were originally restricted to particular islands or areas. Much remains to be done. For example, Dr

J. G. Myers of the then Commonwealth Institute of Biological Control (CIBC) suggested that the possibility should be investigated of effective parasites occurring in Papua New Guinea, which would be adapted to parasitism in thick canes. To our knowledge, this has not been done.

In the Amazon basin, Dr Myers conducted some of the most rigorous and primitive foreign exploration ever engaged in by any entomologist anywhere; on the other hand, he was possibly the first to utilize commercial airlines for the shipment of parasites when in 1932 he took advantage of Pan American's recently inaugurated Caribbean flights to ship parasites from Cuba to Antigua.

Dr Myers had some definite hypotheses regarding biological control ecology that directed his quest, as was the case with Compere and others. He was dealing with a native species that had adopted sugar-cane as a host plant. He based his research and explorations on two main ideas: (1) there were many primitive 'ecological islands' in South America and the Caribbean area, where unknown parasites of *Diatraea* might exist on wild host plants, because they were isolated from sugar-cane areas, and (2) the known parasites of *Diatraea* occurring in sugar-cane areas might not be as well adapted to living in the sugar-cane habitat as was *Diatraea*. According to Myers:

> 'There is surely no valid ecological or practical difference between a pest introduced without its parasites and an indigenous insect which, in a circumscribed biological or geographical island, has learned to live upon a cultivated crop, while its parasites, although perhaps already abundant in the area, even on the edge of the fields, have not yet learned to or are in some way prevented from, attacking it in its new host-plant. The line of control endeavor should here take the course, not of trying hopelessly to establish the parasites in the cultivations, but, in accordance with the older and proven technique, of introducing another and more efficient parasite from outside.'

That he was justified in his assumptions is borne out 'by the fact that the most promising parasite so far discovered, the Amazon fly (*Metagonistylum minense*), was first found, not in sugar-cane, but in a wild host plant, in a primitive plant community' deep in the Amazon basin.

Before importing the Amazon fly, Myers had been engaged in the introduction of a tachinid fly parasite of *Diatraea*, *Lixophaga diatraeae* (Townsend), from Cuba into the Lesser Antilles. The first attempt failed, according to Myers, because it was largely a single-handed effort, lacking a full-time experienced worker on the receiving end. Later, during March to May 1932, Myers, along with his assistant L. C. Scaramuzza and some skillfully selected and trained workers consisting of a Spaniard, a Cuban Negro, a white Cuban, a Portuguese and a Haitian, sent nearly 7000 puparia of the parasitic fly to the entomologists, Mr H. E. Box in Antigua and Mr Mestier in St Kitts, via the rather new Pan American Airways flights. The spread of the fly and the progress of parasitism were speedy, and effective biological control was attained. The next year, this tachinid was sent to St Lucia with similar results, and has since been introduced to many other areas.

Dr Myers spent several years in nearly continuous foreign exploration. Five separate major journeys were made in northern South America. This account will be restricted to the one involving the discovery and importation of the Amazon fly, a major parasite of the sugar-cane borer. His itinerary was roughly as follows: leaving headquarters in Trinidad he proceeded to Para (= Belem), Brazil, from which he crossed one of the mouths of the Amazon to the great island of Marajo and then returned to Para, exploring the River Moju. Thence he proceeded up the Amazon to the Trapajoz River, and up it to Fordlandia and back, continuing up the Amazon to Manaus. From here he branched on to the Rio Negro and Rio Branco, to the Brazilian border of British Guyana. From there, by the headwaters of the Ireng and the Mazaruni Rivers, he proceeded in a circular route to Mt Roraima in Venezuela, returning by the Venezuelan and Brazilian savannahs to the River Uraricuera which flows into the Rio Branco. There, because of lack of communication with Manaus, he returned to the British Guyana border and walked down the cattle trail to the coast, exploring all host-plant associations for sugar-cane borer parasites *en route*. Altogether about 800 miles were covered on foot – according to Myers 'a method of progression which offers the best conditions for entomological work, and one which ought to be adopted more generally were the time available'. Except for this long trek on foot, the entire trip was by water.

Myers had more than his share of the three main difficulties in such South American travel, namely river rapids, disease and shortage of food. On occasion he was deserted under difficult conditions, but of all the varied helpers he had, he considered the Indian aborigines to be especially hardworking, intelligent and efficient as entomological assistants. Myers' wife shared much of his work, but she was invalided to England late in 1931 with severe malaria contracted in the unhealthy delta of the Orinoco.

All of the material of the parasitic Amazon fly was collected near

Santarem on the Amazon in 1933, and essentially all local transport was by water. Virtually all the parasite puparia were collected, by means of boats or canoes, from *Diatraea* infesting the floating grass-beds of *Paspalum repens*, which reach their maximum development in the vicinity of Santarem. Other parasites also were present but were already known elsewhere, and because in this area of the lower Amazon and lower Rio Branco *Diatraea* was scarce on sugar-cane, it seemed possible that the Amazon fly might play an important role in this scarcity.

At the height of the campaign, a small fleet of boats was engaged. Locally a small motor launch, a small sailing boat and 11 dugout canoes were used, employing as many as 40 collectors. A 26-foot launch, which could withstand the very heavy seas of the lower Amazon, was bought to make the round-trip journey, carrying parasites from Santarem to Para every two weeks, as no other reliable transport was available on a regular basis. The entire population of the lower Amazon was amazed to see the launch successfully make trip after trip through the 80 km of treacherous open water before Para.

The arrival of the launch in Para, 750 km from Santarem, was arranged to coincide with the weekly commercial airline flight from Para to Georgetown, British Guyana, where the parasites were to be colonized. This flight took one day, so with the judicious use of ice, the pupal parasites could be kept for up to a maximum of 13 days and still be unemerged and healthy by the time of their arrival in Georgetown.

All told, six shipments were sent during August to October 1933, totaling some 3000 parasite puparia. They were received in Georgetown and colonized by Mr L. D. Cleare, who also cultured many more in the insectary. By March 1934, only a little over six months after the first shipment was made from the Amazon, the tachinid parasite was recovered in some numbers from six different release fields in two localities. It has since become widely established and has effected excellent biological control. (See Myers, 1935.)

Coconut leaf-mining beetle in Fiji, 1933

This beetle, *Promecotheca coeruleipennis* (Blanchard), is a native insect in Fiji that originally, apparently, was well controlled by native natural enemies. However, beginning in about the 1880s, as trade developed and coconuts became extensively cultivated, the beetle gradually became a serious pest. Severe outbreaks became commonplace. The larvae bore within the leaflets, and the adult beetles feed on the leaves externally. In heavy infestations they together caused a reduction of about 75 percent in the leaf surface of every tree over wide areas.

A detailed study of the insect and its natural enemies was not feasible until 1929, when sufficient entomological expertise became available from other duties. This investigation showed that all severe outbreaks had one unusual feature in common; only one stage of the insect was present at any one time, i.e. there was no overlapping of generations. This was remarkable in a salubrious climate like Fiji, where reproduction can continue throughout the year. Such a climatic situation invariably leads to all stages being present simultaneously, yet such was not the case here. The discovery of the reason behind this, and the appreciation of its significance, led to the solution of the problem.

In a nutshell, the principal cause of the outbreak of the beetle was considered to be due to the indirect adverse effect of a predatory mite, *Pyemotes ventricosus* (Newport), on two native parasite species which otherwise were capable of satisfactorily controlling the beetle and were considered to have done so originally. The mite was undoubtedly accidentally introduced some years previously, inasmuch as it is cosmopolitan and feeds on a wide variety of insects. For complex reasons to be briefly explained, the mite caused an 'even-brooded' condition of the beetle, and this resulted in the native parasites being rendered ineffective because they attacked only the larval stages, and these were largely lacking for long periods of time because of the even-brood. Thus the native parasite populations were greatly reduced and rendered ineffective.

The mite exerted its effect thus. It attacked all larval stages and the pupae of the beetle, but not adults or eggs. In the dry season it increased with such incredible rapidity that it would destroy all of the beetles except eggs and adults in a given locality, the adults laid eggs and died, and this resulted later in the single-stage condition consisting principally of beetle larvae, but meanwhile the predatory mite population crashed because of food shortage when only adult beetles and eggs were present. When the wet season came, the beetles increased again very rapidly – essentially without any check – both because parasites were scarce due to the single-brood condition and because the mite was ineffective during wet weather. T. H. C. Taylor concluded that the mite would have replaced the indigenous parasites and done a good job of control, had it not been decimated by wet weather every year. In the wetter parts of Fiji the beetle seldom became a serious pest, presumably because the mite was held down and the native parasites operated without hindrance.

The preceding knowledge showed that a new parasite must be discovered, and led to the formulation of a set of quite definite characteristics that should be possessed by such a parasite in order for it to control the coconut leaf-mining beetle under the single-brooded condition. These requirements mainly included: (1) the parasite should have a more rapid rate of increase relative to the pest than the native parasites, and one generation must require no longer than a month, (2) adults of the parasite should be able to survive the long periods when no individuals of the pest were in a suitable stage for its attack, and (3) it should parasitize all larval stages as well as pupae of the host, so that suitable hosts would be available over a much longer period in each pest generation than for a parasite species which attacked only one stage, and thus the period during which the new parasite would lack suitable hosts would be correspondingly shorter. Also it was desired that the parasite should be internal, very active and capable of rapid dispersal, and tolerant of the climatic conditions of Fiji.

No suitable candidate parasite was known at the time this study was completed, but in 1930 R. W. Paine's investigations in Java resulted in the discovery of a large number of parasites of a related species of *Promecotheca* (*P. nuciferae* Maulik). This beetle occurred throughout Java but was never a pest – a good indication that the parasites might be effective. Taylor followed this up with detailed studies of the Javanese parasites in 1932–3, and concluded that only one of the parasites, *Pediobius parvulus* (Ferrière), met the conditions set down for the new parasite desired for Fiji, even though this species was considered to be one of the least important of the complex controlling the beetle, *P. nuciferae*, in Java.

This small parasitic chalcidoid wasp attacks all stages of larvae as well as the pupae of *Promecotheca*, and many individuals develop in each host individual, giving it a high potential rate of increase with respect to the host. The life cycle is only some three weeks long, as compared to about three and a half months for the beetle in Fiji and, perhaps most important, the adult parasites live for five and a half weeks, which enables them to survive periods when suitable stages of the host are rare or absent.

Taylor transported the parasites from Java to Fiji in the usual manner of the day. Six large, wire-gauze-covered cages, each containing four seedling coconut palms heavily infested with *Promecotheca nuciferae* of all stages, were used. Parasites were included in the cages and were able to reproduce during the voyage. He left Batavia, Java, April 17, 1933 on the S. S. *Van Rees*, which traveled from Singapore to Sydney via Java,

Papua New Guinea (then New Guinea), New Hebrides and New Caledonia. It was necessary to trans-ship at Noumea, New Caledonia, on May 10 to the S. S. *Kareta* in order to go on to Suva, Fiji, where he arrived May 14.

Before arrival in Noumea (four days from Fiji) all palms, soil, containers, etc. were thrown overboard to eliminate any potential pest hazards, after all adult parasites and parasitized larvae and pupae of the host had been collected into glass tubes.

From Suva it was necessary to proceed at once on a chartered motor launch to the Lau group of islands, where outbreaks were raging. They arrived at Nabavatu, on Vanua Balavu, on May 24 (37 days out of Java), where headquarters and an insectary were set up. Some 1200 adults and pupae of *Pediobius parvulus*, the preselected parasite, arrived in excellent condition. To be on the safe side, Taylor also imported the two parasite species considered most important in Java. Only one, *Dimmockia javanica* Ferrière, was successfully introduced, cultured and liberated in sufficiently large numbers, but it failed to become established, thus justifying the predictions in this regard.

The first liberations of *Pediobius parvulus* were made on May 26, 1933 and were continued to April 30, 1934, when so many parasites were present in the field that insectary rearing was no longer necessary. Within a year, it had completely controlled all the severe outbreaks of *Promecotheca* that had raged on Vanua Balavu, Kanacea, Taveuni, Mago, Lakeba and elsewhere. Taylor records that it literally attained 100 percent parasitism on all trees, even though in many outbreaks every tree over hundreds of hectares of land bore about 4000 beetle individuals. At the first peak of the parasite explosion, about 5000 adult parasites were emerging per tree daily, and this continued for about ten days. (It will be recalled that many parasites develop in each host.) Not a single *Promecotheca* individual escaped in many outbreak areas, and the parasite dispersed so well that it controlled the pest even on isolated coconut trees hidden in the forest.

The one minor disadvantage to this parasite was that it was too effective. On small islets or very isolated spots of coconuts it would exterminate the pest and then die out itself. This necessitated reintroduction of the parasite if the beetle happened to reinvade. However, this never happened on larger islands or large coconut estates. Today the coconut leaf-mining beetle remains a rarity, of no economic importance.

This is an amazing case in many respects, but the development in advance of a set of very definite characteristics that the imported parasite would need to possess in order to be successful, and then the deliberate preselection of such a parasite from among a group, at least two of which were much more common and effective in the country of origin, is unique we think in the annals of biological control. Not that the foreign explorer doesn't intuitively employ preselection to a considerable extent, but nearly always the parasites given the most chance to succeed are the ones considered to be most effective in the country of origin. Additionally, this case demonstrates the feasibility of biological control of native insect pests. (See Taylor, 1936, 1937.)

Coffee mealybug in Kenya, 1938

The importance of correct taxonomy to successful biological control is nowhere better illustrated than by this project. The coffee mealybug, *Planococcus kenyae* (LePelley) was, during the early years of its occurrence as a pest in Kenya, first thought to be the citrus mealybug, *Planococcus citri* (Risso), and later to be *Planococcus lilacinus* (Cockerell). Consequently, for 12 years the wrong natural enemies were sought after and imported and unsuccessfully colonized. It is a most interesting story, and the lesson of the necessity for accurate identification still needs to be kept in mind with other projects.

The coffee mealybug was first noticed when a severe infestation on coffee occurred in the Thika district of Kenya in 1923. At first it was assumed to be a native insect but its rapid spread to other nearby districts and ultimately far afield, coupled with its lack of any important parasites in Kenya, made it evident that it was an introduced insect. It soon became one of the notable mealybug plagues of the world. Coffee was the principal cash crop attacked, but severe damage was done to food plants in native gardens and to ornamentals as well. In fact, the cultivation of yams was stopped due to the mealybug, and other cropping practices had to be modified. The prosperity of thousands of small native landowners was severely affected. It was estimated that the financial loss on coffee alone, including cost of control efforts, from 1923 to 1939 (when biological control was achieved) was between £1 000 000 and £1 500 000 sterling, or aproximately US $5 000 000 to $7 500 000, quite a sum for that period.

During the early period in 1925, when the pest was thought to be the citrus mealybug, a parasite of that mealybug, *Leptomastidea abnormis* (Girault), was sent from Sicily to Kenya. Although no parasites survived the trip, it is doubtful if they would have attacked the coffee mealybug.

Another parasite of the citrus mealybug, *Leptomasix dactylopii* Howard, was received later on in large numbers from the University of California at Riverside, but would not successfully reproduce on the coffee mealybug. Later, when the mealybug was considered to be *Planococcus lilacinus*, other natural enemies were sought, not only from other mealybug species but especially from *P. lilacinus*. A major effort was made by Dr R. H. LePelley to obtain parasites of *P. lilacinus* in the Orient, its presumed native home, and to import and colonize these in Kenya. During 1936–7 he explored many countries and, particularly from the Philippine Islands and Java, successfully shipped large numbers of parasites and predators to A. R. Melville in Kenya. However, none of the parasites obtained by so much effort would attack the coffee mealybug when tested in Kenya. Some of the predators were cultured and liberated, but apparently none became established. By now, natural enemies had been imported unsuccessfully from four continents.

At about the time of LePelley's trip it became evident that the coffee mealybug was a species distinct from *P. lilacinus* and, since much of the world had been searched, most probably was native to Africa somewhere outside of Kenya. Here again luck creeps in. Our late colleague Harold Compere was exploring for black scale parasites in Africa for the University of California and in 1937 visited Melville in Nairobi, while LePelley was still absent abroad. Compere recalled that he had seen the coffee mealybug in Uganda, and told Melville that it was rare there.

Following LePelley's return from the Orient, staff became available for a search for parasites in Uganda, and this time A. R. Melville did the exploring, with LePelley in charge of the receiving, quarantine, culture and colonization work. Melville went to Uganda early in 1938, and almost at once found the mealybug and sent parasites to Nairobi. LePelley successfully cultured nine species of primary parasites, destroying several species of hyperparasites that were present. Because of lack of insectary space, the five most promising parasites were concentrated on and the other four dropped, with the idea that they could be reacquired if necessary. However, such never proved to be the case. A species of *Anagyrus* near *kivuensis* was colonized in June 1938, and during September–December 1938, a further 15 000 parasites were liberated. From 1939 to 1941 about an additional 200 000 were liberated each year, and the whole of the mealybug-infested area was colonized.

Of the five species colonized, three were established, but one, *Anagyrus* sp. near *kivuensis* Compere, is given credit for the outstanding results that occurred. According to LePelley it possessed all the chief attributes of an outstanding parasite; 'it is vigorous, hardy, adaptable, mates readily, and is an excellent searcher. It appears capable of maintaining itself on a very low and scattered mealybug infestation. This insect alone . . . has completely controlled infestations in districts where for 10 or 15 years it had been necessary to band large areas at great expense.'

The establishment of the parasites rapidly and remarkably reduced the population of the mealybug. By 1941, losses in coffee plantations were reduced by 92 percent and became less and less over the years. The same remarkable clean-up occurred in native gardens and reserves. In 1938 a motorist could easily notice areas of vegetation along the roadside which were black from sooty-mold fungus growing in the honeydew exuded by masses of mealybugs, and which stank with the mealybug odor. Two years after liberation of parasites there were no such areas, and a detailed search was necessary in order to find any mealybugs at all. When LePelley wrote of the marvelous results in 1943, they had just occurred and he was not quite sure that they would be so completely maintained in the future. He suggested that, if ever required, known additional species of parasites could be introduced. That his doubt proved groundless is shown by Dr W. R. Thompson's report in 1956 that complete biological control was continuing. In 1959 A. R. Melville estimated that this project had saved the coffee industry of Kenya £10 000 000 sterling, at a total expenditure which could not have exceeded £30 000. (See Greathead, 1971; LePelley, 1943.)

Comments and conclusions

As we reflect on the cases just discussed, quite a few points become evident that today are accepted as standard principles and practices in biological control. One thing is clear: highly trained professional entomologists, enthusiastically devoted to their science, furnished the main ingredients for success in case after case. Were it not for men such as Riley, Muir, Silvestri, Taylor, Compere and Myers, to name just a few, how many more years would have elapsed before serious pests were subjugated? The personal equation cannot be given too high a rating in this work, nor can basic research and scientific communication. In several cases, the solution to a problem lay in studies previously conducted by scientists having no connection with the problem. The acquisition of their information was essential, and it emphasizes today's need for a data bank.

It is clear that perseverance may pay off, as with the sugar-cane leafhopper problem in Hawaii, which was completely solved only after 15 years of continuing effort and interest, and just as clear that some projects were prematurely abandoned or insufficiently pursued, such as the citrus whitefly problem in Florida as well as the gypsy moth and sugar-cane borer problems. The former has now been successfully concluded, but there still is a lot that can be done on the latter two.

From the successful results obtained, quite a number of conclusions are evident: generally one, and usually not more than two, natural enemy species are responsible for control, and these effective natural enemies become established and disperse easily, bringing about control rapidly (usually within nine months to two years) and spectacularly. It is not necessary to import massive numbers, when effective natural enemies are concerned. In the case of the spiny blackfly only 20 parasites comprised the initial stock, and only 42 females furnished the stock for the successful citrus blackfly project in Cuba. The importance of correct taxonomic identification of the pest is clearly shown by the coffee mealybug case in Kenya. The same applies to correct identification of natural enemies. The fact that effective natural enemies may be obtained from hosts other than the target pest was established by the sugar-cane beetle borer case, the *Opuntia* cactus case, the coconut moth case and the coconut leaf-mining beetle case, and it is very important to keep this in mind on any project. By inference, this also means that native pests may be controlled by importation of natural enemies from related hosts abroad, and of course this was actually demonstrated with the coconut leaf-mining beetle in Fiji.

The various cases discussed showed that either predators or parasites, acting alone, may produce complete control. It all depends upon the particular pest-natural enemy complex involved, and no real purpose is served by arguments *a priori* as to which type of natural enemy is better. However, parasites were the major factor in most cases. It is also evident that control may be accomplished by the attack of a natural enemy on any one host stage (egg, larval or pupal), and does not necessarily require a sequence of stages to be attacked. Cases in point include the eucalyptus snout-beetle and its egg parasite, and the sugar-cane leafhopper and its egg predator.

Preselection of particular natural enemies for a given job is practiced up to a certain point by all trained foreign explorers, and was eminently successful in the case of the coconut leaf-mining beetle in Fiji, but as has been emphasized by various explorers, the natural enemy they had considered to be best often turned out to be inconsequential and, in fact, the best one even may be overlooked for years, as was *Tytthus*, the egg predator of the sugar-cane leafhopper. It seems that the proof of the pudding lies in the eating, i.e. all likely candidates should be imported and tried. By likely candidates we mean ones that attack the target pest or related species, and that are of more than incidental occurrence in the native home area. Careful quarantine to exclude other, possibly detrimental, organisms must of course always be carried out.

There is no evidence from any of these cases that purposeful importation of any of the natural enemies did any harm, as far as final control of the pest is concerned, regardless of whether the poorest enemies were imported first and the best ones last or vice versa. Neither is there any evidence that complete biological control, once achieved, ever became reduced over the years. If anything, as in the eucalyptus snout-beetle case, control continued to improve for a number of years.

From a financial standpoint, the results were frequently fabulous. The various cases speak for themselves, but several that cost only a few thousand dollars to solve continued to result in savings to agriculture of millions of dollars a year each.

6

Modern foreign exploration and successes

The development of commercial air transport signalled the modern era of foreign exploration. The changes from the earlier years to today have been largely technological ones related to easier, much more rapid means of travel and safer and simpler techniques of shipping and handling natural enemies. Along with this, of course, have come improved quarantine laboratories and insectaries for culture of introduced enemies, better colonization techniques and increased biological, ecological and taxonomic knowledge in general.

The changes have not been abrupt, and there is no clear line of separation. We have arbitrarily chosen the end of the 1930s as the beginning of the modern era. It will be recalled from Chapter 5 that air shipments of sugar-cane borer parasites were made as early as 1932 in the Caribbean, and in 1933 from Para, Brazil, to Georgetown, Guyana. However, as will be discussed shortly, the first long-distance international air shipment apparently occurred in December 1939. Unfortunately, only a few projects were able to take advantage of the increasing air transport facilities before World War II put a stop to such endeavors until 1945.

The modern foreign explorers can often accomplish in days or weeks what previously would have required months or a year or more. Their conditions of travel and living in foreign countries are usually considerably more pleasant and health risks fewer, but off-the-beaten-track conditions are still grim in many places. In the experience of one of the authors (PD), he has had to have armed guards in order to travel and collect in certain localities in Burma; has lived in hovels on stilts where the toilet was a hole in the floor; has had pig-intestine soup for what he later recalled was Thanksgiving Day dinner; has had to have local permission and blessing to enter and collect in bandit-infested parts of the semi-autonomous Tribal Territories of Pakistan (then West Pakistan), and once *en route* received many required penicillin shots from nearly anyone he could find, including airline stewardesses, pharmacists and medical orderlies. On the other hand, he has had some lovely collecting in Rio, Tahiti, Japan and other places.

Today, with a little planning, airmail shipments of live natural enemies can be made from nearly any given country to the home insectary as much as half-way around the world in from four to six days or less. Special arrangements utilizing air freight can reduce this time. One of the authors (DR) even used diplomatic mail services, which are very fast, for this purpose. Thus, the problem of 'bringing 'em back alive' has been greatly alleviated, and the need to grow and maintain host cultures abroad solely for importation purposes virtually eliminated. Heated and pressurized cargo space generally furnishes good conditions for the natural enemies. The proper food and moisture conditions merely have to be provided in the shipment package, which also should be well insulated. Refinements may need to be made to suit particular cases, but there is really no technical obstacle left to prevent our succesfully shipping live natural enemies rapidly from any country to any other country in the world.

Another development that has helped considerably has been the ever-increasing construction of biological control insectaries around the world. These not only may serve as headquarters for visiting foreign explorers, but as foci of knowledge concerning local natural enemies. Sometimes they may have the natural enemies desired already under culture, so that large shipments can be easily made. Many countries, provinces or states, and some agricultural industries, maintain insectaries (see Fig. 6.1). Their lack is the greatest where they are sorely needed – in developing countries. This is partially alleviated by the Commonwealth Institute of Biological Control (currently known as the CAB International Institute of Biological Control), which maintains insectaries in the United Kingdom, Switzerland, Trinidad, Kenya, Pakistan, India and Malaysia, and will carry out special projects for any given country. Also, of course, insectaries with quarantine facilities are necessary in order to preclude the possibility of accidental importation of hyperparasites or other detrimental organisms. Most such insectaries operate under government permit and supervision.

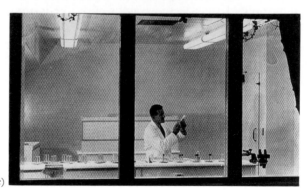

Fig.6.1. (*a*) The insectary of the University of California at Riverside; (*b*) parasite rearing cages in the insectary at Riverside; (*c*) Dr Richard Doutt in the quarantine room of the insectary of the University of California at Berkeley.

Some outstanding modern biological control projects

The 21 successful projects that follow have been chosen from among many, either because they are particularly interesting or because they furnish valuable insights into biological control principles and practices. Many of the other cases that might just as well have been used are included in the summary of worldwide results later in this chapter. The dates given with each project generally represent the time of importation or establishment of the major natural enemies.

Citriculus mealybug in Israel, 1939–40

During the 1920s and 1930s, the Palestine Farmers Federation financed the construction of several insectaries to carry out biologial control projects against such pests as the Mediterranean fruit fly and the common citrus mealybug. In 1938, Israel Cohen was appointed Director of these insectaries, and it is to him that most credit must be given for the completely successful biological control of the citriculus mealybug, *Pseudococcus citriculus* Green, which was an invading pest, first noticed in August 1938, when a severe outbreak occurred in some groves in the coastal plain. In 1938 the infestation (then thought to be the Comstock mealybug, *Pseudococcus comstocki* (Kuwana)) was spreading rapidly, and by 1939 was becoming catastrophic. At a meeting called by the citrus industry in May 1939, the opinion was expressed that 'The intensity of the damage caused and the apparent rapidity of spread threaten to annihilate the citrus industry.' At this meeting it was agreed that spraying experiments should be carried out, but that major stress should be on biological control by importation of parasites from other countries. This was met with strong opposition by the late Professor F. S. Bodenheimer and other professional entomologists, who had noted that native mealybug predators and parasites were already attacking the new invader, and feared that an introduced natural enemy might upset the applecart by competing with them and would do more harm than good. As we shall see, such fears proved groundless.

Attempts to control the pest with oil sprays and other insecticides failed to yield even remotely encouraging results. Although Cohen was not trained as an entomologist, he had studied agriculture at the University of California during the period when another very serious mealybug pest, *Pseudococcus calceolariae*, had been completely controlled there by introduced parasites, and he was convinced that their hope in Israel lay in a similar direction (Cohen, 1975). He sent samples of their new mealybug to various experts, who identified it as *P. comstocki*. He then wrote to various entomologists to ascertain where parasites of this Comstock mealybug might be obtained. Japan was suggested as the native home, but certain parasites were also known to be present in the eastern United States, where *P. comstocki* had become established. About this time Dr I. Carmon, the horticulturist of the Agricultural Experiment Station of the Jewish Agency at Rehovot, was about to journey to the United States, so Cohen suggested that he route his return trip via Japan and search for parasites there. Cohen raised funds to cover the extra expenditures and contacted United States Department of Agriculture (USDA) and University of California entomologists to help and advise Dr Carmon.

With the assistance of a USDA entomologist stationed in Japan, Carmon collected and sent parasites of *P. comstocki* to Israel (then Palestine) in two airmail shipments in December 1939. This apparently represents the first lengthy international transport of natural enemies by air. Additionally, Carmon personally returned by boat in 1940 with many pupae of the various Japanese parasites of *P. comstocki*. Meanwhile, Cohen had a special temporary laboratory-insectary constructed and invited Dr E. Rivnay to direct it.

Dr Rivnay sorted out the various primary parasites and hyperparasites received from Japan, and determined that *Clausenia purpurea* Ishii was the most effective one. The hyperparasites were eliminated, of course.

Fig.6.2. An adult female of *Clausenia purpurea* ovipositing in the citriculus mealybug. (From Rivnay, 1942.)

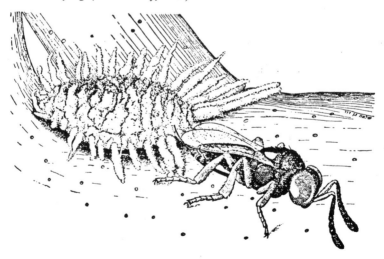

Clausenia purpurea was cultured, and one hundred specimens were colonized under a tent on one tree in April 1940. It became established immediately and spread rapidly. Mass production was started later in 1940, so as to distribute the parasite widely. Trees which received 25 parasites in 1940 showed marked decreases in the mealybug infestation by August 1941. In a very short time control was complete, and has remained so ever since. We can attest from our personal experience that the mealybug is now extremely difficult to find in Israel, and whenever a rare colony is found – usually hidden between appressed leaves – it is invariably heavily parasitized by *Clausenia*. Fig. 6.2 shows *Clausenia* ovipositing in the citriculus mealybug.

One of the most interesting parts of this story is that the pest was later found to be *P. citriculus* rather than *P. comstocki*. This came about because a Japanese parasite, *Allotropa burrelli* Muesebeck, which had been established on *P. comstocki* in the United States, would not develop in the presumed *P. comstocki* in Israel, thereby indicating the Israeli mealybug to be a different species. Thus, the pest was controlled, one might say, with a blend of skill and luck by a parasite of another mealybug species. However, this illustrates very strikingly the point that effective natural enemies may sometimes be obtained where least expected. It also shows that personal initiative, as well as support by government or industry of the biological control effort, is of prime importance. Neither Cohen nor Carmon was an entomologist, yet between them they collaborated to import a parasite responsible for one of the world's great successes. Ironically, Israel could have obtained this parasite from the eastern United States, where it had evidently become fortuitously established before 1939, more readily and cheaply than they did from Japan. (See Rivnay, 1968.)

Klamath weed in California, 1944–6

The weed, *Hypericum perforatum* Linnaeus, of European origin, was first reported in northern California in the vicinity of the Klamath River. It increased rapidly, and by 1944 it occupied over two million hectares of rangeland in 30 counties in the northwestern USA and Canada. Not only were good forage plants greatly reduced, but cattle and sheep lost weight when eating Klamath weed because of its somewhat toxic effect. This resulted in such a great decrease in land values, that it became almost impossible for ranchers to borrow money for improvements. Chemical weed killers were available but not generally practical, because of the cost and the inaccessibility of much of the infested land. Professor H. S. Smith, head of biological control work in California, proposed the importation of insects that fed on the Klamath weed as early as 1922, but the idea of deliberately introducing a plant-feeding insect was not acceptable at that time.

Meanwhile, Australia had been fighting the same problem since 1929 with *Hypericum*-eating insects introduced from England and Europe. Professor Smith, in California, followed the progress there with much interest through correspondence with A. J. Nicholson, Chief Entomologist for the Commonwealth Scientific and Industrial Research Organization (CSIRO). Smith finally obtained authorization in 1944 to import three European species of beetles that showed promise against the Klamath weed in Australia. It was impossible then to consider importations from Europe because of World War II, but rather easier to bring material from Australia through the co-operation of the United States Army Transport Command. The Australian CSIRO, through Dr Nicholson, offered to collect and prepare the material for shipment. Importations started in October 1944, but problems were immediately encountered in changing the timing of the life cycle so the beetles would be in phase with the Northern Hemisphere seasons. Two species of *Chrysolina* that were in the summer resting state of aestivation responded rapidly in California to fine mist sprays of water to become active and lay eggs within three weeks. The third species, an *Agrilus*, was lost. After starvation tests in quarantine on a variety of economic plants, the beetles were ready to be released in the field.

Chrysolina hyperici (Förster) was released in the spring of 1945, and *C. quadrigemina* (Suffrian) in February 1946. Both became readily established, but it soon became evident that *C. quadrigemina* was becoming dominant. Distribution throughout the entire infested areas was rapidly made from the original colonies. From one colony of 5000 beetles released in 1945–6, more than three million beetles were collected for redistribution in California in 1950. They were also sent to Washington, Oregon, Idaho, Montana and British Columbia, where they became successfully established.

The beetle's performance in California was much more spectacular than in Australia. Within ten years after the first release, the Klamath weed was reduced to the status of an uncommon roadside weed in California. Its abundance was reduced by more than 99 percent. Land values immediately increased by three or four times. Loss in weight of cattle and sheep ceased. It has been previously estimated (DeBach, 1964a, p. 13) that at least $20 960 000 in savings accrued to the agriculture industry in California for the period 1953–9, or about $3 500 000 per year. These savings continue to accrue each year, without even taking inflation into account. (See Goeden, 1978; Huffaker, 1967.)

Oriental fruit fly in Hawaii, 1947–51

This fruit fly, *Dacus dorsalis* Hendel, is a native of the Orient. It probably gained entrance to Hawaii in 1945 via military materiel returning from the western Pacific theater of war. It increased rapidly to epidemic proportions during 1946, attacking a wide variety of fruits. The larval infestations not only rendered most fruits worthless in Hawaii, but additionally posed a serious potential problem to the warmer fruit-growing areas of the mainland United States. Chemical control was difficult, expensive, hazardous to health with certain materials, and frequently insufficiently effective. As a consequence, one of the most massive biological control projects of modern times came into being.

The Hawaiian Territorial Board of Agriculture and Forestry initiated exploration for natural enemies in 1947–8 in the Philippines and Malaysia. They actually imported and established the parasites responsible for the success subsequently obtained, but it was impossible to ascertain this during this early stage of the effort. Consequently, in 1948–9 other interested organizations joined into a co-operative effort, including the USDA, the University of California, the Hawaiian Agricultural Experiment Station, the Hawaiian Sugar Planters Experiment Station and the Pineapple Research Institute. Up to September 1951, fourteen foreign explorers had collected parasite material from many fruit fly species in most of the tropical and subtropical areas of the world, in particular from the Philippine Islands, Malaysia, Taiwan, Thailand, Borneo, India, Sri Lanka, Australia, New Caledonia, Fiji, South Africa, Kenya, Congo, Brazil and Mexico. Shipments consisted mainly of parasitized puparia sent by airmail or air freight, and generally reached Hawaii in seven days or less from any given country. All told, more than 4 246 000 fly puparia of over 60 species were sent to Hawaii. About one-third of these were *Dacus dorsalis*, the pest in question. Some 80 species of parasites were obtained from this material, of which 16 or more larval parasites of the genera *Biosteres* and *Opius*, six pupal parasites and one predator were cultured and released in the field. Initially 11 species were recovered in the field, but the most effective one soon became nearly exclusively dominant.

This story is somewhat confused because various of the imported species of *Biosteres* (then known as *Opius*) were very similar in appearance, and some were misidentified at first. Thus, the original material received from the Philippines in 1947 contained the three most important of all the parasites eventually imported, but only one of these, *Biosteres longicaudatus* Ashmead, was correctly identified. The other two, *B. vandenboschi* (Fullaway) and *B. arisanus* (Sonan), were thought to be one species, *B. persulcatus* (Silvestri), which actually was not present. The same applied to the next group of shipments, received from Malaysia in 1948. Parasites from these shipments immediately became established. *Biosteres longicaudatus* increased rapidly in the field after its initial release on Oahu in 1948, but suddenly lost its dominant position during the latter half of 1949 to *B. vandenboschi*, which had been released initially about the same time. In turn, *B. vandenboschi* was replaced during 1950 by *Biosteres arisanus* (then known as *Opius oophilus*), which had first been recognized to be established in 1949. Interestingly, in spite of the competition between these three species, each replacement of one by another was accompanied by a higher total parasitization and a greater reduction in the fruit fly infestation. Both *B. longicaudatus* and *B. vandenboschi* had virtually disappeared by 1951, and this status has since continued.

The final result has been a very substantial reduction in the Oriental fruit fly populations in all of the Islands, estimated to be on the order of 95 percent as compared to the 1947–9 peak abundance. At that time, practically 100 percent of most kinds of fruits were infested. The threat of movement to the mainland has been greatly reduced. Still, some preferred fruit, such as guava and mangoes, which up to 1949 were 100 percent infested, are sometimes infested to the extent of 50 percent, but with many fewer larvae per fruit. Yet, on average, less than 10 percent of the mangoes are now infested. Many kinds of fruits that were once heavily infested are now virtually free of attack.

This project illustrates the basic importance of accurate knowledge of both taxonomy and biology to biological control. *Biosteres arisanus* was mistaken for some time as *B. persulcatus*. Had the latter been imported and established early with some degree of success, it is possible that further work, including the ultimate discovery of the best parasite, *B. arisanus*, might have been dropped. It is of interest that *B. arisanus* has also turned out to be the best parasite of the Mediterranean fruit fly in Hawaii, having displaced *B. tryoni* (Cameron), which previously was well established and fairly effective. Had *B. arisanus* been recognized as a valid species and introduced from Malaysia in 1913–14, when importation of parasites of the Mediterranean fruit fly was being initiated, better biological control would have occurred 40 years earlier.

The biology of *B. arisanus* differs from that of *B. longicaudatus* and *B. vandenboschi*, which deposit their eggs in host larvae. The former is an egg-larval parasite: it lays its egg in an egg of the host, then completes development in the host larva. Had this habit been known in 1935–6, when F. C. Hadden undoubtedly, but unknowingly, imported *B. arisanus* along with other *Biosteres* and *Opius* species from Malaysia and India for control of the Mediterranean fruit fly, it probably could have been cultured and established. However, insectary propagation did not occur, very likely because its habit of ovipositing only in host eggs was not then known, and probably only host larvae were provided in the culture unit. If the biology had been understood so that the parasite had become established on the Mediterranean fruit fly at that time, it not only would have provided better biological control of that fruit fly from 1936 to 1950 (when *B. arisanus* finally became well established), but it would have been present to attack the Oriental fruit fly when it first reached Hawaii in 1945, and could well have made the later massive and expensive project unnecessary. (See Clausen, Clancy & Chock, 1965; Wharton, 1990.)

Red wax scale in Japan, 1948

This case might be likened to 'a gift from the gods'. This omnivorous pest, *Ceroplastes rubens* Maskell, was first found at Nagasaki, Kyushu, in 1897, probably having originated in China. It became a scourge of many economic plants, including citrus, persimmon and tea among others, and soon spread to the main island of Honshu and to Shikoku. No native parasites of any consequence were found attacking it. The introduction of several species of parasites that attacked it in Hawaii and California was attempted in the 1930s, but without success. Thus, chemical control remained necessary, but was not always sufficiently effective because of the protection afforded by the waxy covering of the scale.

About 1942, the red wax scale began to diminish in severity on the southern island of Kyushu. In 1946, after the population of the red wax scale had been reduced noticeably, Professor K. Yasumatsu discovered a new effective species of parasite, *Anicetus beneficus* Ishii and Yasumatsu, that was apparently responsible for the decline. The parasite was then thought to be a race of a closely related species, *A. ceroplastis* Ishii, that parasitized another wax scale, *Ceroplastes pseudoceriferus* Green. Surveys showed that the parasite was not present on Honshu or Shikoku.

The origin of the new parasite remains a mystery to this day. Dr Yasumatsu suggested the possibility that *A. beneficus* may have evolved by mutation from *A. ceroplastis* in relatively recent years. Although no other case of this nature has been demonstrated in biological control, it is not too unlikely a hypothesis because the only major changes would have to have been in host preference and the development of reproductive isolation between the two. They are quite similar morphologically, although distinguishable under high power magnification. They do not interbreed, hence unquestionably are distinct species. Support for Yasumatsu's hypothesis lies in the fact that *A. beneficus* has never been collected outside of Japan, even in the presumed native home of its host, *C. rubens*, in China.

Beginning in 1948, *A. beneficus* was transferred in large numbers from Kyushu to various orchards on the islands of Honshu and Shikoku. It never failed to become established wherever liberated. Results were spectacular, resulting in complete biological control everywhere. One example of the effectiveness of the parasite will suffice. Twenty mated females were released in a heavily infested citrus orchard of 3000 trees. Judging from our own observations in citrus elsewhere, probably at least 100 000 scales were present per tree, or 300 million in the orchard. Within three years of the single release of 20 parasites, it was nearly impossible to find a live scale in the orchard.

This instance of biological control emphasizes the point that it is not always necessary to go abroad to discover effective natural enemies. There are other similar cases where efficient enemies have been transferred from one part of a country to another with great success. (See Yasumatsu, 1958.)

Purple scale in California and elsewhere, 1948 on

Although the purple scale, *Lepidosaphes beckii* (Newman), was one of the world's most serious pests of citrus, it, like many of today's major pests, was never the object of a formalized biological control project. The discovery of the key parasite cannot be attributed just to luck, but it was incidental to the search for parasites of the even more serious California red scale. Had a purple scale project been concentrated on specifically earlier, the parasite could have been discovered years sooner, with consequent prior alleviation of the problem and resultant much greater accumulation of savings to agriculture in many countries. As a matter of fact, the Italian entomologist Filippo Silvestri recorded observing a parasite of purple scale on lemons in Macao, China, during his exploration of South China and Japan in 1924–5 for the University of California. Unfortunately no one followed through on this, because it undoubtedly was the parasite in question, *Aphytis lepidosaphes* Compere, that was to be obtained for the first time years later from nearby Canton, China. We now know it to be the dominant parasite throughout southeast Asia, where it maintains the purple scale at very low population levels.

In November 1948, Dr J. L. Gressitt, who was primarily searching for California red scale parasites, sent live material of *A. lepidosaphes* from Canton to the biological control insectary of the University of California at Riverside by airmail. A total of several thousand live specimens were received from the Canton area and from Taiwan through 1949. It was immediately cultured, and several small colonies were released in the field in southern California during 1949. Subsequent colonizations were made through 1952 to facilitate the rate of increase and distribution. The extremely cold winter of 1949–50 drastically reduced the first colonies, but by the autumn of 1950 establishment was evident in most sites. Over the next few years the parasite became thoroughly established, common, and spread throughout all infested areas. It even moved on its own over 100 miles into Baja California, Mexico, from the nearest point of liberation in California. The purple scale infestations were reduced greatly, but not to the point of completely satisfactory control. However, chemical applications could be reduced by about half because of the effect of the parasite. We were able to determine from field and laboratory studies that the reason for the parasite not being completely successful in southern California was due primarily to mortality caused by low winter temperatures.

Subsequently, from 1952 to 1968, we made shipments of *A. lepidosaphes* to 13 other countries or states having serious infestations of purple scale and, ironically, the parasite was completely successful in subjugating the scale in most of these, from which we either have received reports or have made personal observations. Complete success has been reported for Texas, Mexico, Greece, Cyprus, South Africa and Brazil; Peru reported a substantial degree of control, and Chile partial biological control.

This parasite has an unusually good ability to disperse. We have found it to be established by accident in as many countries as it has been sent to purposely, and it is credited with completely effective biological control in most of these. Florida is a good example. The purple scale was a very serious major pest there for years. The parasite was first recorded there in 1958, and indications were that it probably had not been present more than two or three years. How it arrived is a mystery, but by about 1960 the scale had been reduced to insignificant numbers. This case emphazises again how much loss society suffers from lack of sufficient knowledge and emphasis on the biological control approach. (See DeBach & Landi, 1961; DeBach & Rosen, 1976; Rosen & DeBach, 1978.)

Citrus blackfly in Mexico, 1949; Texas and Florida, 1974–6

This is one of the larger projects and great successes of modern times. It developed as a co-operative effort between the Departments of Agriculture of Mexico and the United States because of the danger of spread into the United States.

The citrus blackfly, *Aleurocanthus woglumi* Ashby, was first found at El Dorado, Sinaloa, on the west coast of Mexico in 1935. In spite of quarantine barriers it spread like wildfire, and by the early 1950s was the most serious pest in most of the commercial citrus areas of the country. Chemical control was difficult, sometimes ineffective and prohibitively expensive. If heavy infestations were not controlled within one year, complete crop failure would result. In general, growers reported approximately 80 percent crop reduction in heavily infested areas. The threat to the citrus industry was so great that in 1951 the President of Mexico decreed the establishment of a National Blackfly Committee. The Spanish name 'mosca prieta' became a household term throughout Mexico.

After *Eretmocerus serius*, the parasite that had earlier controlled the citrus blackfly in Cuba and elsewhere (see Chapter 5), had been imported from the USDA laboratory in Panama and established in 1943 but found to be ineffective except in restricted areas of continual high humidity, other natural enemies were sought in the native home area in the Orient. Herbert D. Smith, of the USDA, went to Malaysia in 1948 to obtain parasites that C. P. Clausen had previously recorded there. He shipped live material the old fashioned way in Wardian cages and also made one air shipment of infested leaves. Nothing came from the Wardian cages and only a few parasites emerged from the airmail shipment but failed to become established. Smith then moved on to India and Pakistan, to search in areas more similar climatically to the Mexican citrus areas. He spent eight months there in 1948–9, conducting a wide-ranging survey; then, from November 1948 to April 1950, made 25 air shipments of parasitized material to the special USDA quarantine laboratory in Mexico City. Of eight parasites and two predators received, four parasite species became established and three – *Amitus hesperidum* Silvestri, *Encarsia clypealis* (Silvestri) and *E. opulenta* (Silvestri) – eventually became dominant in controlling the citrus blackfly. A nationally coordinated program of great magnitude rapidly arranged for the recollection of parasites developing in the field and their distribution to all areas of the country. At the peak period of the campaign about 1600 people were employed, and hundreds of millions of parasites redistributed. By the end of 1953, the pest was well controlled by the parasites in nearly all groves of the states of Morelos, Vera Cruz, Puebla, Oaxaca, Guerrero, Mexico, Colima, Jalisco, Guanajuato, Hidalgo and San Luis Potosi. By 1955, commercially satisfactory biological control occurred in virtually all citrus areas of the country. Now the term 'avispas' for the parasites became a household word.

Amitus hesperidum was the most generally effective parasite, especially in reducing heavy infestations, but is susceptible to lengthy periods of hot weather, so *Encarsia opulenta*, which is quite tolerant of hot weather, became the dominant parasite in states such as Sonora or Nuevo Leon. *E. clypealis* attains optimum performance under humid conditions, and throughout Mexico in general is more effective than *E. opulenta*. Thus, these parasites complement each other very well. Their ecological relationships in Mexico could hardly have been guessed from their distribution and the ecological evidence obtained in India and Pakistan. This affords strong support for the general biological control practice of importing and testing all likely natural enemies, rather than trying to prejudge which one will be the best and thereby excluding some. The idea of preselecting the one 'best' parasite is based on the now-discounted hypothesis that competition between parasite species may reduce overall effectiveness in host population regulation. In spite of the intense degree of competition that occurred between these three species, it was not detrimental to host population regulation but rather enhanced it. (See Smith, Maltby & Jiménez-Jiménez, 1964.)

The citrus blackfly had invaded Florida in 1934 and Texas in 1955, but was then successfully eradicated. In 1971 it again invaded the Lower Rio Grande Valley of Texas, and this time eradication failed and a highly successful biological control project was initiated by the USDA. *Amitus hesperidum, Encarsia clypealis* and *E. opulenta* were introduced from Mexico and established in 1974–5, and have since effected complete biological control of the citrus blackfly throughout the Lower Rio Grande Valley. Then, when an infestation of the blackfly was discovered in Florida in 1976 and chemical eradication failed again, a shipment of *A. hesperidum* and *E. opulenta* was hand-carried there by Dr W. G. Hart from a USDA laboratory in Mexico. Several additional shipments followed, and both parasites became well established. Blackfly populations crashed within one year of parasite releases, and complete biological control has since been achieved. Interestingly, in Texas it was *E. opulenta* that proved to be the dominant parasite, having largely displaced the other two species. In Florida, too, although *A. hesperidum* was responsible for the initial spectacular control, *E. opulenta* increased gradually and has now become the predominant parasite in low-density host populations. *E. clypealis* was released in small numbers in Florida, but did not become established. Again, in both States, competition among natural enemies did not detract in any way from the excellent overall results. (See Dowell *et al.*, 1981; Summy *et al.*, 1983.)

Rhodesgrass mealybug in Texas and elsewhere, 1949 on

This mealybug, *Antonina graminis* (Maskell), formerly known as the Rhodesgrass scale, is a destructive pest of valuable rangeland and lawn grasses. It became extremely serious in Texas following its discovery in 1942, and by the early 1950s had completely destroyed thousands of hectares of excellent stands of Rhodesgrass. The total infested area exceeded 130 000 km² (50 000 square miles). Control with insecticides was unfeasible under rangeland conditions.

It was known that one internal parasite, *Anagyrus antoninae* Timberlake, occurred in Hawaii, where the mealybug was only a minor pest, so this parasite was imported early in 1949. It was cultured and released in large numbers and became established, but proved to be of little value because it could not stand the high temperatures and low humidities of southern Texas. Several parasites were imported from France in 1954–5, but no establishment occurred. Then, in 1957, a new parasite species, *Neodusmetia sangwani* (Subba Rao), was discovered attacking Rhodesgrass mealybug near Delhi and Bangalore, India. It was introduced and established in Texas in 1959. Subsequently, colonies were sent from there to Mexico, Brazil, Bermuda, Arizona, California, Florida and Israel.

The parasite was soon shown to be effective in reducing Rhodesgrass mealybug to subeconomic levels. Substantial host reductions always followed within one year of the establishment of adequate colonies. This was effective in eliminating losses in yield due to mealybug damage. There was one major hitch, however; the parasite female was functionally wingless (brachypterous), and therefore distributed itself slowly from foci of colonization since it had to crawl. Females were also very short-lived, and only dispersed about two meters in their lifetime. To compensate for this, an ingenious program was developed. Parasitized mealybugs were collected from colonies in the field, packaged in suitable containers and mass distributed by dropping from airplanes. This has resulted in a great speed-up in attainment of complete distribution of the parasite throughout the infested area of Texas. Whereas the total cost of this project did not exceed US $200 000, the virtual elimination of the Rhodesgrass mealybug as an economic pest for forage and turf grasses has resulted in benefits to the Texas economy estimated at $177 000 000 a year. This undoubtedly represents more savings in one year than has been spent on biological control research and application over the last 100 years. (See Dean *et al.*, 1979; Schuster, Boling & Marony, 1971.)

Effective biological control of Rhodesgrass mealybug by *Neudosmetia sangwani* has also been reported from Florida, Brazil and Israel.

Olive scale in California, 1951–7

This native of the Middle East, *Parlatoria oleae* (Colvée), became accidentally established on olives near Fresno about 1934. It rapidly spread throughout the Central Valley and into southern California, becoming a major pest of many deciduous fruit crops and ornamentals – some 200 plant species in all – in addition to olives. Monetary losses due to damage and the cost of chemical control were well over US $1 000 000 per year. It not only caused massive damage due to heavy infestations, but was additionally serious on olives because even one scale per olive fruit would cause discoloration and subsequent cullage. This made it evident from the beginning that biological control on olives would have to be especially efficient.

An intensive program of biological control was initiated in 1949. First, a 'strain' of *Aphytis maculicornis* (Masi) attacking olive scale was imported from Egypt, but proved to be of no consequence. In 1951, H. S. Smith arranged for Dr A. M. Boyce to search for natural enemies throughout the presumed native home of the olive scale, from India through the Middle East, North Africa and the Mediterranean. Boyce was abroad about a year and made numerous consignments to California by air from India, Pakistan, Afghanistan, Iran, Iraq, Syria, Lebanon, Israel, Cyprus, Egypt, Greece and Spain. Even during this relatively modern period he had plenty of adventures, including collecting in rebel areas under armed escorts. Follow-up shipments were made by collaborators from India and Pakistan in 1952–3.

Several species of parasites were obtained and colonized in California, including four so-called 'biological strains' or sibling species of *Aphytis maculicornis*. These were practically indistinguishable morphologically, but laboratory studies showed that they possess distinct biological attributes, so they were reared and released separately in the field. It was soon obvious that the 'Persian strain' of *A. maculicornis*, released in 1952, was the only natural enemy showing appreciable promise in the field, so insectary production, colonization and field study plots were concentrated on this parasite. (We have subsequently recognized it as a distinct species, and it is known as *Aphytis paramaculicornis* DeBach and Rosen.) Over 27 million were colonized between 1952 and 1960 at several hundred sites in 24 counties. This illustrates the importance in foreign exploration work of obtaining parasites from all parts of the range of a pest species, and of recognizing that distinct biological races or sibling species may be found, with distinctly different potentials, in diverse parts of the range of distribution.

Soon dramatic results were evident. *Aphytis paramaculicornis* became readily established everywhere and increased rapidly, commonly attaining parasitization rates of 90 percent or more and generally reducing the average olive scale population densities by about 90 to nearly 99 percent. Plant damage *per se* was eliminated. However, even this drastic reduction proved to be economically unsatisfactory on olive in many cases, because, as mentioned, even one scale on a fruit may result in its being culled, and not more than five percent cullage could be tolerated. Since a heavily infested olive tree might have upward of one million scales, a reduction of 99 percent would leave 10 000 scales, which could mean that quite a few fruit might have one or more scales. The parasite was prevented from being more effective because its populations were greatly reduced each summer by the hot dry weather. Otherwise, it undoubtedly would have produced complete biological control everywhere on its own, as it did in certain favorable locations.

The establishment of additional effective natural enemies obviously was necessary. One of the writers (PD), although not assigned to this project, knew of the need through colleagues at Berkeley, and as a result of a fortuitous combination of circumstances was able to obtain and ship to California from Pakistan, the parasite that ultimately made biological control completely satisfactory.

He was searching for parasites of the California red scale in Pakistan early in 1957, and had gone with Dr M. A. Ghani to a remote village in the Tribal Territories where citrus had been reported. It was to be a brief three-day trip with one day at the village for collecting because of continuing travel commitments. Upon arrival there they found that the altitude was somewhat too high and the climate too cool for citrus, so he spent the day looking for other scale insect parasites on deciduous fruit trees and ornamentals. Olive scale was found and parasite activity was noticeable, especially emergence holes of internal parasites in the dead scales. He knew that no internal parasites were established in California. Consequently, he collected as much of this material as he could, and after return to Rawalpindi the next day, packaged it and sent it on to the University of California at Berkeley by airmail.

Two species of parasites emerged from this single shipment, and both were successfully cultured. Only one, however, the then new species, *Coccophagoides utilis* Doutt, became established. By early 1961 it showed great promise of improving the degree of biological control of olive scale in the two groves in which it was first released in 1957–8. This led to mass-culture and colonization of over 4 million of the parasites during 1962–4 at over 170 sites in 25 counties. Widespread, complete biological control soon resulted. *C. utilis* acted as a complementary mortality factor to *Aphytis paramaculicornis*. Although it only added about four to eight percent additional host mortality to what *A. paramaculicornis* would have produced alone, this was sufficient to reduce the equilibrium level of the scale population so considerably that virtually no cullage of olives remained. Twenty years later, these two parasites continue to coexist and to effect complete biological control. Olive scale has become so scarce, that 2500 to 10 000 leaves and their associated twig segments have to be examined in order to obtain five scales. Incidentally, the same *Coccophagoides* had been obtained in small numbers from shipments made by Boyce from India and Pakistan in 1951, but attempts to culture it had failed at the time.

This case provides an excellent illustration of the fruitlessness of trying to evaluate the potential or actual effectiveness of a parasite on the basis of the percent parasitization of the host, and it emphasizes that we should try all available parasites within reason, until completely satisfactory biological control is obtained.

(See Boyce, 1987; DeBach, Rosen & Kennett, 1971; Huffaker, Kennett & Tassan, 1986.)

Comstock mealybug in the Soviet Union, 1954

The comstock mealybug, *Pseudococcus comstocki* (Kuwana), is a native of Japan that was first discovered in Tashkent in 1939 and soon became a serious pest of mulberry (grown for silkworm culture), catalpa and other crops in Soviet Central Asia and the Caucasus. Chemical control was ineffective. Beginning in 1943, *Clausenia purpurea* Ishii was obtained from Israel but did not become established. Then, when parasites introduced from Japan in 1939–41, together with some that were fortuitously established, had effected complete biological control of the mealybug in the eastern United States, several species were introduced from there into the Soviet Union in 1945. One of these, *Pseudaphycus malinus* Gahan, became established in the Tashkent area.

In 1954 a massive program of rearing and redistributing *P. malinus* was initiated. By 1962, more than 400 million parasites were released in the eight Soviet Republics where the Comstock mealybug was a problem. This has been highly successful. The parasite has become established throughout the geographical range of the pest, and has effected complete biological control, keeping the mealybug at very low population levels. In Uzbekistan in 1963, for instance, the pest was effectively suppressed on eight million infested mulberry trees, no chemical treatments were required, and the yield of silkworm cocoons was increased by 25 percent.

Pseudaphycus malinus is continually being colonized in new centers of infestation. In addition, the parasites *Allotropa burrelli* Muesebeck and *A. convexifrons* Muesebeck were introduced in 1962 from Korea and have become established. (See Bartlett, 1978; Beglyarov & Smetnik, 1977.)

Winter moth in Canada, 1955–60

This moth, *Operophtera brumata* (Linnaeus), was looked upon as just a nuisance of shade trees when the project began in Nova Scotia in 1954. It had been found for the first time in 1949, scattered along the southern coast, but apparently was accidentally introduced in the 1930s. It moved slowly, and the only threat seemed to be that it might eventually reach more valuable trees elsewhere. It was not until 1962 that it had invaded New Brunswick and Prince Edward Island. Eventually though, based on its range in Europe, the moth appeared capable of spreading across North America. However, the economics of the situation changed in just a few years. Hardwoods now are a valuable commodity, and it has been stated that had they been so important in 1954, chemical control would probably have been utilized and biological control attempts passed by.

In a ten-year span following the invasion of the winter moth into Nova Scotia from its home in Europe, damage to oak forests in just two counties amounted to some 26 000 cords of wood per year. Currently this would be worth about US $5 million. During this period the moth spread over most of the Province. Other forest and orchard trees also were damaged, and it has been estimated that probably all oaks would have been killed and a minimum of $12 million loss suffered in this area alone had its attacks continued. However, parasites from Europe were imported and effectively colonized from 1955 to 1960 and two species, *Cyzenis albicans* (Fallen), a tachinid fly, and *Agrypon flaveolatum* (Gravenhorst), an ichneumonid wasp, became established, the former first in 1955 and the latter in 1956–7. Eventually they brought about complete biological control for a cost of about $160 000.

Initial increase of the parasites was slower than usual, and it took about six years following liberation before host population collapse occurred at a given release site. In addition to the monetary savings in Nova Scotia, the rate of spread into new areas has been greatly reduced, in proportion to the reduction in host populations by the parasites. The two parasites have been shown to be compatible and strongly complementary. *C. albicans* is the most efficient in reducing high host densities, but *A. flaveolatum* is more effective at low host densities. The original collections in Europe may reflect this indirectly. There, the heaviest infestations possible were sought out in order to obtain the most parasites for shipment. In other words, small outbreaks were sought out. In such

Table 6.1. *Dominance of the two parasites on* Operophtera brumata

Average percent parasitization by			Average host population density‡
Year	*Cyzenis*★	*Agrypon*†	
1959	10	0.1	60
1960	34	4	75
1961	58	7	70
1962	38	40	37
1963⎱ host too scarce to provide data			6
1964⎰ on percent parasitization			0

★Established 1956.
†Established 1957.
‡As measured by percent defoliation of red oak in the vicinity of liberation areas.

dense host populations the dominant parasite may not represent the best regulatory agent, or even a satisfactory one, when host densities are low, but rather the dominant one may be dependent upon high densities in order to maintain itself. These European collections yielded about ten *Cyzenis* to one *Agrypon*. When these two parasites were first well established in Nova Scotia (1959), host densities were high and *Cyzenis* was much the dominant parasite. By the time host densities at the original sites had been considerably reduced in 1962, *Agrypon* was slightly dominant. In 1963, host densities were too low to obtain adequate parasitization data. The data in Table 6.1, from the original sites, show the change in dominance of the two parasite species as the host population decreased.

These data illustrate the point that in searching abroad for new natural enemies, one should not concentrate only on collecting where the host is most common and easily found. To do so could mean missing the best parasite, which may be found where host populations are the lowest.

No releases of the parasites have been made in the Maritime Provinces since 1965, but recent surveys have indicated that they are still effective in residual host populations. When the winter moth eventually reached the West Coast, collections of *Cyzenis* and *Agrypon* were made in Nova Scotia and in Germany, and the parasites were colonized in British Columbia in 1979–80 and in Oregon in 1981.

It is of interest that intensive quantitative population studies of factors affecting the winter moth near Oxford, England, covering some twenty-odd years, did not indicate parasites to be of major consequence. This again raises the question of the value of such studies in the evaluation of the effectiveness of natural enemies in prey population regulation (see Chapters 3 and 7 *re* experimental methods of evaluation). Mathematical model studies in England by G. C. Varley and G. R. Gradwell of winter moth-parasite interactions also have led to predictions by them that in Canada strong oscillations in interacting populations of the winter moth and its parasites will cause damaging outbreaks of the winter moth at nine- or ten-year intervals. This would have been a rather unlikely event, as no such instance has yet been recorded among the many other successful cases of applied biological control. Indeed, nothing of the sort has happened in Nova Scotia, the two introduced parasites having produced stable biological control at very low host population densities. In fact, re-analysis of the Oxford and Nova Scotia data has led to the construction of a more realistic model, which does not predict any such oscillations. (See Embree, 1971; Embree & Otvos, 1984; Hassell, 1980.)

Another interesting point is that this case, as also was true with the eucalyptus snout-beetle in South Africa, is an exception to Clausen's rule. This generally true precept states that if imported natural enemies are destined to become completely successful in host population regulation, they will do so either within three host generations or within three years. In this instance it took six to seven years.

Florida red scale in Israel and elsewhere, 1956 on

The biological control of the Florida red scale, *Chrysomphalus aonidum* (Linnaeus) (= *ficus* Ashmead), in Israel stands out as an illustration of the simplicity, ease and inexpensiveness with which biological control can sometimes be accomplished – given an enthusiast to push it. This outstanding success against Israel's worst citrus pest was achieved principally through correspondence, as was the case earlier with the citriculus mealybug in Israel, and was brought about by the same man, Israel Cohen, head of the Agrotechnical Division of the Citrus Marketing Board of Israel. No foreign explorer was directly involved. Cohen, who, it will be recalled, was not a trained entomologist, wrote to specialists in biological control of scale insects at the University of California at Riverside in 1955, seeking advice as to where natural enemies might be obtained. He was informed that at least two species of parasites were known in Hong Kong, where the scale was indigenous and scarce, and that arrangements might be made with S. K. Cheng, a part-time collector for Riverside, to obtain and make air shipments. Cheng was a biology teacher who had originally been trained 'on the job' by Dr J. L. Gressitt of the Bishop Museum, Honolulu, when Gressitt was doing foreign exploration in the Orient for the University of California at Riverside in the 1940s.

Shipments were begun by Cheng in February 1956, reaching a total of 72 during the next two years. All told, they contained relatively few parasites, with approximately equal numbers of *Aphytis holoxanthus* DeBach (then thought to be *A. lingnanensis*) and *Pteroptrix* (= *Casca*) *smithi* (Compere) being present. Less than 150 live females and males of *Aphytis* were received in 1956, but laboratory culture was immediately successful with the first few specimens tested, and all the others were then released in infested groves in Rehovot, mostly during the period February–May by Dr E. Rivnay and later by David Nadel. These few gave rise to the later millions that reproduced in the field and subjugated the scale. The *Pteroptrix* was not successfully cultured in the laboratory, but about 120 individuals were released directly in one grove from December 1956 to April 1957. *Pteroptrix* was not seen again until one of the authors (DR) recovered it in 1960 on infested date palms in the vicinity of the original release site.

In September 1957, about one and a half years after the first release of *A. holoxanthus*, a homeowner in Rehovot, Israel informed Nadel that for the first time in his memory the citrus trees in his backyard were clean of Florida red scale. *A. holoxanthus* was obviously responsible for this, and an immediate follow-up survey by Cohen and Nadel of the citrus groves around Rehovot showed a considerable area completely clean of the scale. Bordering this area the scale was still abundant but the parasite was extremely common, and furthermore was found to have spread as far as 20 km to the north and ten to the south. Further distribution of the parasite in 1957 was then rapidly accomplished by movement of parasitized scales on twigs and branches to new locations and by release of insectary-reared parasites during 1958–9.

Complete biological control was achieved throughout the main citrus-growing area of Israel, the coastal plain, within two to three years after the first liberation. Establishment in the hot Jordan Valley was considerably slower, but there too satisfactory biological control has eventually been achieved. To this date, except for occasional pesticide-induced upsets, at the most only a few isolated scales can be found where previously there would have been millions. It has been conservatively estimated that savings of well over $1 million per year can be attributed to this successful project that cost only a few thousand dollars. In 1956, before the parasites were exerting any effect, 3000 tonnes of oil were used to spray for the Florida red scale; in 1957 this was halved to 1500 tonnes; by 1958 – two years after the first release of parasites – it was reduced to 500 tonnes, and in 1959 only 10–20 tonnes were used – a reduction of over 99 percent.

Although *Pteroptrix smithi* was established at the same time as *Aphytis holoxanthus*, it did not have any appreciable effect in this spectacular success. In fact, its presence was not even detected until the project was practically over. However, over the years it has slowly extended its range along the coastal plain of Israel (but not to the interior valleys), and has become the dominant parasite in some very low populations of the Florida red scale. In certain coastal groves it now outnumbers *Aphytis* as much as ten to one. This illustrates how difficult it may be to pre-select the best natural enemy for a given project. A foreign explorer, observing the current situation in most coastal groves in Israel without any knowledge of what had gone on previously, would not be likely to consider *A. holoxanthus* as the most promising candidate for importation against the Florida red scale. Yet it was this species that effected complete biological control in Israel, and subsequently in several other countries.

Regardless of the closed border between Israel and Lebanon, *Aphytis holoxanthus* soon dispersed into that country and produced similar results. It has been purposefully introduced subsequently into Florida, Mexico, Brazil, Peru, South Africa and Australia, with outstanding results and at a cost of only a few dollars to cover the air shipments in most of these cases. Additional millions of dollars in reduced pest control costs and damage are accruing annually from these successes.

Historically and technically, there are several points of interest in this case. They all revolve around the facts that inadequate knowledge of the taxonomy of *Aphytis* species and lack of financial support for biological control work prevented the discovery of *A. holoxanthus* and its successful use in the countries mentioned, as well as in some others where it still remains a pest, for at least 80 years.

William Ashmead of the USDA, stationed in Florida to work on citrus insects, attempted by correspondence as early as 1879 to ascertain the native home of the Florida red scale in order to obtain its natural enemies for use in that State. He was unsuccessful, but had funds been available and knowledge sufficient, the biological control of the Florida red scale in Florida could have been the first great successful project rather than that of the cottony-cushion scale in California, which occurred ten years later. Thus, the Florida red scale remained a serious pest in Florida 80 years longer than it needed to.

Another chance was missed when Alfred Koebele reared an *Aphytis* from Florida red scale in Hong Kong, apparently in the late 1890s, and sent the specimens to the United States National Museum in Washington, DC. They were misidentified as *Aphytis mytilaspidis* (Le Baron) and subsequently ignored, probably because *mytilaspidis* was common in the United States, but as far as can be told today (they are poorly prepared) they are *A. holoxanthus*. Then, in 1924–5, while Professor F. Silvestri was collecting in China for the University of California at Riverside, he recorded noticing differences between the *Aphytis* parasitizing the California red scale and the Florida red scale. He thought the parasite of the former to be *A. chrysomphali* (Mercet) and designated the one from Florida red scale as *Aphytis* sp. Again, this was probably *A. holoxanthus*, and had Silvestri formally described it as a new species, as he did with several other important parasites, the resulting publication would have undoubtedly come to the attention of Florida entomologists or others and the successful importation made in this case some 30 years before it finally occurred in Israel. Yet another chance to 'discover' what apparently was *A. holoxanthus* was overlooked following the clues published by T. H. C. Taylor in 1935 in connection with the coconut scale project in Fiji. He recorded a 'form' of *A. chrysomphali* as a parasite of Florida red scale in Java. The pupal stage he pictured is obviously not *chrysomphali*, but is identical in appearance with pupae of *A. holoxanthus*. These missed opportunities emphasize, as we do in Chapter 7 and elsewhere, the great need for taxonomic research and for increased foreign exploration and importation of new natural enemies. With many of today's major pests we have done no more than William Ashmead was able to do against the Florida red scale in 1879. (See DeBach, Rosen & Kennett, 1971; Rivnay, 1968; Steinberg, Podoler & Rosen, 1986.)

California red scale in California, 1957

This scale, *Aonidiella aurantii* (Maskell), is of Oriental origin. It became established in California by the 1870s and for years was the number one pest of citrus, being responsible for losses of millions of dollars annually from damage to trees, loss in fruit quality and cost of insecticidal treatment. Attempts at biological control began in 1889, when Albert Koebele sent in some predatory beetles from Australia, and work has continued with varying degrees of emphasis up to the present. The campaign against the California red scale is the longest in the history of biological control. A detailed and valuable history of the earlier efforts in California has been given by Compere (1961). His account shows how a combination of confused taxonomy and lack of adequate knowledge of biology and ecology prevented the discovery and importation of the most important parasites of the California red scale for 50 or 60 years. The biological and ecological problems involved in working toward a successful solution have been so complex that a large book would be necessary to cover the entire story, which is still continuing. Only the briefest résumé can be given here.

Up to 1941, several predatory beetles and one parasite, *Aphytis chrysomphali* (Mercet), attacked the scale in California. They were generally of no practical significance, although we were able to demonstrate later that *A. chrysomphali* was potentially capable of controlling red scale populations very effectively in mild coastal locations, if not interfered with by chemical treatments or by honeydew-seeking ants. Rarely were both of these factors absent from any given grove in the restricted climatic zone favorable to the parasite, so it had been considered to be of no consequence over the years, after it became accidentally established in California about 1900. Inasmuch as virtually all *Aphytis* reared from the California red scale abroad up to 1948 were considered to be *A. chrysomphali*, no attempt was made to import any of these, which, as we shall see, ultimately included the two best parasites now established. Thus, the inadequate state of the taxonomy of the genus *Aphytis* caused a delay of some 50 years in the introduction of the most valuable parasites into California, and subsequently elsewhere.

The degree of success in biological control of the California red scale in California can now be considered as substantial. It was considered to be a failure for nearly 60 years, however. Completely satisfactory control occurs in some groves year after year, and this is essentially the case in entire districts. In groves under good biological control the scale is so scarce as to be difficult to find, even by an expert. Nonetheless, a large amount of chemical treatment is still practiced. We consider the real need for much of this to be questionable or definitely unnecessary, especially in coastal and intermediate climatic zones, but there are certain climatic areas where the parasites currently established are rendered ineffective by cold winter weather, hot summer weather, or both. In general, the degree of unfavorability increases with distance from the coast. Thus, in the desert areas of the Imperial Valley, the extent of biological control is reduced to a negligible level. In any event, the California red scale has lost the dubious distinction of being California's major citrus pest. Many thousands of hectares now go untreated each year, with enormous savings to the citrus industry.

Four species of parasites are responsible for the degree of biological control now achieved. All were established between 1941 and 1957, with the last one being the most important. Two species of *Aphytis* are in general the most effective of the natural enemies. Each of these alone can regulate California red scale populations at extremely low levels if habitat conditions are favorable to them. *Aphytis melinus* DeBach, imported by one of the authors (PD) from India and Pakistan in 1956–7, must be rated number one. It is completely dominant in interior areas and generally in intermediate and some coastal climatic areas, but is complemented in its work by the red scale race of *Comperiella bifasciata* Howard, which was imported from southern China in 1941. The latter was established from an original stock of only five mated females. *Aphytis lingnanensis* Compere, imported from southern China in 1947–8, remains dominant in some coastal districts, but is strongly complemented by the red scale race of *Encarsia perniciosi* (Tower), imported from Taiwan in 1949.

These four parasite species are the last natural enemies to become established, and are by far the most effective of all the many species of predators and parasites imported and liberated since 1890. This illustrates the value of continued research and indicates that there probably never is any certain point in time when a given biological control project can safely be pronounced to be a failure. That 'best' parasite may still remain undiscovered.

The discovery and importation of the last and currently best one, *Aphytis melinus*, was a direct result of increased taxonomic knowledge and of previous ecological field studies with *A. chrysomphali*, beginning in 1946, and with *A. lingnanensis* starting in 1948. These studies for the first time proved experimentally in the field that species of *Aphytis* can be highly effective in host population regulation. The investigation also demonstrated that parasites can be precluded from exerting adequate control by adverse climatic factors. As a consequence, we deliberately sought another species of *Aphytis* from climatic zones similar to the harsher interior zones of southern California. Areas of northwest India and Pakistan were some of the few that were known to be similar, hence were searched. This led to the discovery of *Aphytis melinus*. We still hope that another species will be discovered that will be fully adapted to the most severe weather conditions where citrus is grown in southern California and in the San Joaquin Valley. It may exist in the interior of south China, which has been unavailable for foreign exploration during most of the modern era. *Aphytis riyadhi* DeBach, obtained in 1979 from a related host species, *Aonidiella orientalis* (Newstead), in Saudi Arabia does not seem to offer much hope in this respect. Meanwhile, *A. melinus* has been sent to many other countries for use against the California red scale, including Argentina, Australia, Chile, Cyprus, Greece, Israel, Morocco, Mexico, Sicily and South Africa. It has become established in all of these, effecting substantial control, and in certain instances has spread of its own accord into neighboring countries.

An interesting sidelight of this project has helped clarify the basic role of competition between parasite species, as it is related to host population regulation and the biological control policy of multiple species importations. It has been hypothesized by some that, because multiple importations will necessarily involve competition between parasite species, this will result in poorer regulation of host populations. Therefore, they have proposed that only the single best natural enemy should be introduced, although how this could be determined *a priori* has never been ascertained. Virtually all biological control specialists disagree with the latter idea as being both impractical and ecologically unrealistic. Results from the California red scale project support this, and further demonstrate strikingly that even the most severe competition between parasite species does not interfere with host population regulation but rather enhances it. The story and reasons are as follows.

Admittedly, much trial and error has been involved in the eighty-odd-year campaign against the California red scale. About 50 or more species of natural enemies have been purposely introduced; others became accidentally established, and certain native ones adopted the red scale. A highly complex and drastic succession of changes has occurred over the years among the various predators and parasites of this scale. Natural enemy species have become established, increased in abundance, sometimes to become dominant, and then later on, as additional species were imported, decreased drastically or even disappeared completely. Meanwhile, there has never been any observation or suggestion made that all this competition between natural enemies had ever proved favorable to population increase of the California red scale at any time. In fact, the opposite is true: California red scale populations have continued to decrease.

The most intense and interesting competition has occurred during the past 40 years, principally between species of *Aphytis*, *Encarsia*, *Comperiella* and several predatory ladybeetles. The interactions now have nearly run their course and homeostasis between the remaining species is being approached. The current situation is that briefly mentioned previously, with *A. melinus* dominant and *Comperiella bifasciata* complementary in interior areas and with *A. lingnanensis* dominant and *Encarsia perniciosi* strongly complementary in some coastal areas.

This competition between natural enemies over the years has caused the virtual disappearance by the newer parasites of the predaceous ladybeetles, which up to about 1950 used to reduce heavy outbreaks of the scale periodically but were unable to maintain it at low levels. Following the establishment of *Aphytis lingnanensis* in 1948, we have documented its role in causing the virtual, if not complete, extinction of *Aphytis chrysomphali* which was the dominant, generally distributed enemy of California red scale until the late 1940s and which produced excellent biological control in certain favorable habitats. We subsequently have followed the complete elimination of *A. lingnanensis* from interior citrus areas, where it was exclusively dominant in the mid-1950s, by competition provided by *A. melinus* which was first established in 1957. *A. melinus* is more tolerant of climatic extremes than *A. lingnanensis* and is capable of utilizing smaller host scales for the production of female progeny. Meanwhile, we have seen that the apparently less efficient parasites, *Comperiella* and *Encarsia*, continue to coexist with one or the other species of *Aphytis* and to complement them, because they attack different host stages and are not strict ecological homologs.

In spite of this intense and continuing competition between natural enemies for a single host species, the extent of biological control has concurrently increased until today California red scale populations are much lower in general than they ever have been. From careful study of the large mass of biological and ecological data accumulated on this project, we see no possibility that the observed results could have been predicted with any accuracy beforehand. Too many variables are involved, even in what might appear to be a simple ecosystem such as citrus. We have to conclude that we can knowledgeably restrict the number of promising natural enemy candidates for importation to a relative few, but we cannot predict in advance which one, or which combination, will be the best in any given habitat, much less in the variety of habitats usually inhabited by a major pest. (See DeBach, Rosen & Kennett, 1971; Luck & Podoler, 1985.)

Dictyospermum scale in Greece, 1962

This co-operative pilot project, designed to add impetus to biological control research in Greece, was initiated in 1962 when one of the authors (PD) was invited to spend his sabbatical leave as a Senior Fulbright Fellow at the Benaki Phytopathological Institute in Kiphissia, near Athens. It was chosen because it could be carried out rapidly, at little expense, and with good probability of success because the parasites to be tested were known to be effective in other countries. At that time the dictyospermum scale, *Chrysomphalus dictyospermi* (Morgan), was the number one pest of citrus in Greece, the California red scale was second in importance, and the purple scale third. Parasites were imported for all three, but this discussion will cover only the results obtained against the dictyospermum scale. Favorable results were obtained in each case, however. Chemical treatment was generally practiced, but damage from scale insects often occurred in spite of it. Heavily infested and damaged citrus trees were commonly seen in the principal citrus areas in the Peloponnesus and Crete, as evidenced by leaf-drop, killed twigs and scale-encrusted fruits.

Three established species of parasites were found to attack the dictyospermum scale, of which *Aphytis chrysomphali* was the most abundant and potentially the most effective. However, as was the case earlier in California against the California red scale, *A. chrysomphali* was apparently precluded by adverse environmental conditions from exerting any appreciably reliable dègree of biological control.

At that time, several species of *Aphytis* were being mass cultured in the insectary at the University of California at Riverside, all of which had been reared from the California red scale in different countries and imported for trials against that scale in California. The three that were considered most likely candidates for testing in Greece against both the dictyospermum scale and the California red scale were *A. melinus* (originally from India and Pakistan), *A. lingnanensis* (originally from southern China) and *A. coheni* DeBach (originally from Israel). The former two were known to be effective parasites of the California red scale in certain locations in the field in California, as well as in their native homes. *A. coheni* was new, and its status unknown.

None of these parasites was known with certainty to occur on dictyospermum scale anywhere in the field (this scale is not a pest in California), but laboratory tests at Riverside had shown that *A. lingnanensis,* and to a lesser extent *A. melinus* and *A. coheni,* did very well on dictyospermum scale.

Inasmuch as no scale cultures for rearing of the parasites were then available at the Benaki Institute, direct field colonizations of imported pure cultures of adult *Aphytis* parasites were indicated. The main problem, even with air shipment, is that the adult parasites are quite short-lived and suffer considerable mortality within a few days, especially if subjected to temperature extremes *en route.* The first air shipments from California in October 1962 evidently were transferred *en route* to Athens and held to await airplane connections, hence took several days. Many of the parasites were dead on arrival, but the survivors were colonized. Therefore, special arrangements were made with Trans World Airlines for the next shipment to receive preferential treatment. As a result, the next and final air freight shipment of November 1962 arrived within 24 hours after leaving Riverside, California. This is probably a record for such long-distance shipment of natural enemies. Fig. 6.3 shows this shipment being received by one of the authors (PD) at the Athens airport. No mortality occurred among these parasites. They were colonized in the field, even in Crete, within the first or second day after arrival. Substantial numbers, ranging between about 8000 to 20 000 of each species, were liberated.

The following winter of 1962–3 was unusually cold, but recoveries were made the following spring in low numbers, with *A. melinus* being the only one of any consequence taken and ultimately the only one to persist. In spite of the fact that about 18 000 additional *A. lingnanensis* were liberated in June 1963, a very favorable period, *lingnanensis* failed to become established.

Aphytis melinus increased and spread rapidly during 1963, both on dictyospermum and on California red scale. It was recovered from four of five sites sampled in October–November 1963, one year after release, and already comprised 79 percent of the total parasite population as compared to 21 percent for *A. chrysomphali.* It had become dominant at the points of liberation within one year; within two years it had completely replaced *chrysomphali* at these sites. After about two and a half years, *A. melinus* had successfully dispersed from the original sites some 200 km across country which is cultivated only in scattered areas and is generally quite barren. Other data bearing on this point indicate an effective dispersal rate of about 100 km per year. This far surpasses any of the other species studied which attack *Chrysomphalus dictyospermi* or *Aonidiella aurantii*. This ability to disperse may be related to searching ability, and thus may indicate why *A. melinus* is such an effective parasite. By the end of 1966, *A. melinus* was strongly dominant and had virtually replaced *A. chrysomphali* throughout the Peloponnesus and in all areas surveyed in Crete. It had also dispersed to the islands of Lesbos and Chios (see Fig. 6.4). Soon thereafter it was found for the first time on the mainland of

Fig.6.3. Paul DeBach receiving a shipment of several species of adult *Aphytis* parasites in Athens, Greece, only 24 hours after they had been sent air-freight from the University of California insectary at Riverside, California. The parasites were released directly in the field and brought about complete biological control of the two worst scale-insect pests of citrus and substantial control of two others.

Turkey at Izmir, and, aided by further releases, has since spread all over the citrus areas along the Aegean and Mediterranean coasts.

Biological control of the dictyospermum scale has been outstandingly complete. At the original sites in the Peloponnesus near Trejenia it was virtually impossible to find scales on trees that were encrusted three years before. As *A. melinus* spread and became abundant, the dictyospermum scale became everywhere reduced to insignificant levels. When one of us (PD) revisited Greece in 1968, he was unable to find anything but a rare specimen of this scale on trees which previously he knew to be chronically encrusted. The entire cost of this project, as far as the practical results are concerned, was only a few hundred dollars, including the cost of importation of the parasites and their field colonization in Greece. There are any number of projects in the world today that could be conducted just as cheaply and easily, if the people responsible for decision-making merely had the necessary knowledge that parasites are available. (See DeBach & Argyriou, 1967.)

Fig.6.4. Map of southern Greece and Crete, showing the four original localities in which the scale-insect parasite *Aphytis melinus* was colonized in October–November, 1963. The shaded area shows approximate dispersal of *Aphytis melinus* from these sites by December, 1966. It became strongly, if not exclusively, dominant over the originally established *Aphytis chrysomphali* by that time.

Green vegetable bug in Australia, Hawaii and elsewhere, 1933–62

This plant-sucking pest, *Nezara viridula* (Linnaeus), known also as the southern green stink bug, is of cosmopolitan distribution, being found in North and South America, Africa, Australia, New Zealand, various Pacific islands, the Middle East and Japan. Its feeding habits are very broad, including a wide range of fruit and ornamental trees, field crops and vegetables. Such omnivorous feeding characteristics have made it an important pest nearly everywhere it invaded from its native areas in Africa or the Mediterranean.

Australia was the first country to start a biological control attack against the green vegetable bug, by using the imported egg parasite *Trissolcus basalis* (Wollaston) from Egypt. Large releases of *Trissolcus* were carried out in various States, lasting from 1934 to 1943. Results were not always monitored consistently, so most reports circulated abroad were that the results were of little consequence, even though *Trissolcus* was well established in all infested States. Consequently, this project might justifiably be termed the 'case of the forgotten natural enemies', except in Western Australia which claimed outstanding results early in the game. Otherwise, for some 15–20 years, not much credit was given to the activities of the parasites, especially in areas having relatively cold winters. By 1960, the benefits from the establishment of *Trissolcus* were considered variable – what we would term a partial success. As we have seen, the program was rated an outstanding success in Western Australia. Excellent results also were reported from South Australia and satisfactory control from New South Wales, except in inland areas. This was true of other inland areas.

Meanwhile, later introductions of other races or strains of *T. basalis* from various countries were made, but only the Pakistan race is known to have become established. More recent reviews of the situation show that the pest is now under excellent control in most inland areas, with most credit for this going to the Pakistan race. Then, when the bug was first recorded in northern Australia in 1974 and heavy infestations occurred, *T. basalis* was introduced from southwestern Australia and complete biological control was attained. Thus, with continuing research efforts, Australia achieved an outstanding success against this serious pest. One remaining problem is that the efficacy of *T. basalis* is reduced on certain crops, such as soybean and sesame, and additional strains are being sought in order to overcome this. This project is an excellent illustration of the value of so-called races, strains or ecotypes in biological control, as well as of the pay-off that may come from continued research. (See Ratcliffe, 1965; Waterhouse & Norris, 1987.)

In Hawaii, the green vegetable bug was first discovered in late 1961, and an immediate program of parasite introductions began early in 1962. The success achieved in Australia was already known, so parasites were obtained first from there. Two species of tachinid flies that parasitize adult bugs were also flown in – *Trichopoda pilipes* (Fabricius) from Antigua and Monserrat in 1962 and *T. pennipes* (Fabricius) from Florida in 1963. Large-scale mass production of the imported parasites in 1963 yielded 767 525 egg parasites and 15 712 *Trichopoda* for field release. As a result, by 1964 *Trissolcus basalis*, the egg parasite, *Trichopoda pennipes* and especially *Trichopoda pilipes* were thoroughly established.

Trissolcus basalis was first found to be established in August 1962, and thereafter increased and spread with great rapidity. Rearing records from ten sites on Oahu during the last half of 1963, i.e. about one and a half years following first releases, revealed an average egg parasitization of 94.9 percent. This figure continued over the ensuing years, during which period egg clusters of the green vegetable bug became difficult to find in the field. The tachinid *T. pilipes* also increased and spread rapidly on Oahu, showing an average parasitization of about 41 percent for the same period. Both of these parasite species later became abundant on the islands of Hawaii, Kauai and Maui. Complete economic control was achieved first on Oahu and Kauai, and has been general on all islands since 1965. (See Davis, 1967.)

Results in New Zealand, where the bug was discovered in 1944, were achieved by the importation of *Trissolcus basalis* from Australia in 1949. Serious damage was soon nullified, and 80 percent parasitization was reported by 1950. In 1952, 90 percent parasitization was recorded in certain districts. Overwintering bug populations were greatly reduced that year. Apparently some fluctuations still occur, but there has been a gradual decline in the severity of plant damage, and the current situation is satisfactory (see Cumber, 1964). The opportunity apparently remains for New Zealand to import and establish one or more other species or strains of parasites.

Trissolcus basalis was imported into Fiji from Australia in 1941. Establishment occurred readily. Also *Trichopoda pennipes* was imported from Florida, but establishment is uncertain and results are unknown

(O'Connor, 1960). Similarly, *Trissolcus basalis* was introduced into a number of other Pacific islands including Tonga (1941), New Caledonia (1942–3), Samoa (1953) and Papua New Guinea (1978), with establishments recorded in each case, but biological control results are unavailable. (See Clausen, 1978; Waterhouse & Norris, 1987.)

Walnut aphid in California, 1959 and 1968

This aphid, *Chromaphis juglandicola* (Kaltenbach), is a native of the Old World that invaded California at about the beginning of this century. It soon covered all the walnut-growing areas of the State. In spite of attacks by several native natural enemies, the aphid frequently literally covered the trees with exuded honeydew and the resultant black sooty-mold fungus that grows in it. Thus for years, routine insecticidal treatments were necessary, beginning with nicotine sulfate in the early days and progressing to a variety of synthetic organophosphorus insecticides and chlorinated hydrocarbons in more recent years. This posed the usual problems of upsets, resurgences, development of resistance and drift of toxic materials outside the area being treated.

A change of tactics to biological control obviously was desirable, but there has been a long-standing idea prevalent among entomologists that aphids may not be particularly amenable to biological control, because they can multiply so rapidly in the spring that damage is done before their natural enemies can catch up with them. Dr Robert van den Bosch, who was in charge of this project, suspected that this applied principally to the less effective, high-density, omnivorous natural enemies, but perhaps not to more host-specific forms. He therefore concentrated his search on the latter, and in southern France in 1959 he found a parasitic wasp, *Trioxys pallidus* (Haliday), having a high degree of host specificity. It was successfully imported into California and became eminently effective against the aphid in the coastal plain of southern California, spreading rapidly and destroying a very high percentage of the host population. However, the problem was not completely solved because this parasite was not well adapted to conditions in northern and especially central California where, in spite of large-scale releases, it apparently never became permanently established. Evidently, the French strain of *T. pallidus* lacked genetic characteristics enabling it to reproduce and survive to any extent in areas of extreme summer heat and low humidity.

Recalling that *Trioxys pallidus* had been observed and collected during 1960 on the hot, dry central plateau of Iran, van den Bosch obtained live specimens from there in 1969. They were cultured and colonized in central California that summer and autumn, with spectacular results. Small releases made in June and July gave rise to abundant parasite populations, which had dispersed significantly by autumn. Further releases were made in the Central Valley in 1969 from insectary cultures. Surveys of release plots in October 1969 showed the parasite to be established in all new release plots, and to have spread to a distance of at least eight miles. By 1970 it had colonized all major walnut-growing areas of California. By 1971–2, biological control of the walnut aphid could be said to be complete throughout California, except where upsets occurred due to the adverse effects of insecticides used against the codling moth or the walnut huskfly, and to a lesser extent from the Argentine ant, interfering with the parasites.

The biological control impact of the Iranian *T. pallidus* was closely evaluated from 1969 to 1974. It was shown that, in spite of rather heavy hyperparasitism, the parasite essentially eliminates the springtime peak of the walnut aphid, which formerly was the most injurious because it coincided with rapid walnut growth. In one grove, population indices showed an average reduction of from 2550 aphids per sample in mid-May 1969 to only 8 per sample in mid-May 1970. Some population peaks may subsequently develop during summer or autumn, but their economic impact appears to be low or negligible. The benefit to the California walnut industry from this successful project has been conservatively estimated at from half to one million dollars per year.

Had taxonomy remained in the 'alpha' or descriptive stage, and the existence of ecotypes not been appreciated, the Iranian *Trioxys pallidus* probably never would have been imported, because it is indistinguishable morphologically from the French form imported and established in 1959. It would have been considered to be already established in California. Instead of a complete success, only a partial success would have been achieved. How many more of these opportunities remain undiscovered? We are sure there are a great many, but only strong support for, and rigorous dedication by, well-trained bio-ecologists thoroughly familiar with biological control will supply the ultimate answers. (See van den Bosch *et al.*, 1970, 1979.)

Cereal leaf beetle in the midwestern United States, 1966–72

This European species, *Oulema melanopus* (Linnaeus), probably invaded the United States in the late 1940s. It was first recognized in Michigan in 1962 and, in spite of attempts to place the original sites of infestation under quarantine, has since spread over most of the midwestern and northeastern States and into Canada. It is a migratory, univoltine pest, overwintering as adult beetles. Both the larvae and adults feed on the leaves of various cereal plants, including wheat, barley, oats, rye and corn, as well as certain forage grasses and weeds. Losses as high as 55 percent in spring wheat and 23 percent in winter wheat were reported early on. Eradication by blanket sprays was attempted for seven years, but was unsuccessful.

In 1963, the USDA initiated an extensive search for parasites in Europe. From 1964 to 1971, some 230 000 parasites were collected in numerous locations in 14 European countries and shipped by air to the USDA quarantine laboratory at Moorestown, New Jersey, where they were screened and forwarded to researchers in the Midwest. *Anaphes flavipes* (Förster), a mymarid egg parasite, was established in Michigan in 1966. By 1972, three larval parasites – the eulophid *Tetrastichus julis* (Walker) and the ichneumonids *Lemophagus curtus* Townes and *Diaparsis temporalis* Horstmann – were also established, and all four were distributed throughout the infested area, which was still expanding. Fig. 6.5 shows the cereal leaf beetle and two of these parasites.

The Animal and Plant Health Inspection Service (APHIS) of USDA established a laboratory in Niles, Michigan, for mass rearing and distribution of the parasites. Each year, farmers and extension officers were invited to 'biological control days' and provided with 'bouquets' of grain plants bearing parasitized eggs and larvae of the beetle, grown in special 'field insectaries', for distribution in infested fields.

The project has been rated as a complete or near-complete success. The parasites have generally reduced beetle populations by 60 percent and grain losses to less than one percent, and chemical control is virtually no longer necessary. Only 8000 hectares were sprayed against the cereal leaf beetle in 1981, as compared to more than 640 000 hectares in 1966. It has been estimated that the quick importation and dispersal of the parasites has prevented the cereal leaf beetle from ever becoming a major pest in the United States. Such *prophylactic* control is just as valuable, perhaps more

(a)

(b)

(c)

Fig.6.5. (a) An adult cereal leaf beetle feeding on a barley leaf; (b) *Anaphes flavipes*, an egg parasite imported from Europe, ovipositing in a cereal leaf beetle egg; (c) one of the larval parasites imported from Europe, *Lemophagus curtus*, ovipositing in a cereal leaf beetle larva. (Photos published in *Agricultural Research*, **31**, No. 5 (November 1982); courtesy USDA.)

so, than the more spectacular results that would have occurred had the pest originally been more entrenched before parasites were imported. (See Dysart, Maltby & Brunson, 1973; Haynes & Gage, 1981.)

Woolly whitefly in California and elsewhere, 1967 on

This whitefly, *Aleurothrixus floccosus* (Maskell), of tropical American origin, was detected in San Diego, California, in 1966 and soon proved to be very damaging to citrus. Populations increased rapidly to high densities, averaging 100 live larvae per leaf or more and causing defoliation, twig death, reduced fruit set and poor fruit quality. The vast amounts of honeydew produced by the larval stages served as a substrate for an extensive black growth of sooty mold fungi, hastening defoliation and tree decline, and attracted the Argentine ant, *Iridomyrmex humilis*, which induced outbreaks of various scale insect and mite pests through interference with their natural enemies. By 1972, potential losses caused by the whitefly were estimated at $22 million per year.

Biological control efforts began as early as 1967, when one of the authors (PD) hand-carried two species of parasites, *Amitus spiniferus* (Brèthes), a minute platygasterid wasp (see Fig. 2.18, page 59) and *Eretmocerus* n. sp., an undescribed aphelinid, from Mexico to California and established them in San Diego. However, in 1969 a multi-million dollar chemical eradication program was initiated in the San Diego area and all biological control activity had to be stopped. This created a rather unique circumstance, because infestations of woolly whitefly had been discovered in both California and Baja California Norte, Mexico, at about the same time. Thus, only a political border separated the US whitefly populations, which came under chemical eradication, from infestations in Mexico where essentially no pesticide treatments were undertaken. Therefore, the parasites were collected from San Diego colonization sites and re-colonized in Tijuana, just across the border. Only 30 *Amitus* and 169 *Eretmocerus* were initially colonized there, but co-operation with Mexico's Dirección General de Sanidad Vegetal during 1969 and 1970 resulted in further importations totaling some 12 000 of these parasites from Sinaloa, Mexico.

The parasites were successfully established in Tijuana. Large reproducing populations of both species were maintained on selected 'field insectary trees' that were kept as free as possible of ants, dust, chemicals and other adverse factors. Parasitized whiteflies were transferred to new infestations on cut twigs, thus eliminating the need for handling delicate adult parasites and costly insectary operations. By the end of 1970, more than 27 000 *Amitus* and *Eretmocerus* were redistributed in this manner. *Amitus spiniferus* and a second *Eretmocerus* n. sp. from El Salvador were also successfully colonized during 1970, as was *Cales noacki* Howard, another aphelinid parasite imported from Chile.

Meanwhile, in San Diego, the eradication program that had been in operation for more than two years was judged a failure. Consequently, biological control was reinitiated in California in 1971 within a so-called 'quarantine area' of approximately 500 km², and more than 20 000 adults of the parasites established in Baja California Norte were transferred to some 100 sites there. Outside of this area chemical eradicaiton – later termed 'containment' – was continued. *Amitus spiniferus*, *Cales noacki* and the two *Eretmocerus* species were all successfully colonized in California. Field insectary trees were again selected, adverse factors were eliminated, and infested twigs bearing all species of parasites were distributed within the quarantine area. Further explorations were conducted by Paul DeBach and Mike Rose, and by the end of the importation phase of the project – 1976 – a total of 32 species of parasites had been imported from Mexico, Central and South America and the Caribbean.

Both in Baja California Norte and in San Diego, complete biological control was achieved in the initial colonization sites within two growing seasons. Two of the parasite species, *A. spiniferus* and *C. noacki*, proved particularly effective in reducing whitefly populations and regulating them at very low levels. On average, woolly whitefly populations were reduced by more than 95 percent, and on demonstration trees that were kept free of interference from ants, dust and insecticides, reductions of 99.95 percent were measured, even during periods of peak whitefly activity. So much so, that by 1975 it was necessary to inspect 1000 leaves or more to find a living whitefly on the study trees. However, in new infestations where parasites were still absent, or wherever they were interfered with by pesticides, woolly whitefly populations increased exponentially and reached extremely high densities.

While this was going on, chemical 'containment' failed and the woolly whitefly continued to expand its range in California. By 1972 the known infested area more than doubled and extended to Fallbrook; by 1973 the whitefly reached Orange County, and by 1974 it existed in virtually all the coastal and several key inland citrus areas as far north as Los Angeles

County. By 1981 it was established in all southern California citrus growing areas except the southern desert. As the range of the whitefly expanded, so did the biological control program. Insectary trees were established in each infested area, and the various parasite species and biotypes were colonized separately on them and redistributed. In this manner, prophylactic biological control was often achieved so rapidly that most citrus growers were saved from the need to apply pesticides against the whitefly. This method also assured that the different parasites were established throughout the range of the pest, so that competition would eventually determine the best natural enemy composition. Complete biological control was achieved by *A. spiniferus* and *C. noacki* throughout the entire infested area, and woolly whitefly populations were reduced by 95 to more than 99 percent. Perhaps of equal importance, the natural enemies that regulate other important pests on citrus were conserved. Based on earlier dollar estimates, savings of about $40 million per year to the California citrus industry are attributed to the biological control of woolly whitefly. The cost of the entire program was less than $1 million.

Following this success in California, complete biological control has also been achieved in the Mediterranean region. The citrus-growing regions of Spain and France were invaded by the woolly whitefly at about the same time that it was discovered in Baja California Norte and California. In 1970, Mike Rose collected adult *Cales* and *Amitus* from field study sites in Baja California Norte and San Diego and hand-carried them on the same day to San Francisco, where DeBach immediately took them to Spain and released them on infested citrus trees in the Malaga cemetery. Some of that material was forwarded to France, and several additional shipments were made. *Cales noacki* has become thoroughly established throughout the infested areas in Europe and North Africa, with great success. Early benefits from biological control of woolly whitefly in Spain have been estimated at about $100 million. (See DeBach & Rose, 1976. A full account of the woolly whitefly project is currently in preparation.)

Rhinoceros beetle in the South Pacific, 1967 on

This large beetle, *Oryctes rhinoceros* (Linnaeus), is a serious pest of coconut, oil palm and various other palms. A native of southeast Asia, it has spread during the present century into many islands of the Pacific and Indian Oceans. Preferred breeding sites include manure and compost heaps, sawdust pits, standing dead palms, stumps and fallen trunks. The grubs are therefore of little economic importance, but the adult beetles bore into the crown of palms, causing severe damage to the developing fronds, reduction of yield and, in heavy attack, death of palms. Their feeding also provides points of entry for deadly attacks by pathogens and other pests.

Early attempts at biological control in the South Pacific involved the importation of numerous natural enemies. One of these, the larval parasite *Scolia ruficornis* F. from Zanzibar, was widely established but of little economic consequence. However, investigations of parasites and predators virtually ceased when a baculovirus was discovered in rhinoceros beetle larvae in Malaysia in 1963 and its potential as a biological control agent was realized. This pathogen, named *Rhabdion-virus oryctes* Huger, was subsequently found to occur also in Indonesia and the Philippines, and appeared to be responsible for the fact that damage caused by the pest in southeast Asia was less severe than in the Pacific.

This virus develops in the nuclei of infected cells but normally does not produce inclusion bodies (see Chapter 2), and is therefore more difficult to diagnose and handle than other insect viruses. It is lethal to larvae, but multiplies also in adult beetles and may be disseminated by them to breeding and feeding sites. Thus, infected adults may infect other adults, as well as larvae. The most effective method of establishing the virus is through release of laboratory-infected beetles. It is entirely safe to vertebrates.

The virus was first introduced from Malaysia into Western Samoa in 1967, and its establishment there was followed by a marked decline in beetle damage to the palms. Similar results were reported from various other islands, where the virus was established from 1970 to 1978. In Fiji, the extent of palm damage fell from 75–90 percent in some areas in 1971 to 5–15 percent in 1974. Damage to coconuts was reduced by 60–95 percent in Mauritius, by 82 percent on Wallis Island. In Papua New Guinea, the rhinoceros beetle ceased to be a problem six to seven years after the release of the virus. In some countries, the use of baculovirus has been supplemented by application of the fungus *Metarrhizium anisopliae* to breeding sites and by such cultural methods as sanitation of plantations, destruction of breeding sites, or growth of dense cover crops to conceal decaying logs.

Although the results have been less spectacular than those of some of the other projects reported here and in Chapter 5, this project is significant because it is the only one known to us where an imported insect pathogen has alone effected substantial biological control of an important pest. (See Bedford, 1980, 1986; Waterhouse & Norris, 1987.)

Alfalfa blotch leafminer in the eastern United States, 1977

This European leafmining fly was first reported from Massachusetts in 1968 and has since spread over 16 northeastern states from Michigan to North Carolina in the USA and into Quebec, Ontario and the Maritime Provinces of Canada. Enormous populations developed, causing alfalfa fields to turn light brown or gray from the abundant mines. Direct losses to the first two cuttings of alfalfa were compounded by the threat that extensive chemical control would upset the alfalfa weevil (*Hypera postica*) and the pea aphid (*Acyrtosiphon pisum*), both of which had been brought under effective biological control. Thus, when the pest was definitively identified in 1972 as *Agromyza frontella* (Rondani), a species common but of insignificant importance in its native Europe, a biological control project was initiated.

Fourteen species of parasites were collected in France, Denmark, West Germany, Switzerland, Austria, Luxembourg and Lichtenstein by the USDA European Parasite Laboratory at Sèvres, France, during 1974–8 and shipped to the Beneficial Insects Research Laboratory at Newark, Delaware, USA. Early importations were unsuccessful, but when comparative studies revealed that, unlike in Europe, there were no effective parasites in the United States that attacked the host larva and emerged from the puparium, the effort was redirected to obtain species that would fill that vacant niche. Some 30 000–40 000 puparia of the fly were received, and three species of parasites became established. Two of these – *Dacnusa dryas* (Nixon), a braconid that prefers to oviposit in early-instar hosts, and *Chrysocharis punctifacies* Delucchi, a eulophid ovipositing in late instars – proved to be highly effective. In all, only 5207 *D. dryas* and 3307 *C. punctifacies* were released during 1977–8 in the original release sites in Delaware. They were 'dribble released' in small numbers throughout the season, from early May to mid-November. By 1978 they became so abundant there, that they could be collected by sweeping and distributed to new locations. By 1985 they were firmly established in 13 states, and complete biological control was in progress.

The parasites have required about five years to exert effective control.

In Delaware, levels of parasitism rose to 71–2 percent by 1981, and infestations were reduced from pre-establishment maximum levels of 10–25 mines per stem to 0.05 in the first alfalfa cutting, and from 24–42 to 2 in the second cutting. Not more than one percent of mature alfalfa leaflets were infested, well below the economic injury level. Annual savings of $13 million were reported in 1983 for the infested US area, as compared to a $1.1 million total cost of the entire project.

This rather simple, straightforward project emphasizes the point that direct importation work, unencumbered by peripheral research, is still the best approach to biological control. (See Drea & Hendrickson, 1986; Hendrickson & Plummer, 1983.)

Arrowhead scale in Japan, 1980

The arrowhead scale, *Unaspis yanonensis* (Kuwana), was until recently the most destructive pest of citrus in Japan. A native of China, it was first detected in Japan near Nagasaki in 1907, and by 1930 had spread all over the citrus-growing areas of that country. Severe infestations often resulted in the death of trees, but chemical control with organophosphorus insecticides was a mixed blessing because tremendous outbreaks of the citrus red mite resulted.

Biological control work – augmentation type – on the arrowhead scale started as early as 1954. Mass releases of a native coccinellid predator, *Chilocorus kuwanae* Silvestri, were tried in the 1960s and showed some promise. Then, in 1972, *Aphytis lingnanensis* Compere, a parasite of the California red scale that also attacked the citrus snow scale, *Unaspis citri* (Comstock), was introduced from Hong Kong. It was found to attack the arrowhead scale but would not overwinter in Japan, so a program of mass rearing and periodic releases of the parasite was developed, which by 1979 was rather effective. Such programs are, of course, much more costly than the permanent establishment of an exotic natural enemy.

In 1980, one of the authors (PD) visited the People's Republic of China, with the request from Japanese colleagues to keep an eye open for arrowhead scale parasites. In an experiment station near Chongqing, he was shown a chart depicting a parasite that had been reared from the arrowhead scale. A search was made in nearby citrus groves and two parasites were discovered – *Physcus* sp., an aphelinid endoparasite, and *Aphytis* sp., an ectoparasite. A shipment was immediately made to Japan, but it did not arrive until two weeks later and all the parasites had died *en route*. However, on his return flight to California, DeBach had a prearranged meeting at Narita Airport, Tokyo, with three Japanese colleagues who were soon leaving for China to search for arrowhead scale parasites. Thus, he was able to inform them of the species he had found and to direct them to the exact locality where they could be collected. This proved to be very useful, and the Japanese explorers – K. Furuhashi, M. Nishino and K. Takagi – made several successful shipments of the two parasites. The *Physcus* was determined as *P. fulvus* Compere and Annecke, a species that had been recorded from arrowhead scale in China in 1948. The *Aphytis* was sent to the authors for identification, and we found it to be new to science. We subsequently described it as *A. yanonensis* DeBach and Rosen. Both species were mass-reared and released, and have become firmly established in most citrus areas of Japan. They coexist and complement each other's work, *Aphytis* producing 11 generations per year and *Physcus* 5. By 1981, parasitism in the original release plots in Shizuoka Prefecture had reached 70–80 percent, and the population density of the arrowhead scale was reduced to 10 percent of its former level. By 1986, effective biological control was achieved throughout Japan wherever the parasites were not decimated by broad-spectrum pesticides. This has also made biological control of the citrus red mite feasible in Japan. (See DeBach & Rosen, 1982; Furuhashi & Nishino, 1983; Tanaka, 1982, 1989.)

Cassava mealybug in Africa, 1981

This outstanding project, of unprecedented geographical scope, illustrates the truly international nature of biological control. Cassava (*Manihot esculenta*) was introduced into Africa from South America in the sixteenth century. It is the staple food crop of more than 200 million people living in the African 'cassava belt', an area more than twice that of the United States extending from 15°N to 20°S. An unknown mealybug was discovered infesting cassava in Congo and Zaire in 1973, and in 1977 it was described as a new species, *Phenacoccus manihoti* Matile-Ferrero. Spreading up to 300 km per year, it soon became the major pest of cassava in Africa, stunting the growth points of plants, defoliating entire plants and causing severe tuber yield losses. Some crops were totally destroyed, and the poor quality of cuttings from infested plants prevented the replanting of cassava in certain areas. By the end of 1986 the cassava mealybug had invaded 25 countries and covered about 70 percent of the cassava belt.

Biological control efforts started in 1977, when the CAB International

Institute of Biological Control (CIBC) began searching for natural enemies in the Caribbean, Venezuela, the Guyanas and northeastern Brazil. Then, in 1980, the Africa-wide Biological Control Project (ABCP) was organized, with European, American, Canadian and United Nations support, by the International Institute of Tropical Agriculture (IITA) at Ibadan, Nigeria, and a search was made in the southern United States, Mexico, Central America, northern Colombia and Venezuela. Unfortunately, these early efforts were hampered by inadequate systematics. A mealybug infesting cassava in northern South America and causing similar symptoms was at first thought to be *P. manihoti*, but its parasites failed to reproduce on the cassava mealybug in the Congo. In 1981, when it was realized that this mealybug differed in some morphological characters as well as in its mode of reproduction (it is biparental, whereas *P. manihoti* is uniparental), it was described as another new species, *P. herreni* Cox and Williams. When this confusion was cleared, the true cassava mealybug was found in Paraguay in 1981 by the Centro Internacional de Agricultura Tropical (CIAT). Only then were several parasites and predators collected by CIBC and sent to IITA through the CIBC Quarantine in London and the Nigerian Plant Quarantine. Further explorations were made by IITA in 1983–6 in Paraguay, Bolivia and Brazil, and the cassava mealybug was discovered in several localities. Populations there were so low, that laboratory-infested plants had to be exposed in the field in order to attract natural enemies. Several parasites and predators were thus obtained and shipped to Africa.

One of the South American parasites, the encyrtid *Epidinocarsis lopezi* (De Santis), proved to be an immediate success. It was first released in Nigeria in November 1981, and less than a year and a half later was found to be established up to 170 km from the original release sites. Exhibiting an amazing rate of dispersal, by the end of 1984 it was found in 70 percent of all cassava fields on more than 200 000 km² in southwestern Nigeria. It has since been sent to many other countries, where it has been extensively released from the ground and from the air. By the end of 1986 it had been established in 16 African countries, over a total area of more than 750 000 km². About 1 percent of that area (750 000 ha) is planted to cassava, under diverse ecological conditions ranging from rain forest to savannah to East African highlands.

Currently, as the cassava mealybug continues to spread in Africa, so does the parasite, effecting complete biological control wherever it becomes established. In southwestern Nigeria, in spite of hyperparasitism and competition with other natural enemies, cassava mealybug populations collapsed following the introduction of *E. lopezi*, and they remained low for the next five years. The proportion of cassava plants showing damage symptoms declined from 87.6 in 1983 to 23.4 in 1984, and the average mealybug population was only 11 per plant tip, as compared to hundreds in other areas. The cassava mealybug is no longer the serious pest it used to be in southwestern Nigeria, and similar reductions in its populations have been reported from many other African countries. African farmers are again growing cassava where mealybug damage had previously devastated the crop. Already, this project has become one of the most spectacular successes in the history of biological control. (See Neuenschwander & Hammond, 1988; Neuenschwander & Herren, 1988.)

Lesser successes and their value

Most, but not all, of the 21 cases just discussed have been considered complete successes. Obviously, there are all degrees of success possible in applied biological control importation projects, but for convenience they can be divided into complete, substantial and partial degrees of success. These categories may be defined as follows: (1) *Complete* biological control refers to the successful reduction of a major, widespread pest to a non-pest level and its permanent maintenance below the economic threshold, so that insecticidal treatment becomes rarely, if ever, necessary. (2) *Substantial* biological control includes cases where economic savings are somewhat less pronounced, because of the pest or crop being less important or the crop area being less extensive, or the control being such that occasional insecticidal treatment is indicated. The latter should mean a reduction in chemical treatment of 75 percent or more. Also, control can be complete over most of the pest area but only substantial or partial over some smaller portion of it. (3) *Partial* biological control includes cases where the pest is definitely reduced but not consistently so, or is not maintained regularly below the economic injury level. Thus, regular chemical treatment may remain necessary but may be reduced by approximately 50 percent; intervals between treatment may be lengthened and outbreaks occur less commonly. Partial control can easily mean a saving of 50 percent in reduced treatment costs and a concurrent reduction in environmental pollution. When evaluating published records of successes for purposes of tabulation, we also include under this category cases where complete or substantial biological control is obtained in only a minor portion of the infested area, or where we are dealing with minor pests or incipiently established pests, as well as insufficiently documented cases which infer a greater degree of success. Partial successes tend to be overlooked or discounted by the average entomologist because they are not as obvious as complete or substantial successes, but they do represent important pest reductions and often may be the key to being able to reduce insecticidal treatment enough to make integrated control programs feasible. Even though partial successes tend to be overlooked or discounted, nonetheless they often represent a considerable saving as measured by reduction in damage, lessened need for treatment or reduced hazard of spread.

It might be thought that partial successes constitute the majority of results obtained in biological control. This is far from the case. Over 80 percent of the successes in control of insect pest species have been either complete or substantial, whereas only one-sixth or so have been partial. This is discussed in more detail in the second section following.

Summary of worldwide results against weeds

Since 1865, numerous species of weeds have been studied as possible objects of biological control importation projects. Research has progressed to the stage of natural enemy importation and colonization against 89 of these weeds, and 214 exotic natural enemies have been introduced into 53 countries for their control up to 1985. In addition, 38 native natural enemies have been used against 44 weed species. Since in several cases both exotic and native enemies were used against the same weed, in all 125 weed species, representing 69 plant genera in 37 families, have served as targets for biological control. This reflects a rapid expansion of this field during the last 30 years or so, particularly in North America. Biological control of weeds is currently practiced all over the world, but primarily in the United States (including Hawaii), Canada, Australia and other British Commonwealth countries.

Most of the weeds considered for biological control until about 1960 were alien, perennial species infesting relatively undisturbed terrestrial habitats, predominantly grazing lands. However, more recently, an increasing number of annual and biennial, aquatic and semiaquatic, native as well as alien weeds have been objects of biological control research.

The successes achieved in biological weed control have utilized insects as natural enemies for the most part, but there have been also mites, nematodes and fungi. The plant-feeding insects involved have been a diverse lot, representing 8 orders and 42 families, with about 75 percent of the species belonging to the Coleoptera and Lepidoptera. The biology, form and habitats of the natural enemies involved in successful results are widely diverse. Between them they variously attack all of the major vegetative or reproductive parts of these plants, considered as a whole. Up to now, most successes have been registered against perennial alien rangeland weeds. Perhaps this reflects a concentration of effort on this type of weed until recent years.

Judging by the published results available, a rather high degree of success has been achieved relative to projects attempted. As mentioned, attempts to utilize weed-feeding natural enemies have been made against 125 weed species. There have been 267 projects worldwide, and 48 percent of them have achieved a measurable degree of success. Most of them – 228 projects – involved the importation of exotic organisms, and 103 of these have been rated as successful, resulting in effective biological control of 49 species of weeds. Of 701 separate natural enemy importations on record, 398 – more than half – have resulted in establishment, and the results of many recent cases are still pending. Sixty-four percent of the natural enemies introduced in these projects have so far become established, and 26 percent were rated as effective. Thus, 40 percent of those that became established (55 of 136) have been effective in biological control. When successful projects are analyzed in terms of degree of success, some two-thirds of them fall in the complete or substantial categories. Also, of 39 projects involving native organisms, 25 have been successful, and 33 weed species were controlled. Altogether, nearly two-thirds of the target weed species (80 of 125) have been brought under biological control in at least one project. This must be considered an impressive achievement, warranting increased emphasis on this ecological approach.

In addition to all these, nine species of exotic fish have been utilized in 38 countries against a variety of aquatic weeds and algae.

(Data from Julien, 1987. For further discussion and descriptions of numerous projects, see also Andres *et al.*, 1976; Andres & Goeden, 1971; Goeden, 1978; Julian, Kerr & Chan, 1984; Rosenthal, Maddox & Brunetti, 1985.)

Summary of worldwide results against insect pests

The number and extent of successes obtained from biological control importation projects against insect pests have been tabulated and analyzed on a world basis by several authors (see DeBach, 1972; Laing & Hamai, 1976; Luck, 1981). The following account is based on these sources and on BIOCAT, a computerized database compiled by the CAB International Institute of Biological Control from published and unpublished information (David J. Greathead, personal communication, 1988), with some additions from our own records.

By 1988, the importation method alone had resulted in the permanent control (either complete, substantial or partial) of at least 164 species of pest insects in eight major orders of insects. Of these, 75 pests were completely controlled, 74 species were substantially controlled, and 15 fell in the partial control category. Additionally, various of these pest species were subsequently controlled in other countries following the first success, so that altogether there have been at least 384 successful projects worldwide up to 1988. Numerous additional projects are currently going on, but for the most part the results from these are still not sufficiently complete for inclusion. This is by far the greatest number of successes achieved by any one non-chemical method, with the possible exception of cultural control, which to our knowledge has not been tabulated. The use of resistant plant varieties or irradiation-sterilized insects each apply thus far only to control of a few pest species. All other non-chemical methods combined would add only a few more.

Our tabulation shows that natural enemies had been imported against some 416 species of insect pests, and some measurable degree of success obtained with the 164 pest species mentioned earlier. This represents some degree of success with nearly two-fifths (39.4 percent) of the species involved, and includes completely successful control of 75 species. Thus, these formerly serious and major pests no longer require chemical treatment in the country where the natural enemy was introduced. A total of 149 species are included in the combined categories of complete and substantial control. The latter category means that only very occasional treatment is needed, or perhaps that complete control was not achieved over the entire range of the pest. Only 15 (9 percent) of the pest species controlled fell in the category of partial success. The authenticity of these successes (aside from the cases where formal experimental proof was presented) is vouched for by the fact that nearly one-third of the pest species controlled (50) were again successfully controlled in other countries following the first success.

The preceding data show that the chances of obtaining some significant extent of success against any given insect pest species from importation of natural enemies are about 4 in 10. In view of such truly outstanding results, it seems a remarkable indictment against our profession that only about 416 pest species have been subjected to the natural enemy importation method, whereas there are approximately 10 000 species of insect pests recorded worldwide. In other words, over 95 percent of the world's insect pests have had no biological control importation project directed against them.

Our analysis reveals that the chances of obtaining a degree of control (i.e. either complete or substantial) that would nearly or completely eliminate the need for further insecticidal treatment included more than one-third of all target pest species, or 35.6 percent. This is a very impressive success ratio, and the economic savings have been correspondingly great. The dollars returned in savings resulting from successful biological control for each dollar invested have been carefully estimated as being about 30 to 1. In California alone, major biological control projects carried out during the present century have resulted in total savings of well over $500 million until 1985. This is a conservative estimate, based on calculation of annual savings due to reduced crop losses, elimination of chemical control treatments, etc. (see DeBach, 1964a; Huffaker, Simmonds & Laing, 1976) and adjusted for inflation. It does not even include the cottony-cushion scale project, which has literally saved the entire citrus industry from destruction.

These results belie the opinions sometimes expressed in the literature by the casual writer, or by entomologists unacquainted with the field, to the effect that biological control is the best method of all when it works but it is seldom successful. Such statements obviously do not stand up against the information just presented, and it appears that they are based on a misconception of what constitutes success in biological control. They mistakenly interpret lack of success as being the proportion of imported natural enemy species that fail to become established. Records at hand show that out of a total of 4226 natural enemy importations worldwide, 1251 have become established, 2038 failed, and the fate of 932 is still unknown. Thus, 29.6 percent of all attempts, or 38 percent of those with known outcome, have resulted in establishment – a rather impressive record in itself. However, this proportion cannot be equated with the ultimate success of a project, because only one or two natural enemies out of a large number imported need to be successful. In fact, as we have seen in Chapter 3, the most effective enemies tend to displace or exclude the less effective ones, so a high establishment ratio is not to be expected. In the long run, it is not the proportion established, or the proportion established to successes achieved, but rather the cost-benefit ratio that is really important, and we have already shown this to be a very good 30 to 1. With insecticides it is estimated to be only about 3 to 1 (see Pimentel *et al.*, 1981).

The research and development cost of one new chemical pesticide in the market place is now at least $20 million. The research and development cost to obtain one new natural enemy will be only a small fraction of this, ranging from less than $100 to several hundred or even occasionally tens of thousands of dollars, but we believe the average would not be more than a few thousand dollars per entomophagous species.

The analysis thus far has been concerned only with the proportion of target pest *species* against which successful results have been obtained. However, successful results have been repeated with quite a few of these species in other countries. The cottony-cushion scale has been controlled in over 50 countries following the first success in California, and the woolly apple aphid in 25. These represent the extra, easy dividends in biological control. The original project may cost from a few hundred to many thousands of dollars, but the repeat or transfer projects often cost only the amount of an airmail shipment and some correspondence.

The total of all successful *projects*, from original importations as well as from subsequent transfers of natural enemies to other countries, is over 384. This includes 220 more successes than if the original projects alone were considered, i.e. the number of successes was more than doubled. Over 75 percent of these additional successful projects fall in the substantial or complete control category. The value of these repeats should not be minimized. They are just as important to the country in which they occur as was the first success in the original country.

If all attempts to import natural enemies against a given pest in a given country are counted as one project, there have been 1279 importation projects in at least 153 countries to date, and 384 of them have resulted in a measurable degree of success: 156 projects have been rated as complete successes, 164 as substantial, and 64 as partial successes. Thus, 30 percent of all projects have been successful; however, 83.3 percent of all *successful* projects have been complete or substantial, whereas only 16.7 percent have been partial.

There are 12 leading countries or states in the world that account for nearly half of the successful importation projects thus far obtained: Hawaii with a total of 31 recorded successes, California with 24, the rest of the continental USA with 20, Chile with 16, Australia and New Zealand with 13 each, Canada, Israel and France with 11 each, Mauritius with 10, Peru with 9 and South Africa with 8. It has been shown that it is not latitude, longitude or climate, the island or continental status, or the type of crop or pest, but rather simply the amount of support and research effort that has been directed toward the importation and colonization of new natural enemies, that has accounted for the pre-eminent status of these countries in the biological control field. In some of these leading countries, one or a few devoted individual scientists have been largely responsible for the progress made.

It has often been assumed that biological control only works well in tropical or subtropical areas; however, worldwide results show that the majority of successes have occurred in temperate areas. This is again interpreted as being related to the greater amount of research effort being expended in temperate areas, rather than to a direct correlation with climate. Certainly, tropical areas are quite favorable to the activity of natural enemies, and there is some indication that the proportion of complete successes has been greater in such areas. The important thing to realize is that there doesn't appear to be any climatic zone where significant biological control of agricultural pests may not be achieved.

There appears to be little doubt that importation of new natural enemies from abroad should generally receive first priority in all biological control efforts (including conservation and augmentation), as well as in all other non-chemical approaches to integrated pest management. It is obvious that much greater research emphasis and financial support is needed. The probability of success of the method has been shown to be very good; it has general application; results are permanent; the cost-benefit ratio is highly favorable, with initial costs being relatively low and recurrent costs essentially nil; problems relative to the adverse effects of chemical insecticides are reduced or eliminated by its successful use, and no adverse effects on the ecosystem occur from biological control. No other method approaches this totality of favorable aspects.

As long as we ignore, anywhere in the world, an effective natural enemy capable of controlling one of our major pests, we are postponing inexpensive, reliable and permanent control until we import such enemies; yet we continue to do just this with 95 percent of the world's insect pests. The cost of other treatments and crop loss in the interim may justly be charged against the entomological profession. This was strikingly illustrated by several of the successes discussed earlier in this chapter and in Chapter 5, where it was shown that, but for lack of effort and adequate research, various serious major pests could have been subjugated years before they finally were. (See DeBach, 1972.)

7

Maximizing biological control through research

Applied biological control can be viewed as a discipline or profession like the field of medicine. It concerns problem solving which, nowadays especially, is achieved through research. Such research is immensely diversified in medicine, and likewise in biological control. Both include an imposing gamut of more discrete and specialized biological disciplines, and the research involved must cover a broad spectrum of biological principles, practices and procedures. In the case of biological control we are concerned with all of the varied disciplines involved in ecology, and this does not leave much science out. This chapter emphasizes research on biological control of insect pests, but similar sorts of research are just as important with biological control of weeds or other organisms. Principles of biological control of weeds have been discussed by Huffaker (1964) and by Schroeder & Goeden (1986).

As a broad phase of ecology, and depending on the project, biological control is most likely to encompass research on: (1) biology, including behavior, developmental cycles, physiology, genetics, reproduction, nutrition and culture; (2) systematics, phylogeny and biogeography; and (3) population biology and ecology.

Despite the substantial progress that has been made in the field of biological control, there is no doubt that its application could be greatly increased. Applied results largely depend on basic taxonomic, biological and ecological research, and the amount of such research remains much too small considering the enormity of today's problems. The paucity of support is evident from the fact that the majority of state universities or their experiment stations in the United States have departments of entomology, but only a handful have even one person specializing in biological control research. Aside from their internal budgets, all receive unsolicited donations for research, principally from chemical companies. The following figures reported for one of the largest entomological departments (including a large biological control research group) in the United States are quite revealing in this connection. In the three-year period, 1984–7, US $17 833 were donated for biological control research, whereas $1 008 825 were donated for other entomological research, predominantly on chemical pesticides. Similarly, in West Germany DM 200 million are spent annually on research for new pesticides, whereas the funding for development of biological control methods in that country amounts to DM 2 million per year (van Lenteren, 1987). Of the problems that biological control could help materially in solving, we refer principally to: (1) protection of the food supply from pests, (2) protection of humans, wildlife and general environmental quality from toxic chemicals used in pest control, (3) protection of the public health and well-being by biological control of vectors of disease and of nuisance pests, and (4) economic savings to growers and consumers.

From the analysis of worldwide successes in classical biological control of insect pests discussed in Chapter 6, it is abundantly clear that the single most important characteristic of the 12 leading countries or states achieving the most successes have been their amount of research support for dedicated, well-trained scientists assigned to biological control importation projects. In this respect Hawaii, a tiny land area, leads the world and California follows closely. Both have vigorously supported biological control from approximately 1890 on. They, plus the other leading countries or states, are so widely distributed geographically, and the projects so varied, that it is obvious that neither latitude, climate, island or continental status, nor type of pest or crop, can be assigned a major influence on successful achievement of biological control. Here we have been referring to an analysis of importation projects, but it should go without saying that the two other main phases of applied biological control, conservation and augmentation, require as much and probably more research effort. The complexity of adequate field studies necessary for a really scientific understanding of how to manipulate established natural enemies to increase their effectiveness offers a formidable research challenge. Currently, the People's Republic of China appears to be one of the world leaders in this respect.

Systematics and biological control

The broad dependence of classical biological control importation projects on systematics – more accurately nowadays on biosystematics – furnishes an excellent example of the absolute necessity for basic research. True, some projects would be successful on an empirical basis, but the failures would be, and have been, greater because of the lack of adequate knowledge of systematics. Perhaps only the need for detailed studies in population ecology, in order to solve problems in the conservation and augmentation phases of biological control, is quite so critical.

In biosystematics, the taxonomist goes much further than is the case with so-called alpha-taxonomy, which relies principally on morphological characteristics to distinguish between species. If two specimens look the same to a given taxonomist, he will call them the same species. However, the morphological and coloration criteria used as distinguishing characteristics differ greatly between taxonomists, so that one person might include several valid species under one species name, whereas another might name as different species several races or variants of what was actually one species. The biosystematician uses other aids to help in determining what actually constitutes a valid species. This may include data on biology, physiology, habits, sound-production, type of pheromone produced, etc., but the ultimate test (with bisexual species at least) is whether or not one population interbreeds with another population and produces the normal number of viable and fertile offspring. Failure to do so is, with very few exceptions, proof that distinct species are involved. With parasitic and predatory groups, data on host or prey relationships, host plant, and microhabitat preferences are often useful taxonomic tools, and obviously of utmost importance in biological control. The adequate taxonomist also acquires data on strains or races, geographical distribution and phylogenetic relationships with other species or groups.

What we have been leading up to is that, more frequently than not, a natural enemy cannot be used in biological control importation projects until it is discovered and recognized as a distinct species. This usually means receiving a formal scientific description, but even this is not always enough because as Compere (1969) has pointed out, 'of the named species, few have been adequately described; and of these, fewer still classified with regard to natural relationships. The great majority of described chalcidoids cannot be identified with any degree of certainty on the basis of existing descriptions.' Also, if a species is misidentified, the error may be made that it is already established, or else that it is a parasite of another host species, with the consequence that it would be ignored. Of course, accurate identification of the host or prey (pest) insect also is critical, especially where highly specific natural enemies are concerned; but generally speaking, pest insect taxonomy receives more adequate research support. However, if the host is misidentified, a parasite may be imported that comes from another host species and will not attack the one we are striving to control.

Just how serious is the lack of biosystematic knowledge affecting applied biological control? It is far worse than anyone, even most experts in the field, realize. Comprehensive reviews and discussions of this question will be found in Delucchi, Rosen & Schlinger (1976) and Rosen (1986).

The problem can be analyzed from the standpoint of: (1) undiscovered or unnamed species, (2) species misidentified because of poor taxonomy, (3) unrecognized species, because they are so similar morphologically to already known species that only careful biosystematic studies will reveal them, and (4) the occurrence, detection and value of races or strains of species. Before going on, however, there is a corollary problem nearly as critical. The applied biological control specialist must be enough of a taxonomist to easily keep up with the literature, so as to keep abreast of all new taxonomic advances of importance to him. A perfect parasite for his purpose may be described from Afghanistan, but if he does not learn of this, he is postponing its usefulness for an indefinite period.

The undiscovered (i.e. uncollected) and unnamed (i.e. collected but not recognized as new) species of parasitic Hymenoptera (the order which contains the most effective and most numerous species of natural enemies) constitute by far the vast majority of the species that actually exist. The English specialist on parasitic Hymenoptera, G. J. Kerrich, estimated (1960) that if the parasitic Hymenoptera had been studied to the same extent as have the Coleoptera, there would be about 500 000 described species, whereas there are less than 50 000. In other words, and because not all Coleoptera are yet known, possibly considerably less than 10 percent of the parasitic Hymenoptera are known to science and 'available' for use in biological control. Another estimate he used supports this overall figure; that is, of the Ichneumonidae of tropical America, tropical Asia, Africa, and Australia, only about 10 percent have been described. In this regard we should note that this group includes many of the larger, more readily seen and collected parasites. Finally he makes the amazing, but not unjustifiable, 'guesstimate' that as many as one million species of the superfamily Ichneumonoidea may exist in the world. This figure is a total commonly used for all insects described so far.

Dr Henry Townes (1971), a world authority on the Ichneumonidae, made the somewhat more conservative estimate that about 70 percent of the parasitic Hymenoptera are still undescribed species, and that there is not even a scrap of biological information concerning 97 percent of the species. In other words, not even the hosts are known for 97 percent, so these are of greatly reduced usefulness for employment in biological control projects. In any event, if the unknown species of parasitic Hymenoptera constitute somewhere between 70 and 90 percent of the species actually in existence, it is obvious that we have just begun to scratch the surface in the search for insect natural enemies, and are even much further behind in the absolutely essential knowledge of their host or microhabitat preferences. Townes points out that the Ichneumonidae comprise 20 percent of all parasitic insects and that, from the biological control viewpoint, the most important thing to know is their microhabitat preferences. They tend not to be host-specific but specific to certain situations – for example, to borers of various species in certain twigs, or to cocoons of several species under the bark of a particular tree. Thus, with many ichneumonids we should look for species that occur in a specific microhabitat, not necessarily for a species that attacks the host we wish to control. This knowledge is the product of good biosystematic research.

The discovery, naming and adequate description of hitherto unknown natural enemies may be regarded as placing new weapons in the arsenal of biological control. Thus, for instance, a systematic world revision of the genus *Aphytis* (Rosen & DeBach, 1979) has made some 35 new species available for biological control projects against armored scale insect pests.

The problem of misidentified species because of poor taxonomy has several causes, but the results are self-evident. A natural enemy species, perhaps undescribed and new to science, but wrongly given the name of a described species, may be as effectively hidden as if it were never collected (DeBach, 1960). Expert systematists know that this occurs commonly. It stems directly or indirectly from the great shortage of professional systematists. The scientists in our national and other museums are spread thinly, and are often called upon to identify specimens in families or genera with which they have only superficial acquaintance. The problem may be compounded when the field worker, unable to obtain identification of his submitted specimens rapidly, undertakes his own identification with an insufficient background. Publication of mistakenly identified species, or worse, of inadequately or erroneously described new species, may result. This is not to discourage the field man from such activity, because his contribution is often necessary to help close the gap. It is urged, however, that he develop considerable expertise, preferably in groups of restricted size, before publishing descriptions of new species.

Misidentification of an exotic natural enemy as one already present in the target country may delay its utilization in biological control programs. Even worse, perhaps, is the possibility of misidentifying an undesirable organism as a useful natural enemy. Certain hyperparasites, for instance, are closely related to primary parasites – they belong to the same family, or even to the same genus – and their misidentification may lead to their inadvertent importation. For this reason, all imported parasites are reared in quarantine on a primary host for at least one generation.

A considerable, but unknown proportion of the undescribed natural enemies may be so because they are so-called sibling or cryptic species. Such species are valid ones, but are so similar morphologically to one or more others as to be indistinguishable from them. They can only be detected by the biosystematic type of investigation mentioned earlier. Few systematists have the time, facilities or financial support for such work, inasmuch as it entails culturing the host and parasite, or predator and prey, and studying living specimens. Probably most of such studies have been done in connection with biological control projects. A glimpse of the potential from such research and of what lies hidden may be obtained from studies of the genus *Aphytis*. The species of this genus include the most important parasites of armored scale insects throughout the world, hence have received more than usual emphasis.

We have found that what only a few years ago was considered to be one species, *A. chrysomphali*, was in fact a complex of sibling species. Actually, each of the originally described, more common species in the genus is now known to include a group of sibling species. All told, at least 20 percent of the 100 known species fall into this sibling or near-sibling category (DeBach, 1969; Rosen & DeBach, 1979). The importance of recognizing such cryptic species is that each may attack different hosts, which is essential knowledge in biological control, and even if they attack the same host species, they will have different inherent capabilities for biological control. Hafez & Doutt (1954) found that the *Aphytis maculicornis* imported for control of the olive scale included four so-called biological strains or sibling species, all indistinguishable from one another but possessing distinct biological attributes. The 'Persian strain', now known as *A. paramaculicornis*, was the only one that proved to be effective in biological control.

Subspecific categories (so-called races, strains, ecotypes, biotypes, subspecies and semispecies) are of real importance in biological control. Again, biosystematic studies are necessary for their detection. Races with distinct host preferences do exist and, as such, are the practical equivalent of distinct species. Two races of the parasite *Comperiella bifasciata*, imported into California, represent an excellent example. They are morphologically indistinguishable and cross readily in the laboratory, with resulting normal progeny production and sex ratios. The Chinese race may prefer the California red scale as a host, but it develops well in the yellow scale; the Japanese race strongly prefers the yellow scale and virtually never completes development in the California red scale. They occur together in the field in the San Joaquin Valley of California where, somewhat surprisingly because they hybridize readily in the laboratory, they maintain their distinctness. The Japanese race is always the one reared from yellow scale there, and the Chinese race from the California red scale. Apparently the hybrids are not as well adapted as either parent is to their particular host – hence, natural selection continually operates to eliminate the hybrids and to maintain the integrity of the two races.

Semispecies exhibit considerable reproductive isolation, whereas races by definition do not. In biosystematic studies of imported and cultured *Aphytis*, the species *lingnanensis* was found to contain a complex of semispecies as is shown by Fig. 7.1. Although none of these can be distinguished morphologically, differences in host preferences and other biological characteristics make certain of them the practical equivalent of different species. Take, for instance, *A. lingnanensis*, a parasite of the California red scale (*Aonidiella aurantii*) from China; '2002', a parasite of the coconut scale (*Aspidiotus destructor*) from Puerto Rico; and 'R-65-23', a parasite of the citrus snow scale (*Unaspis citri*) from Florida. Careful crossing tests have shown that, whereas *lingnanensis* and 'R-65-23' are only partially reproductively isolated from '2002', they are completely isolated from one another. In other words, *lingnanensis* and 'R-65-23' are good biological species in relation to one another, but both are semispecies in relation to '2002'. Legner (1983) did similar work with house fly parasites of the genera *Muscidifurax* and *Spalangia*.

Many examples could be cited, wherein lack of adequate systematics has precluded successful biological control for a number of years, as well as a few where good systematics has helped, but space limitation necessitates restricting these to several only. Reflection will tell us that the cases we know of represent only the few that have finally been solved. The ones chosen here are associated with the successful projects treated in Chapters 5 and 6.

The final outstanding biological control of the sugar-cane leafhopper in Hawaii was delayed about 15 years because its predator, *Tytthus mundulus*, was so poorly known as to be thought to be a plant-feeding insect. Success with the coffee mealybug in Kenya was delayed a dozen

Fig.7.1. The crossing relationships of some geographical forms of the *Aphytis lingnanensis* complex. These forms are indistinguishable on a morphological basis one from the other, yet all degrees of reproductive isolation are evident. (From Rao & DeBach, 1969.)

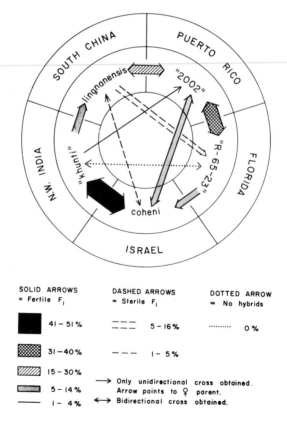

years or so because of misidentification of the mealybug as another species. As a result, an unnecessary wild-goose chase was conducted in the Orient to obtain parasites from the wrong mealybug. It was not until the mealybug was correctly identified that its native home was found and effective parasites imported. On the positive side of the ledger, the successful biological control of the citrus blackfly in Cuba and other Carribean countries, and eventually in Mexico, stemmed directly from an earlier systematic study by Filippo Silvestri of certain parasitic Hymenoptera in Tropical Asia, in which he described as new species the parasites that later accomplished the job.

The successful establishment of the effective *Aphytis paramaculicornis* on the olive scale in California doubtless was delayed a number of years, because of the failure in those days to duly appreciate the existence of sibling species. Failure to determine that the parasite of Florida red scale in Hong Kong was not *Aphytis lingnanensis* but actually a new species, *A. holoxanthus*, precluded the importation of that parasite into Israel for an indeterminate number of years. Following its importation in 1956, it gave complete control of Israel's then worst citrus pest. Adequate biosystematics could have made it available many years earlier. Although it has since been sent to several other countries, each with similar excellent results, certain ones – apparently because of lack of knowledge of the literature – have not yet taken advantage of what would be a 'free ride' today. Florida did so in 1959 with spectacular results, but it is most apropos of our discussion to note that 80 years previous to this, in 1879, William H. Ashmead, who was an excellent taxonomist, was attempting to learn of the native home of the Florida red scale in order to obtain its natural enemies for Florida. Had systematics been far enough advanced then, Florida would have had biological control 80 years earlier, with a consequent savings of the many millions of dollars that were spent over the years for insecticidal control. Another chance was missed about the end of the last century, because Albert Koebele reared an *Aphytis* from the Florida red scale in Hong Kong which was sent to the National Museum in Washington, but mistakenly identified as *A. mytilaspidis* and subsequently ignored. One of the authors (DR) examined these specimens in 1970 and determined them to be *A. holoxanthus*, the principal parasite of the Florida red scale. Still another opportunity was missed in the 1920s during T. H. C. Taylor's work on the coconut scale in Fiji. An *Aphytis* reared from the Florida red scale in Java was misidentified as a form of *A. chrysomphali* – a cosmopolitan species. From Taylor's figures of the pupa it is definitely not *A. chrysomphali* but very likely *A. holoxanthus*.

The California red scale story in California includes a long history of confusion of the scale with related species, as well as misidentification of parasites. Inadequate systematics was responsible for the failure for many years to successfully import and establish *Comperiella bifasciata* parasitic on the California red scale, as well as *Aphytis lingnanensis* and *A. melinus*. The latter two are the principal red scale parasites in California, but were not imported earlier in the mistaken belief that they were *A. chrysomphali*, a species already present in California and thought to be of little consequence. We now know from biosystematic studies that what was once thought to be the single species, *Aphytis chrysomphali*, parasitic on the California red scale in the Orient and elsewhere and accidentally established in California, includes at least seven species having different adaptations.

The final example, previously discussed in Chapter 6, also comes from California but illustrates an improving trend in appreciation of the use of biosystematics in biological control research. A monophagous parasite, *Trioxys pallidus*, of the walnut aphid, a major pest of walnuts in California, was discovered and introduced from France about 1959. It was colonized in all principal walnut-growing areas, quickly becoming established and effective in control in coastal San Diego County, but it never became established in the commercially important San Joaquin Valley area, despite massive colonizations. Studies indicated that the parasite was not adapted to the hot, arid climate of the San Joaquin Valley. This led to the likely presumption that other ecotypes, sibling or closely related species might occur in climatic zones similar to the San Joaquin Valley. Finally, an Iranian race or ecotype – possibly a sibling species – indistinguishable from the French form, was discovered and imported in 1968. It readily became estabished and rapidly spread, with the result that the walnut aphid is now under complete biological control in the San Joaquin Valley.

Biology and biological control

Basic research on the biology of a natural enemy, or of the host or prey insect, can be the key to success in any given project. Several examples will suffice to illustrate this.

For years, researchers had been importing parasitic species of Aphelinidae in the genera *Pteroptrix* (= *Casca*), *Physcus*, *Coccophagus* and others, for use against scale insects and mealybugs. Some of these parasites appeared very promising in the native home, yet time after time culture attempts following importation were unsuccessful. Finally, it was discovered that their very unusual and complicated developmental biology was the reason for previous failures. The usual type of reproduction and development in parasitic Hymenoptera, it will be recalled, is for females to mate soon after emergence, store the sperm from the male in the spermatheca and then, depending on external stimuli, either release sperm as eggs are being laid to fertilize the eggs and produce female offspring, or retain the sperm in the spermatheca so that the eggs remain unfertilized and produce male progeny. Both males and females develop as primary parasites.

The unusual type, mentioned briefly in Chapter 2, has been known as hyperparasitic male development. There are variations, but generally in such cases the unfertilized, and therefore male-producing, eggs develop only on the immature stages – a larva or a pupa – of another parasite, either of another species or of their own. In extreme cases, known as adelphoparasitic, males develop only on a female larva or pupa of their own species. When such a species is imported and culture attempted, the usual procedure would result in starting with all fertilized females (because in shipments both sexes are present, and the males are quite efficient) which would lay only fertilized female-producing eggs, because the host culture would not contain parasite larvae or pupae to trigger the deposition of male-producing eggs. Hence, no males would be available in the next generation, and the culture would be lost. Often the same trials were repeated over and over. The key to the discovery of this phenomenon lay in biological studies done in Fiji with coconut scale parasites. Final clarification was accomplished by Dr S. E. Flanders at the University of California, Riverside, in the 1930s. As a result of these meticulous studies, it has since been possible to successfully culture and establish several species which previously would have been, or actually were, lost. One striking example involves *Coccophagoides utilis*, which has been the final 'straw' to effect successful biological control of the olive scale in California. This species has hyperparasitic males, which develop on female pupae or prepupae of their own species. Even so, it was successfully cultured from a single shipment, yielding only a few specimens from an isolated area in the Tribal Territory of Pakistan (then West Pakistan). Had it not been known that this species was in a group likely to have hyperparasitic male development, the chances are great that culture would not have been achieved and that the olive scale would remain a much more serious problem today.

Means of breaking obligatory diapause of natural enemies are critical to certain projects, when the species are imported from the Southern to the Northern Hemisphere or vice versa. If one imports them during the best season for colonization in the Northern Hemisphere, spring or early summer, and they are a species in winter diapause from Australia, they will not complete development under usual insectary conditions and the culture will be lost. Studies on the effect of photoperiod on natural enemies have enabled the successful breaking of diapause in certain cases; in others it was more complicated, involving an indirect effect operating through the host plant or host insect.

Knowledge of the biological or nutritional requirements of natural enemies may furnish the key to their successful establishment or effective operation. In connection with ichneumonids, the most numerous parasitic family, Dr Henry Townes has stressed that host species that do not occur in the microhabitat of the exploring female parasite are never attacked, even though in the laboratory they may be suitable hosts. Thus, much effort could be wasted on futile mass rearing and field colonization of an imported species that looked promising in the laboratory on the host desired to be controlled. On the other hand, within its own microhabitat, consisting, say, of pupae occurring in leaf rolls, it may attack a large variety of host species. Thus, as already briefly mentioned, rather than trying only to obtain new ichneumonids that attack the particular host we wish to control, success may likely be achieved by importing ones attacking other hosts occurring in similar or identical microhabitats.

Another biological characteristic of adult ichneumonids that is essential to establishment of new species and to effectiveness of native ones is that, like other insects, they have a high ratio of evaporative surface to body water, hence lose water easily. Since they do not feed on plant tissue, they must in general drink free water daily. Thus, the availability of moisture in the form of dew or rain is the dominant feature in the bio-ecology of adult ichneumonids. Such knowledge may better enable us to manipulate them in certain cases, and in others not to waste time trying to establish them in areas lacking in frequent dews or rain.

Some cryptic relationships between parasites or predators and host plants are not only biologically interesting, but affect the success of certain natural enemies. Several attempts to import the California red scale parasite *Comperiella bisfasciata* in the early days failed because, before air transport, infested plants had to be taken to the Orient to have the scales parasitized there and then returned to California. Citrus could not be used because of quarantine restrictions. Therefore, the sago palm, *Cycas revoluta*, an excellent and very hardy host of the scale, was used. Although the parasites oviposited freely in the scale on *Cycas*, none completed development and for some time it was thought that the wrong 'strain' of the parasite had been obtained. Actually, it was finally found that the *Cycas* plant provides the California red scale, when developing on it, with immunity to *Comperiella*. A similar effect has been noted with the famous vedalia beetle. It will not develop on cottony-cushion scale growing on the ornamental *Cocculus*, and does poorly on others such as maple and scotch-broom. Host-plant-induced immunity of a host to a parasite or predator could lead to false conclusions as to its effectiveness, or even presence, in a particular area.

Recent discoveries of various chemical cues, known as kairomones, that affect the searching behavior of natural enemies may also have practical implications in biological control. As we shall see in a later section, such chemicals may be employed to augment the performance of natural enemies.

Ecology and biological control

To a large extent, biological control is dependent on ecological research. In the discussion of the three major phases of biological control that follow this section, the need for, and value of, ecological research and knowledge will be covered as it applies to each. Here discussion will be limited to only one of the most important illustrations of how to maximize biological control through ecological research: the development of experimental methods for evaluating the impact of natural enemies in the field. This is necessary for all phases of biological control, because it enables us to determine whether we need to emphasize importation of new enemies, conservation, or augmentation of established enemies, or all three.

Precise methods of evaluation of the effectiveness of natural enemies in pest population regulation were lacking for years. Without such methods, one cannot determine whether effective natural enemies of a given pest are actually absent from a habitat, or whether they are present but rendered ineffective by some sort of adverse environmental condition. If it can be shown that they are lacking, then obviously research should be started on the importation of new ones; if they can be shown to be present, but in some manner rendered ineffective, then other research approaches to alleviate the situation are indicated. Although today the general methods are known, research on refinements and on application to particular pest–natural enemy complexes lags badly, and even the experimental approach has been disregarded by some ecologists. This important subject has been discussed in some detail by DeBach & Bartlett (1964), DeBach & Huffaker (1971), DeBach, Huffaker & MacPhee (1976), and Luck, Shepard & Kenmore (1988).

The first experimental technique for the measurement of the effectiveness of a parasite in biological control was developed in the early 1940s. Others followed, until today there are both alternative and complementary techniques. Experimental methods were developed because quantitative statistical methods, although valuable for other purposes, do not provide the rigorous proof of control (or its lack) by natural enemies that is needed. Periodic census and life-table data furnish useful information, but such quantitative methods, including modeling techniques and regression analysis, are inadequate for the measurement of the regulatory impact of natural enemies on host or prey populations. A major weakness is that they do not distinguish between cause and effect. For example, does the host population drop because of the activity of parasites, which then necessarily follow suit, or does it drop from climatic effects and the parasite population *then* follows suit?

One absolute essential to any adequate evaluation method is that the study be conducted in test blocks of sufficient size and quality, so that natural enemies will have had the opportunity to express their full effectiveness and will not have been repressed by artificial adverse conditions. Experimental evaluation plots are then established within the main study block. This will be discussed further in the section on research on conservation of natural enemies.

Experimental methods of evaluation may be divided into two categories, namely (1) exclusion methods and (2) interference or repression methods. In the former, natural enemies are excluded from a 'plot' (anything from a part of a leaf to an entire tree or a group of plants), usually after being artificially removed and also without otherwise significantly affecting the pest population.

Cages of one sort or another are usually used for exclusion, but various kinds of barriers may be employed for certain predators. These tests must always involve a series of paired comparisons, starting with comparable population densities, so that one series of cages completely excludes enemies, whereas another series utilizes the same sort of cages but includes enough means of ingress and egress for enemies, for them to operate normally within the 'open' cages. The open cages are designed so as not to change the microhabitat significantly more than the closed cages, so they furnish an evaluation of any effect of the cages *per se*. The time involved depends on the organisms, but may run from weeks or months to a year or more. Cages are only suitable for certain types of insects, mainly ones tending to be restricted in their movements. The technique employed in each case, of course, must be designed with the habits of the organisms involved in mind. Fig 7.2 shows a pair of sleeve cages, one open and one closed, used for evaluating the efficacy of the natural enemies of the California red scale on citrus in California.

If the host or prey population increases after a period of time in the exclusion cages, but not in the non-exclusion cages, it can be concluded that enemies were responsible for the previously observed prey population density. If there is no change, the conclusion is that the enemies are

Fig.7.2. Mechanical exclusion of natural enemies: a closed (left) and an open sleeve cage (right), enclosing California red scale populations on citrus. The closed cage was initially impregnated with DDT to eliminate natural enemies. (From DeBach, Dietrick & Fleschner, 1949.)

lacking or completely ineffective. Once designed properly, the tests are simple; only a 'before' and 'after' count actually is necessary.

Interference methods similarly involve paired-plot comparisons, but differ in that natural enemies are not completely eliminated or excluded, but rather are seriously decimated and rendered inefficient for a lengthy period of time from one series of plots. The check plots are not interfered with in any way. Again, initial prey population densities in each must be comparable.

Long-lasting insecticides, such as DDT, which are highly toxic to the natural enemies involved but relatively non-toxic to the prey population, have usually been used. Comparison of prey population densities between the treated and untreated check plots after a sufficient period of time has elapsed, will reveal whether or not the natural enemies were responsible for the originally observed population density in the untreated study plots. If there has been a striking increase, this is evidence that natural enemies were responsible for the original population density on the untreated trees. This is because the toxic chemical makes a poor natural enemy out of an effective one and causes the prey population to rise to a higher average density (see Chapter 3). If there is no change, some other factor, such as climate, was responsible for the natural control observed.

In some cases, especially with phytophagous mites, it has been stated by some but denied by others that insecticidal residues may have a stimulatory effect on reproduction, in addition to any adverse effect on enemies. In employing an insecticidal interference method, this possibility should be kept in mind and checked out, preferably by using some other type of experimental method simultaneously. Hand removal of natural enemies from certain plots over a period of time, but not from others, is one such. It seems to be an ideal method, but is time consuming and expensive. The so-called 'biological check method' is another. It utilizes honeydew-seeking ants in certain plots to interfere with natural enemies while having no effect on the prey, and at the same time it excludes ants from the check plots. It has no known disadvantages, except that it is of restricted application to plants or crops having honeydew-seeking ants present.

The 'trap method' provides another alternative. It involves the regular treatment with toxic insecticides of a barrier zone surrounding a central untreated block. With care and plots of sufficient size, chemical drift is kept out of the central block. However, natural enemies moving in and out through the barrier zone are killed, and so become reduced in efficiency in the central block. An increase in prey population density in the central block shows that natural enemies were responsible for the original density. In general, the 'trap method' tends to work best with rather sessile prey species and with ones not affected by the insecticides used to trap the natural enemies, in case some accidental drift of the insecticides should occur into the central untreated block.

Perhaps the best method for sedentary pests and active natural enemies is to use a pair of open sleeve cages, one of which is impregnated with a long-lasting pesticide. Since both cages are left open, the microclimate remains the same in both of them. Natural enemies present in the impregnated cage, or entering it during the experiment, eventually come in contact with the pesticide, whereas the pest population remains unaffected by it.

By the use of such techniques, it has been shown in several cases that what were considered to be serious pests actually represented upsets from one cause or another. It was also possible to prove that *Aphytis chrysomphali*, which was thought to be an ineffective parasite of California red scale in California, actually was highly effective in all coastal citrus areas if not interfered with by ants or chemicals. The tests likewise showed that *A. chrysomphali* was ineffective in interior areas because of the adverse effects of climate. This knowledge led to further foreign exploration for related, but better adapted, species of *Aphytis* and resulted in the discovery and importation of the most effective parasites obtained during 70 years of effort in California.

Importation of new natural enemies

The importation and establishment of new exotic enemies encompasses the single greatest opportunity for increasing and maximizing biological control. Foreign exploration, especially with exotic invading pests, is the classical and first approach in applied biological control, but the applicability of this method against native pests bears re-emphasis.

A substantial number of native pests have been controlled by the importation of natural enemies from abroad, including such well-known cases as the coconut moth and the coconut leaf-mining beetle in Fiji, discussed in Chapter 5, the sweet-potato leaf-miner (*Bedellia orchilella*) in Hawaii, the coconut hispid (*Brontispa mariana*) in the South Pacific Marianas and Carolines, the sugar-cane beetle (*Oryctes tarandus*) in Mauritius and the red coconut scale (*Furcaspis oceanica*) in the Western

Carolines. Most of the successful natural enemies came from closely related host species, but not all.

Much indirect evidence also supports such possibilities (see Carl, 1982, for a recent review). A number of introduced pests have been controlled by parasites and predators imported from abroad, where they attacked related species. It is a well-known fact that native natural enemies commonly adopt accidentally introduced hosts, and may exert satisfactory control. There are also instances where a purposefully imported parasite has achieved control of both the target pest and another pest, as well as cases where the parasite did not provide satisfactory control of the target pest but did of a related pest. This all means that many such parasites exist that would be suitable candidates to introduce against native pests. The possibility of biological control of native pests also seems especially promising with the use of introduced ichneumonids, or other types, that select a particular kind of microhabitat in which to parasitize a host, rather than selecting a particular host species.

On the other hand, the idea should not be carried to extremes. In a recent paper, Hakkonen & Pimentel (1984) suggested that 'new associations', i.e. natural enemies that had not evolved with the target pest, were more likely to effect biological control than did 'old associations', and should therefore be preferred. This is simply not true. Analysis of available records shows that, both with insect pests and with weeds, 'old' associations have accounted for a higher proportion of successes than 'new' ones. Thus, although the parasites and predators of related species should certainly not be ignored, those that have evolved with the target pest in its land of origin should continue to be the first choice, whenever available, in importation projects. (See Goeden & Kok, 1986; Waage & Greathead, 1988.)

It will be recalled from Chapter 6 that, up to 1988, importation of new natural enemies alone resulted in the permanent control of 164 pest species. Seventy-five were completely controlled, 74 substantially controlled, and 15 partially controlled. Inasmuch as various of these pest species have been controlled in more than one country, the grand total of successful projects worldwide is 384. No other non-chemical method of pest control can compare with this record, but we consider that the surface has just been scratched. This seems obvious, when we remember that somewhere between 70 and 90 percent of all parasitic Hymenoptera remain to be discovered. It seems obvious when we find that most pests, even in the countries leading in biological control, have not had any biological control research directed toward them. It seems even more obvious, when we know that certain countries have serious pests, yet fail to import a recorded effective natural enemy of that pest, when they could do so merely by a written request to an insectary in a neighboring country. These examples, and many more, emphasize that there are multitudinous opportunities being neglected, mainly because of a lack of trained personnel, research support and administrators who are able to assess correct priorities.

An indication of how much more can be done is provided in a recent review by Wharton (1990) of the biological control of fruit-infesting tephritid flies. He shows that out of at least 82 parasite species reared from these flies in foreign exploration programs, only 44 have ever been colonized in the field and only 22 are known to have become established. He recommends that 'more effort should be made to reintroduce some of these species, particularly since many of the early failures may be attributed to transportation and rearing difficulties'.

A good example of what new leads adequate foreign exploration can uncover is provided by the world-wide exploration and in-depth study by E. F. Legner of the parasites and predators of house flies and other filth-breeding fly pests. It had been assumed that the geographical distribution of the four major parasitic species was largely complete; however, he found, for example, that one of the best ones, *Muscidifurax raptor*, was either absent or infrequent in East Africa, Mauritius, Australia, New Zealand, Samoa and Hawaii, while it was common in other similar climatic areas. Obviously, it should be imported where absent. It was also demonstrated that what had been thought of as one species included various strains and reproductively isolated sibling species having differing biological characteristics, and hence could represent valuable introductions into new areas. Another species, *Tachinaephagus zealandicus*, was found only in the Southern Hemisphere, where it ranges through diverse climatic conditions. (It is a parasite of filth flies, not of beneficial Tachinidae as the name implies.) Because it attacks fly larvae rather than pupae, it occupies a niche that is largely unutilized in other parts of the world, hence is a prime candidate for importation into many countries. It has indeed been introduced and established in California. (See Legner, Sjorgen & Hall, 1974; Legner, 1983.)

How should we determine priorities for research in biological control and each of its three main phases, as compared to other aspects of pest management? The past records, advantages, disadvantages and relative probabilities of success of other non-chemical methods will be covered in Chapter 9. It will be shown that no single one, and probably not all combined, have the breadth of opportunity and application that biological control does.

The priority importance of the three phases of applied biological control research – importation, conservation and augmentation – must be determined by ecological studies of the faunal complex of an agro-ecosystem in chemically untreated study blocks. Included in these are the utilization of experimental methods of evaluation (see previous section), in order to determine which of the pests present constitute 'real' ones and which are man-made upsets resulting from insecticides or other artificial practices. If certain pests are shown to be real ones, in the sense that they lack effective established natural enemies, then the first priority with respect to them is importation of new exotic natural enemies. If, however, the 'pest' is demonstrated to be a man-made one, resulting from an upset of a non-target species, or is a magnified one resulting from resurgence of a target pest, then in the first case research on conservation should be stressed, and in the latter case both research on conservation as well as importation of new natural enemies should be emphasized concurrently.

If, after thoroughly adequate foreign exploration has been conducted and the problem of a real pest or of a resurgent one has not been solved, then, and generally only then, research on augmentation of natural enemies should be emphasized. All research on foreign importation should not be dropped, however, because rarely do we know that the search really has been adequate. For example, just recall the number of years before the finally successful natural enemy was found in connection with many of the famous projects.

From the preceding, it is clear that foreign exploration and importation should always receive first priority in biological control research, with the one exception that if the 'pest' has resulted from an upset, then research on conservation should be stressed.

The number of successes that have occurred in recent years (see Chapter 6) along with the other reasons presented, clearly show that biological control by importation has not reached a point of diminishing returns, but rather needs greatly increased research emphasis. Every example of fortuitous biological control (see Chapter 3) represents a case where man could have achieved the same results earlier if research had been adequate.

Some people have been critical of the probability of achieving success in biological control, and have usually based this on the ratio of the number of enemies imported to those established. As pointed out in Chapter 6 (see also DeBach, 1972), the establishment ratio actually is relatively good (about 1 in 3 with natural enemies of insect pests, more than 1 in 2 with natural enemies of weeds), although it should – and could – be improved, as will be discussed in a little while. But in any event, this is not the really important criterion, nor is the proportion of natural enemies established to successes achieved. Rather, the cost-benefit ratio is the critical measure, and it is very good. We have shown benefits to be at least 30 times as great as costs in California over a period of about 60 years. This probably represents a good international average: Hawaii would probably be better, but some others might be poorer. The cost-benefit ratio for insecticides has been rated at about 3 to 1. It really does not matter if nine natural enemies fail to become established, if the tenth one produces complete biological control and saves millions of dollars. It merely means that a certain amount of time and money was wasted on the first nine. It now costs from US $20 million to $45 million to research, develop and market one new pesticide, and only one in about 15 000 to 20 000 candidate chemicals ever reaches the marketing stage (Hollingworth, 1987). Even then, it has to prove itself in competition with others. It is, in a sense, comparable to a newly imported natural enemy. Both remain to be finally tested in the field. However, the research and development costs for one new natural enemy will amount to only an infinitesimal fraction of that of a new insecticide. Several years ago, one of the authors (PD) discovered and imported two effective new parasites of the woolly whitefly at a cost of less than $500 (see Chapter 6). He has also been involved in the importation of quite a few already known natural enemies into countries which lacked them, where the cost has been considerably less than $100 per species. Such transfers have resulted in several outstanding cases of complete biological control. Other projects will cost considerably more, but we estimate at the most, generally less than a few thousand dollars per entomophagous species.

An analysis of all worldwide importation projects to 1988 shows that some significant degree of success has been achieved with about 40 percent of all target pest species. Substantial or complete success was obtained with about 35 percent of all target pest species, which means that a degree of control was attained that essentially eliminated any further need for insecticidal treatment. To us this is such an outstanding success ratio, that it is difficult to understand why every single pest of any consequence does not have a team of biological control specialists assigned to do research on it. However, there is one hitch that would have to be straightened out first. There are currently many more recorded pests than there are adequately trained specialists in biological control.

If, instead of talking about the degree of success obtained against target species, we consider it from the standpoint of all successful projects, we then find that the number of worldwide successes more than doubles, because certain pest species have been subjugated by enemies – usually, but not always, the same ones – in more than one country. Of these successful projects, more than 80 percent were in the substantial or complete category. Thus, success in one country may have reverberations around the world, so that many benefit. If all countries were doing as much research in this area as the leading 12, the number of problems solved would jump enormously.

Although the ratio of establishment of imported natural enemies should not be viewed as a *criterion* for the success or failure of biological control, it goes without saying that improving this ratio would increase the *probability* of success. Several authors have recently analyzed and discussed some possible reasons for failure of natural enemies to become established (see Beirne, 1985; Hoy, 1985*c*). Apart from obvious causes such as inadequate systematic, biological or ecological knowledge, as discussed in the preceding sections, most *avoidable* problems seem to be related to careless or faulty procedures: importation and release of inadequate numbers, release of biological material in poor condition, inappropriate choice of release sites, etc. Although certain effective natural enemies have been successfully established from very small initial colonies, obviously importation and release of larger numbers would assure a broader genetic spectrum. There appears to be a broad correlation between the numbers released and the probability of establishment. In Canada, for instance, the rate of establishment was much higher when more than 30 000 of a species were released (or when more than 800 were released in each site) than when these numbers were considerably lower. Use of release sites contaminated with pesticide residues should of course be avoided. The statistics of biological control 'failures' are probably full of projects that were so badly handled that they did not have a fair chance to begin with!

How can research in discovery, importation and colonization of new natural enemies be maximized, aside from what has already been said? Assume that the priority has been determined to be importation. It should go without saying that the work should be handled by one or more trained specialists, with good quarantine facilities available. It has just been pointed out that there is a shortage here that needs correction. Preferably, one person should be in charge of the entire project who has overall responsibility for foreign exploration, importation, quarantine, culture, colonization and evaluation of results. The number of personnel necessary might range from two to five or more, depending on the importance and urgency of the project.

Certain procedures should be followed in order to best determine: (1) where to go, (2) how to discover the enemies wanted, (3) which species among many to choose (if some choice is required for practical reasons), (4) which are likely to be better – parasites, predators or pathogens, (5) which habits and characteristics are signs of an effective enemy, and various others. This subject has been covered by Bartlett & van den Bosch (1964) and Zwölfer, Ghani & Rao (1976).

Obviously, a first step is to consult the literature to determine where the pest occurs, and especially where it is not considered a problem, as well as what natural enemies are recorded and any evaluations of them. Sometimes the latter can be obtained by correspondence, but generally foreign exploration is necessary and desirable in order to achieve a really thorough check of natural enemies occurring abroad.

Based on ecological principles, an analysis of successful and unsuccessful projects, personal experience and the experience of many other biological control specialists, a series of conclusions and recommendations have been developed (DeBach, 1972) regarding biological control importation principles, policies and practices, as follows:

(1) Exotic pests are obviously suitable subjects for biological control by importation of natural enemies, but native pests also have been controlled by imported enemies and should receive serious consideration. Effective natural enemies may be obtained abroad from either the target pest or from species related to it, or from the same type of microhabitat.

(2) The ultimate success of a given natural enemy candidate for importation *cannot be accurately predicted in advance*, but we can find and select good probabilities. The effectiveness of natural enemies can be evaluated – experimentally or otherwise – in the country where they occur, and their potential estimated. An effective natural enemy can be deduced to have the following characteristics: (*a*) high searching ability, (*b*) high degree of host specificity or preference, (*c*) good reproductive capacity relative to the host, and (*d*) good adaptation to a wide range of

environmental conditions. The most essential characteristic is high searching ability. It should be borne in mind that a really effective enemy may be scarce in its native home, because it regulates the host population at low levels.

(3) A really effective natural enemy becomes established easily and rapidly, *if correctly handled*, usually within the first year. Intensive and continued research effort should not be wasted on ones that do not. Rather, additional species or strains of species should be sought. Complicated and expensive programs are not necessarily required to achieve successful results.

(4) Multiple importation, either simultaneously or sequentially, of diverse natural enemy species is the only practical manner of obtaining the best natural enemy for a given habitat, or the best combination of natural enemies for this habitat, or the best combination for the entire host range.

(5) Multiple importation of similar species of enemies is not ecologically unsound, because competition between natural enemies normally is not detrimental to *host population* regulation; in fact, the displacement of a fairly effective established enemy species by another imported species means that the second one is even more effective, and will produce even better host population regulation. Although competing natural enemy species may affect the individual efficiency of each other, their *combined* effect in host population regulation is greater.

(6) There usually is one best enemy for each pest species *in a given habitat*, and one frequently is sufficient for complete biological control. Often, however, a second or third enemy species may add to host population regulation, and may in fact be necessary to achieve satisfactory biological control. The best enemy species may differ for different host habitats; hence there is generally no single best natural enemy extending throughout the range of a pest species. Therefore, all natural enemy species of any promise should be imported and each one colonized in all of the different habitats within the range of the pest species.

(7) The best natural enemy may not be found until all natural enemies are known; hence, basic studies in biosystematics are of the utmost importance and need increased emphasis and support. Current knowledge is woefully inadequate.

(8) The genetics of natural enemy populations needs to be taken more into consideration. Collections should be made over the range of a species, with the view of obtaining either distinct races or ecotypes which would

best fit the host and the new environment, or of obtaining sufficient genetic variability among the imported stocks, so that natural selection would result in the best fit becoming established.

(9) From a *practical* standpoint, long-term basic ecological research on a 'real' pest species need not precede importation of new natural enemies. Such studies are not likely to help a *really effective* enemy, and obviously the enemy cannot effect control unless it is imported. Any delay in discovery, importation and colonization of a really effective natural enemy merely lengthens the period that the pest species remains a problem.

(10) No geographical area or crop or pest insect should be prejudged as being unsatisfactory for biological control attempts. The wide variety of successful results now obtained from importation of natural enemies suggests that nearly anything is possible.

(11) Direct pests (i.e. ones that at low population densities directly damage the marketable item, such as fruit) are suitable objects for biological control, but the probabilities for success may be somewhat lower than with indirect pests, which can exist at higher population densities without causing economic damage.

(12) Basic research on importation policy, and on the population ecology and genetics of colonizing species, should receive more emphasis but should not reduce in any way continued and increasing emphasis on current procedures for importation of new enemies.

We have no doubt that if trained people could be assigned to this research area in significant numbers and with adequate support, more problems in applied entomology could be permanently and rather rapidly solved than by any other approach.

Conservation of natural enemies

Effective conservation of established natural enemies is absolutely essential if biological control is to work at all. The process involves manipulation of the environment to favor natural enemies, either by removing or mitigating adverse factors or by providing the lacking requisites. A general discussion of this subject may be found in van den Bosch & Telford (1964), Rabb, Stinner & van den Bosch (1976) and van Lenteren (1987).

The efficiency of established effective natural enemies can be so greatly affected by adverse factors, that they will be rendered incapable of controlling the pest satisfactorily. The same factors can prevent a newly imported enemy from being effective, or even becoming established. Although there are a number of such factors, the major overriding ones today, as we have seen, are pesticides. At the risk of repetition, it is imperative to restress this. The upsets of non-target insects caused by the effects of pesticides on natural enemies may last for up to three or four years after a given pesticide application has been made. Obviously, if they are used often, the normal balance will never be reattained. The toxicity of a pesticide, its residual life, as well as the frequency and type of application, all interact to determine the degree of adversity to natural enemies. There is no question but that the indiscriminate use of pesticides will preclude the general possibility of obtaining satisfactory biological control. An international working group of the International Organization for Biological Control (IOBC) in Europe, has put a lot of effort in recent years into developing and standardizing methods for testing the effects of pesticides on a variety of natural enemies (see Hassan, 1986).

The only acceptable approach to preserving effective biological control of non-target organisms is to employ selective pesticides, at the least injurious time and in the least disruptive formulation. Even then they should only be used when it is found to be absolutely necessary, based on scientifically established economic injury thresholds. A successful example of the pay-off of research in this area applies to fly control. It was shown conclusively that the practice of applying insecticides to the larval breeding sites in dung in commercial dairy, poultry, feed-lot and stable environments destroyed almost 100 percent of the natural enemies. These natural enemies were shown to be able to cause more than 95 percent destruction of the fly population. If only the resting sites of the adult flies on walls etc. were sprayed, the parasites were not interfered with. Application of poisoned baits to these preferred sites improved the method. When this practice was combined with staggered removal of manure deposits, again in order to preserve enemies, fly control of maximum effectiveness at minimum cost was attained. (See Axtell, 1981; Legner, 1986.)

Once the phenomenon of suppression of effectiveness of enemies has occurred and biological control either has been reversed or prevented from occurring, the real ability of the enemies is hidden and generally forgotten, if it was ever known. It can only be brought into the light by ecological studies in chemically untreated plots, using experimental check methods in the manner previously discussed. This constitutes the main foundation for research to develop better conservation.

In this regard, the key lies in the selection of study sites and the design of appropriate ecological experimentation and observation. There has been a massive neglect of such research. Study areas must be so selected that cryptic interference phenomena do not mask or preclude biological control unbeknownst to the investigator. One cannot learn about the true potential of natural enemies in an inadequate size plot which, even though unsprayed directly, may be subject to heavy drift of insecticides from adjacent plots. By the same token, such a plot may magnify adverse effects of weather or of emigration or immigration of pests or natural enemies. To be adequate, a study plot must be large enough to be completely representative of the faunal complex in a much larger surrounding area. It should go without saying that no chemical pesticides or other natural enemy disruptive practices should be permitted unless absolutely necessary.

Various other adverse cultural practices should be eliminated, if at all possible. Then the plot must be studied and experimental checks conducted for a sufficiently long period, to be certain that balance has been achieved. This can require two, three or four years depending upon how thoroughly natural enemy–prey interactions were disrupted initially. It all boils down to making the study site as ideal for natural enemies as possible. Only then can their true potential be determined, and this is the only logical starting place to develop future corrective procedures for maximizing natural enemies in any given crop. The fact that what is done in the study site may not be practical in the crop should be disregarded. It is first necessary to find out which natural enemies are inherently effective and which are not, and to isolate the factors responsible for nullifying the effectiveness of the potentially good ones.

The opportunity of solving problems through research is virtually as great with conservation as through foreign importation, because so many pest problems are man-made but so few are recognized as such. However, we have but to recall the great extent of natural biological control known to occur, and the increasingly large numbers of upsets and resurgences plaguing us today, to realize the need for research on conservation.

Of course, various other factors are adverse to natural enemies and must be determined for each particular case. Cultural practices used in cropping, as well as cultural pest control techniques, may be detrimental from the overall viewpoint. For example, regular disc harrowing, which produces a lot of dust, may lead to substantial increases in mite and certain insect populations. The reason is that dust kills small predators and parasites, while having little effect on the pests. We have demonstrated that by washing the dust from citrus trees at weekly intervals, populations of the California red scale will be reduced by about one-half due to increased efficiency of parasites. Cultural pest control techniques, such as burning or plowing under stubble or harvest residues, may sometimes do more harm than good. Burning can destroy only parasitized larvae remaining in the stems, while the healthy ones escape because they have entered the soil to pupate. Each case needs to be investigated separately, both with respect to the target insect and to other members of the faunal complex.

The adverse effect of honeydew-seeking ants on natural enemies has already been mentioned in connection with experimental check methods. Thorough ant control may be an absolute necessity if satisfactory biological control is to be attained on various fruit tree and other crops. However, application of this knowledge lags. It should be remembered that certain ant species, such as the Argentine ant (*Iridomyrmex humilis*), are very aggressive and detrimental to natural enemies, whereas others are less so. Again, each case in each crop should be investigated separately.

Weather is another environmental factor known to have strikingly adverse effects on natural enemies and on their ability to produce satisfactory biological control. In general, not much can be done to diminish the ill-effects of weather extremes, but research may indicate some possibilities, as well as indicate what is needed in the way of new natural enemies. The weather that the insect experiences in its microhabitat can be modified by manipulation of irrigation, by use of cover crops, by modification of pruning techniques, or by changes in mass-harvesting practices. In alfalfa it has been shown that harvesting large acreages all at one time greatly decimated the exposed natural enemies, whereas if the fields were harvested in alternate strips on a chronological basis, predators and parasites would be conserved and natural biological control greatly improved.

Aside from adverse factors, natural enemies may be neutralized by lack of certain necessities. The reliance of most adult ichneumonid parasites on a daily drink of water has been mentioned. Sources of food such as pollen, honeydew or nectar are essential to many adult parasites and predators, but are often periodically in short supply. The most outsanding development in solving this problem stems from Dr Kenneth Hagen's research with supplementary diets. He and his associates have shown that dusting pollen on crop plants may serve to increase the populations of certain predatory insects and mites, either directly, as a supplementary food source for the natural enemies themselves, or indirectly, as food for their alternative prey. Similarly, spraying 'artificial honeydew' preparations, consisting mainly of yeast hydrolysates, sugar and water, with some tryptophan added as an attractant, may attract adult green lacewings to a given field and increase their fecundity. By using such treatments they have demonstrated important increases in predation of pests on alfalfa, cotton, bell peppers and other crops (see Hagen, 1986). More on the applicability of these methods in Chapter 8.

Increasing vegetational diversity in an agro-ecosystem may also serve to solve this problem. Experiments in the Soviet Union have demonstrated this, and as a consequence nectar-producing plants are planted in orchards to provide food for adult parasites of the codling moth (*Cydia pomonella*), the San Jose scale (*Quadraspidiotus perniciosus*) and others. Planting a ground cover or 'green manure' may also favorably affect the microclimate. Thus, in China, a green manure crop (*Ageratum conyzoides*) planted in citrus groves not only provides pollen as food for the phytoseiid predators of the citrus red mite (*Panonychus citri*), but also lowers the air temperature in the grove by an average of 5 degrees and raises humidity. This has proved very important, as the predators are rendered ineffective by high temperature and low humidity. (See Knutson & Gordon, 1982; Shumakov, Gusev & Fedorinchik, 1974.)

Availability of nesting sites for predatory *Polistes* wasps, as well as for insect-eating birds, limits their efficiency. F. R. Lawson and R. L. Rabb increased mortality of hornworms on tobacco by providing nesting boxes for *Polistes*. There are major projects in Spain and other European countries to control forest pests by providing nesting sites and drinking water for insectivorous birds. Similarly, straw bundles are used in China to provide shelter for spiders in rice fields, and this practice alone has reportedly reduced the use of insecticides by 50 to 60 percent. (See Biliotti, 1977; Sparks, Ables & Jones, 1982.)

Alternative hosts or prey may be necessary for certain natural enemies. Doutt & Nakata (1973) found that *Anagrus epos*, the principal parasite of the grape leafhopper (*Erythroneura elegantula*) in California required an alternative host species in order to overwinter in the vineyards, and that this other leafhopper (*Dikrella californica*) occurred on wild blackberries. However, efforts to enhance biological control by establishing blackberry 'refuges' near commercial vineyards have not been successful, because blackberry bushes grown under such conditions were less attractive to the leafhoppers than wild bushes. Further studies have shown that French prune refuges, harboring yet another leafhopper species (*Edwardsiana prunicola*), may be much more effective than blackberry in enhancing this parasite's activity, especially if planted upwind from the vineyard and protected by a windbreak (Wilson *et al.*, 1989). This example represents a type of habitat diversification, and there are many others that have been shown to favor natural enemies and perhaps act as trap crops at the same time. E. J. Hambleton and J. E. Wille in Peru increased the effectiveness of natural enemies generally by crop diversification, and specifically by planting strips of corn in cotton fields to attract the earworm, *Heliothis*, so as to more rapidly increase the supply of parasites and predators. Raul Castilla Chacon in Mexico planted corn plots adjacent to cotton for the same purpose, and further studies in Mexico have shown that intercropping cotton with corn did improve the biological control of *Heliothis*, but not of other pests (Jiménez & Carrillo, 1978). V. M. Stern and his colleagues planted strips of alfalfa in cotton to attract *Lygus* bugs and increase predation. Various other examples are discussed by Altieri & Letourneau (1982). However, as pointed out in Chapter 3, careful experimental field studies are necessary in order to determine the most favorable type of diversification. For example, Burleigh, Young & Morrison (1973) found that biological control in Oklahoma cotton was improved by adjacent plantings of sorghum, but somewhat reduced by adjacent plantings of corn. Planting rows of sorghum around cotton fields is currently an accepted practice in some parts of Mexico. In very few instances has adequate research been done in this area of conservation. From what little has been done, it appears that the opportunities for natural enemy management may be just as promising as certain programs developed for wildlife management.

Augmentation of natural enemies

Augmentation of parasites or predators to increase their effectiveness involves their direct manipulation, either by mass production and periodic colonization, or by some type of planned genetic improvement, or by employing chemical cues that affect their behavior (see DeBach & Hagen, 1964; Rabb, Stinner & van den Bosch, 1976; Ridgway & Vinson, 1977*a*; Vinson, 1986). Although research on means of augmentation seems to strike a popular chord, it should in general be given the lowest priority in biological control research endeavors, and not resorted to until it has been determined that the solution does not lie in foreign exploration and importation of new natural enemies, or some type of conservation. The latter approaches can solve the problem permanently, and often very cheaply. Augmentation techniques are more expensive to develop and, with the possible exception of the release of genetically improved strains, they are also more expensive to apply. As implied by the name, periodic colonization in particular is a repetitive process, which can be just as costly as insecticidal control. This is not to imply that augmentation of certain natural enemies may not be the ultimate key to complete biological control in a given faunal complex; rather, it is meant to put research priorities in perspective.

Augmentation attempts should usually be restricted to those natural enemies which have been demonstrated by research to be inherently effective in prey population regulation, but are prevented from doing so principally because they are not adequately adapted to weather extremes, are not synchronized with the necessary stages of the host, or are otherwise rendered ineffective by periodic environmental unfavorability. At the present stage of our knowledge, there is little hope of accomplishing anything practical by trying to change an *inherently* ineffective enemy into a good one. In other words, we should not work with an enemy primarily because it is easy to culture.

Periodic colonization

Periodic colonization of natural enemies may involve either inoculative or inundative releases. Inoculative releases consist of relatively small numbers of natural enemies, intended to propagate in the target habitat so that their progeny would effect control for several subsequent generations. Inundative releases, on the other hand, consist of very large numbers and are aimed at achieving 'immediate' control by the released natural enemies themselves, not by their progeny. Either mass-produced or field-collected natural enemies may be used for both types of releases.

Inoculative releases may be appropriate when the population of an otherwise effective natural enemy is severely decimated periodically by adverse weather conditions, cultivation practices, seasonal lack of hosts, or pesticidal treatments. Small numbers, released at the onset of the favorable season, may solve the problem. Thus, for instance, Croft & McMurtry (1972) have shown that release of a phytoseiid predator, *Metaseiulus occidentalis*, on apple at the rate of 128 per tree in early summer may reduce the populations of the McDaniel spider mite

(*Tetranychus mcdanieli*) by 90 to 99 percent and effect economic control for the rest of the season.

A large-scale inoculative program has been developed in the eastern United States against the Mexican bean beetle (*Epilachna varivestis*), a serious pest of soybeans and other beans. *Pediobius foveolatus*, an inherently effective parasite introduced from India, is unable to over-winter in the United States because it lacks suitable winter hosts and does not diapause. It is, therefore, being reared in the laboratory and released annually, first into snap-bean 'nurse plots', where beetle populations develop early in the season, and then in soybean fields. Complete biological control results on a regional basis, and economically the program is more than competitive with chemical control (see Stevens, Steinhauer & Coulson, 1975; Schroder, 1981).

A similar program has been carried out in the Soviet Union against the woolly apple aphid (*Eriosoma lanigerum*), a serious pest of apple. The parasite *Aphelinus mali*, introduced in 1926, is generally highly effective, except in places with cold damp springs and cloudy summers and certain other situations, which serve as foci for new infestations. Small branches and shoots bearing parasitized aphids are collected in autumn in apple orchards where the parasite is well established, and cold-stored during winter. At the onset of warm weather, they are hung on trees in newly infested orchards, at the rate of 1000 parasites per hectare. Parasitism of 95–98 percent may result during the first year after release, and usually no other control measures are required (Beglyarov & Smetnik, 1977).

Another approach is to make small periodic releases throughout most of the year, to augment an existing natural enemy population that is rendered ineffective by prevailing weather conditions. *Aphytis lingnanensis*, for instance, was an important natural enemy of the California red scale on citrus in California and elsewhere, but was incapable of effecting complete biological control in extreme climatic zones (see Chapter 6). Inoculative releases of about 1 million wasps per hectare (= 400 000 per acre) per year, in 10 monthly increments, were found to provide effective control in some such areas in California (DeBach *et al.*, 1950; DeBach, Landi & White, 1955). Incidentally, the discovery, importation and establishment of the better adapted *A. melinus* has made such releases unnecessary.

Whenever the augmentation of an established natural enemy population by periodic inoculative releases is considered, attention should be paid to the significance of relative numbers. Even a seemingly residual population of natural enemies may be rather large. In a heavy infestation of scale insects on citrus, for instance, a low incidence of parasites of, say, 1 or 2 percent may in fact represent several hundred thousands, or even millions, of live parasites per hectare. The addition of several thousand insectary-reared parasites may, in that case, be no more than a drop in the bucket. The efficacy of such releases should therefore be clearly demonstrated by large-scale field experiments.

Contrary to one's intuitive feelings, when initial pest infestations at the beginning of the season are very light, patchy and irregular, or lacking in suitable stages, it may sometimes be advisable to inoculate the pest along with, or even before, the release of natural enemies, in order to create a more suitable light infestation on which the natural enemies may successfully increase. This 'pest-in-first' method was first tried success-fully on strawberries by C. B. Huffaker & C. E. Kennett (1956), who made inoculative releases of the cyclamen mite (*Steneotarsonemus pallidus*) and then followed with the release of phytoseiid predators. This method is especially useful against various glasshouse pests, as will be discussed in some detail in Chapter 8.

F. D. Parker (1971) and colleagues conducted some very impressive experimentation along these lines in Missouri that successfully controlled the serious imported cabbageworm (*Pieris rapae*). Although most cabbage growers in Missouri considered three species of aphids and three Lepidoptera to be major pests, the studies by Parker, F. R. Lawson and R. E. Pinnell of the United States Department of Agriculture's Columbia, Missouri Biological Control Laboratory indicated that only one, the imported cabbageworm, is a real pest, except for occasional outbreaks of the others. The principal manipulative methods were based on studies showing primarily that the main established parasite, *Apanteles glomera-tus*, of the imported cabbageworm was not well synchronized with early spring populations of the host; the host density of the first two generations was too low for the parasite to increase satisfactorily, and overwintered parasite densities were too low because of winter mortality and attacks of hyperparasites. Two new species of parasites, *Trichogramma evanescens* and *Apanteles rubecula*, were imported to partially solve the problem. Then, mass releases of *A. rubecula* and *T. evanescens* were devised to synchronize and increase parasite density and additionally, early spring host density was brought to a level satisfactory for early increase and continued synchronization by artificially releasing cabbage butterflies in the field. The combined effect of releasing the two parasite species as well as the host resulted in substantial increases in host mortality of the cabbageworm, and the production of nearly all Grade A, No. 1 cabbage heads in a series of test plots. None of the check plots produced marketable cabbage (see Fig. 7.3). Seven other pests or potential pests were present. The cabbage looper (*Trichoplusia ni*) was controlled in the experimental plots by releases of *Trichogramma evanescens* and by spraying with a virus. The other species were held under natural biological control. Later studies (Parker & Pinnell, 1972) showed that control could be obtained either by managing the pest population with releases of both the pest and its parasites, or by releasing only parasites at timely intervals. Releases of *T. evanescens* alone controlled the host population during favorable weather; if extended periods of cold and rainy weather persisted, especially when the plants were beginning to mature, releases of *A. rubecula* at a rate of about 12 500 adults per hectare controlled the escaping host population. This is an outstanding example of what can be accomplished by appropriate research.

A similar approach has been adopted on a large scale in the lac industry of Guangdong Province, China. The purple lac insect (*Kerria lacca*) has long been cultured on various forest trees, but lac production often suffered great losses from attack by the predatory larvae of *Eublemma amabilis*, a noctuid moth. *Bracon greeni*, a larval parasite, was introduced into Guangdong in 1972 from Hainan Island and produced satisfactory biological control, but usually needed help because of periodic lack of suitable host stages. In order to optimize the parasite population, branches bearing larvae of the pest are hung in the lac forest at times of scarcity. This practice has increased markedly both the yield and quality of lac, and no further control measures have been required (Coulson *et al.*, 1982; Huffaker, 1977).

Inundative releases of natural enemies may, in a sense, be regarded as application of 'biotic insecticides'. By their very nature, such large scale releases are considerably more expensive than either importation or inoculative releases. They are, therefore, most suitable against univoltine pest species, or against pests that cause economic damage only during a limited period of the year. However, a natural enemy utilized in inundative releases need not possess all the attributes desired of a species imported for permanent establishment, or even of a species used in inoculative releases. Thus, inasmuch as only the immediate performance of the released natural enemies is of direct concern, their ability to reproduce, disperse or overwinter in the field may be relatively unimportant. On the other hand, the ability to cause high host/prey

(a)

(b)

Fig.7.3. (a) Typical damage caused to cabbage by the cabbageworm in plots not having biological control augmented, as compared to (b) Grade A, No. 1 cabbage produced in plots receiving mass colonizations of the parasites *Trichogramma evanescens* and *Apanteles rubecula*, as well as purposeful early field releases of the host in order to increase host–parasite synchronization. (Photos courtesy of Frank D. Parker.)

mortality rapidly, and the rearing and 'application' costs are of outstanding importance in an inundative release program.

Inundative releases have been attempted with quite a few natural enemy species, but by far the most extensive use and most encouraging results have been with egg parasites of the genus *Trichogramma* (see Fig. 2.8, pp. 46–47). Much of the early field work was empirical and did not include experimental checks. Good results have been claimed in many cases; others were admittedly unsatisfactory. The main reason for controversy stems from the inadequate research basis for many of the projects, including the misidentification of species, so that completely unsuitable stocks were cultured and released. It was thought for some years that only one species, *T. minutum*, occurred in the United States. Recently, biosystematic research has demonstrated the existence of 11 biparental species in the continental United States, several of which include biologically distinct semispecies and sibling species. Whereas in the past all the species of *Trichogramma* were believed to be very polyphagous, recent studies have shown that some of them are rather highly host specific. Such basic research, along with numerous accurate field trials, have increased the efficacy and reliability of the method, and today there seems to be little question but that inundative releases of *Trichogramma* spp. are achieving good results in the United States, especially in cotton. Some 3.5 billion *Trichogramma* are produced annually, with more than 2 billion of them reared for commercial release on some 200 000 hectares of crops (see Ridgway *et al.*, 1981; Ridgway, King & Carillo, 1977; Stinner, 1977). Even more impressive results have been reported from Mexico, the Soviet Union, and the People's Republic of China.

In Mexico, there are 20 government *Trichogramma* insectaries scattered throughout the country. A good example is one at Torreon, Coahuila, that started in 1963 under the direction of Raul Castilla Chacon of the Direccion General de Sanidad Vegetal. They produce hundreds of millions of *Trichogramma* annually for release against the bollworm (*Heliothis* spp.) and the pink bollworm (*Pectinophora gossypiella*), the two worst pests of cotton in this large area. When they started, massive applications of insecticides were being made, yet losses and costs were so high as to be virtually prohibitive; since 1963, losses have declined substantially each year, coincidentally with significant reductions in insecticide usage. The results are attributed to *Trichogramma*, which parasitizes a high proportion of the eggs of each pest species. It is interesting that the pink bollworm and the *Heliothis* bollworm are still feared major pests in certain areas of the United States and elsewhere, where *Trichogramma* mass releases have not been adequately tested, if at all.

The USSR and China lead the world in research and application of mass releases of species of *Trichogramma*. The All-Union Plant Protection Institute of the USSR studied 17 ecotypes of four species, both in the laboratory and field, to determine their biological characteristics and their relative potentialities with respect to 12 important pest species. As a result, 15 intraspecific forms of three species were selected, and today are in common use in pest management in the USSR. In 1976, more than ten 'biofactories' were in operation in various parts of the USSR, with a total productive capacity of over 50 billion (i.e. US billion = thousand million) egg parasites per season, and the total area covered by inundative *Trichogramma* releases exceeded 7.5 million hectares, in various crops. *Trichogramma evanescens*, which has four biological races exhibiting different host preferences, is used against 15 species of field and vegetable crop pests, including the cutworm (*Agrotis segetum*), the cabbage white butterfly (*Pieris brassicae*) and the European corn borer (*Ostrinia nubilalis*). *T. embryophagum* and *T. cacoeciae pallida* are used to control the codling moth (*Cydia pomonella*) on apples in regions where the pest has only one generation. In general, pest numbers have been reduced by about 80 to 85 percent and substantial increases in yield resulted with wheat, corn, sugar-beets, cabbage and apples.

Basic research was responsible for this. First, taxonomic studies of natural populations clarified the various species and ecotypes involved, along with their host preferences. Then, ecological studies pinpointed the preferred conditions under which the several forms performed best. Various *Trichogramma* species and forms show different specializations according to pest species and to habitats. Some prefer open fields and live near the ground. Those attacking the cabbage moth (*Mamestra brassicae*) eggs prefer moist conditions, while those attacking corn borer eggs in corn fields are drought-resistant. *T. cacoeciae pallida* and *T. embryophagum* are forest-dwellers, capable of populating the tree tops. In orchards they prefer eggs of the codling moth to those of other moths. *T. c. pallida* populates the whole tree, whereas *T. embryophagum* prefers the upper parts of the tree. *T. pini*, another forest dweller, populates the upper parts, even of high trees. Consequently, in the USSR, the use of the several species and forms of *Trichogramma* has been arranged according to zones, regions, crops and pests in relation to the biological and ecological qualities of the parasites. This is an elegant system that has evolved to its present state from the early, more or less isolated, empirical, use of *Trichogramma*. It is now the main basis for integrated control, which also stresses the use of microbial bioinsecticides and the minimization of chemicals (see Beglyarov & Smetnik, 1977; Ščepetilnikova, 1970).

In China, several *Trichogramma* species are utilized in inundative releases against various lepidopterous pests on more than one million ha of crops including corn, rice, sugar-cane and cotton and on one million ha of pine forests. In rice, up to 750 000 wasps are released per ha, effecting control at half the cost of chemical control. At one experimental station in Guangdong Province, 7 to 8 billion *Trichogramma* were produced in 1979 against three species of sugar-cane borers and released over an area of about 4300 ha. As a result, sugar-cane yields in the Province have increased by 10 to 15 percent. (See Coulson *et al.*, 1982; Huffaker, 1977.)

Our knowledge of the systematics and biology of *Trichogramma* has improved greatly in recent years. With modern taxonomic and biosystematic techniques, including scanning electron microscopy, computerized biometry, electrophoresis and crossing tests, the number of described species has risen dramatically, from 20 in 1970 to more than 110 in 1985 (Vogele, 1988). Improved methods of mass-rearing would, of course, allow for more cost-effective use of *Trichogramma* in inundative releases. The ultimate in this respect would be *in vitro* rearing on inexpensive artificial media. Attempts at artifical rearing have been made with several natural enemies, but the most intensive efforts have been with species of *Trichogramma*. The problem is two-fold: (1) developing an artificial diet providing adequate nutrition for complete development and high fecundity and fitness of the parasites, and (2) inducing the adult female parasite to accept the artificial 'host' and oviposit in it. Early studies in China and the United States in the 1970s made use of insect hemolymph as a rearing medium. More recent research has developed media devoid of insect additives and has identified suitable 'egg membranes' as well as various chemicals that act as ovipositional stimulants. Droplets of the diet are encapsulated in tiny egg-like spheres coated with wax and plastic, and these 'artificial eggs' are readily accepted by the ovipositing parasites. It is now possible to rear *Trichogramma in vitro* in large numbers continuously for 50 generations, and release the parasites effectively in the field. What started as basic research on insect nutrition now holds great promise for applied biological control. (See Thompson, 1986; Xie *et at.*, 1986.)

Various natural enemies other than *Trichogramma* have also been used in inundative releases. In the United States, R. L. Ridgway & S. L. Jones (1969) have experimentally demonstrated excellent control of the bollworm (*Heliothis*) on cotton in Texas by the mass release of the green lacewing *Chrysoperla carnea*. These releases reduced bollworm larvae by 96 percent and resulted in a three-fold increase in yield of seed cotton. The difference in yield is shown in Fig. 7.4 by the plots white with cotton as compared to the darker ones that received no green lacewing releases. Subsequent trials were also successful, and a mechanized system has been developed for mass-releasing *Chrysoperla* larvae, using sawdust as a carrier (Ridgway, King & Carillo, 1977). These general predators appear promising for use in a number of mass release programs. They are currently being produced in a commercial insectary in California for use against a variety of pests (for more detail, see Chapter 8).

In Guangdong, China, the univoltine stinkbug *Tessaratoma papillosa* is the main pest of litchi. An egg parasite, *Anastatus* sp., is mass-reared in the eggs of an alternative host and released at the rate of 600 per tree, achieving up to 95 percent parasitism and complete control. Since they have a winter diapause, the developing parasites are cold-stored from November to March for spring release. Recently, *Anastatus* has been reared on an artificial medium, with equally successful control results. (Cock, 1985; Liu *et al.*, 1988.)

Fig.7.4. Six plots of cotton, showing increases in yield in the three plots white with cotton that received releases of green lacewings to control bollworms (right foreground, left center, and right background), as compared to the three darker plots that received no natural enemies. (From Ridgway & Jones, 1969.)

The use of insect pathogens in inundative releases appears especially applicable. The literature in this field is voluminous but a good cross section of it, as well as an outline of research needs, is contained in articles by Falcon (1985), Ignoffo (1985), and Surtees (1971). Active interest in the use of pathogens for controlling vectors of disease is being shown by public health entomologists. Demands are becoming more urgent and frequent, to completely curtail the persistent use of chemicals for vector control. Additionally, resistance of mosquitoes and flies to various insecticides has become serious, so that other approaches are indicated. With the backing of the World Health Organization and the co-operation of many insect pathologists, a strong research program has developed, stressing mainly the use of *Bacillus thuringiensis israelensis* (*Bti*, known also as *Bt* H–14) and certain fungi and nematodes.

In agricultural pest control, *Bacillus thuringiensis* has come to be commercially produced and utilized on a wide scale, mainly against lepidopterous pests. It is somewhat questionable whether it should be considered as a natural enemy or an 'insecticide', because the killing action comes from the toxic crystals included in the bacteria (see Chapter 2), and usually the organisms do not reproduce in the field. In any event, it constitutes an excellent selective mortality measure with no known ecological drawbacks.

The successful use of mass releases by spraying of insect viruses has been demonstrated in large-scale tests with the cotton bollworm, the tobacco budworm, the cabbage looper, the imported cabbageworm, the citrus red mite, several pine-infesting sawflies and others, and offers a great deal of promise. A breakthrough is needed to achieve cheap commercial production of viruses on artificial media. An interesting example of research leading to promotion of virus epizootics in populations of the Swaine jack pine sawfly (*Neodiprion swainei*) in northeastern Canada has been reported by Smirnoff (1972). Periodic outbreaks of this sawfly cause defoliation of jack pine stands and many affected trees die. To prevent these losses, a particularly virulent strain of a nuclear polyhedrosis virus, *Baculovirus swainei*, shows promise through manipulation. The objective is to introduce the infectious disease into healthy sawfly populations. Various techniques of accomplishing this were investigated and the most practical one involves introduction of the virus by the dissemination in the forest of infected cocoons which give rise to infected adults. These adults then fly to the trees and move about, spreading the disease by laying diseased eggs. By the time the first larvae from these eggs have died from the disease, indigenous larvae are infected. The third and fourth instars of the indigenous larvae all succumb, but the fifth instars that are simultaneously present proceed to spin cocoons after becoming infected, and the adults from these cocoons contribute to the spreading of epizootic centers in the population.

The induced viral epizootics spread naturally into surrounding infested areas at the rate of only several hundred meters per year. Acceleration of spread could be achieved by wide distribution of the infected cocoons. To obtain such cocoons for dissemination in the field to initiate epizootic centers, pine trees infested with numerous fourth and fifth instar sawfly larvae were sprayed with a virus concentration of two million polyhedra per ml. Branches bearing inoculated larvae were then cut and the larvae allowed to spin cocoons. These cocoons were then put in containers, from which later emerging adults could escape, and distributed in the field to create epizootic centers by dropping them in infested stands from a helicopter, throwing them from a car or transporting them deep into the forest by horse. This method eliminates many difficulties associated with mass production of the virus.

Several viruses have been successfully mass produced *in vivo*, in host larvae, and some have been registered in the United States and elsewhere as microbial insecticides. They can be applied with the same equipment that is used for chemical pesticides, either from the ground or from the air, and may even be mixed with certain chemical pesticides. High cost of production, timing of application and a very short residual effect are the main problems. *In vitro* production in cell cultures has been successful in certain cases but has not reached the industrial stage yet. In Canada and northeastern United States, serial applications of nuclear polyhedrosis and granulosis viruses have been quite effective against various forest pests. (See Entwistle & Evans, 1985.)

An interesting approach has been tried by L. A. Falcon (1985) and his colleagues. They used light traps to attract insects, which were then coated with a microbial insecticide and released, automatically, to spread the disease.

Inasmuch as most crops may be attacked by several pest species at the same time, excessive specificity of a virus can sometimes be a drawback. The recent discovery of certain baculoviruses with a broader than usual host range has therefore attracted considerable attention. An example is an NPV from the alfalfa looper (*Autographa californica*), which is known to infect 43 species of Lepidoptera in 11 families (Payne, 1988).

Genetic improvement

When a potentially effective natural enemy is deficient in some crucial adaptive attribute, genetic improvement of that attribute may be attempted. Rather than wait for natural selection to take its slow course, selective breeding and hybridization may be carried out in the laboratory to speed up the process, and even to obtain results that might never occur in nature. Arthropods are certainly amenable to genetic manipulation. Many adaptive races are known to exist, and there usually is considerable genetic variation in natural populations. In fact, certain domesticated or semi-domesticated insects such as the silkworm or the honey bee have been genetically improved, with remarkable success, for centuries (see DeBach, 1958a; Hoy, 1976b).

Almost any desired trait can be selected for. Various parasites and predators have been successfully selected for improved climatic tolerances, sex ratio, host-finding ability, host preferences and pesticide resistance. Until recently, however, the field effectiveness of such selected strains had either not been evaluated or had not been significantly improved. One of the authors (PD), for instance, selected a strain of *Aphytis lingnanensis* with considerably higher tolerance of both high and low temperatures, but its field performance was never tested because a better adapted species, *A. melinus*, was discovered, imported and established in California at about the same time (White, DeBach & Garber, 1970). Another interesting approach was tried by E. F. Legner (1972, 1988). He made reciprocal crosses between imported strains of widely geographically separated populations of synanthropic fly parasites. Expressions of hybrid vigor were found in F_1 through F_3 generations of crosses from climatically similar but geographically isolated areas. This indicated the possibility of producing superior synthetic strains through directed heterosis for use in mass-release programs, and perhaps enhancing the establishment of imported natural enemies, but again there was no outstanding field success.

The recent development and successful utilization of pesticide-resistant strains of the phytoseiid mite predator, *Metaseiulus occidentalis*, has been a major breakthrough. Field populations of this species were found to have acquired resistance to organophosphorus (OP) pesticides and sulfur, and OP-resistant strains from Washington and Utah were introduced into apple orchards in several countries, where such resistance had not occurred. Strains resistant to carbaryl (a carbamate) and permethrin (a pyrethroid) were then developed by selective breeding and were released in pesticide-treated deciduous fruit tree orchards and vineyards in California. These artificially selected strains have now been successfully established and have demonstrated their efficacy in the face of commercial pesticide applications. Similar work has now been done with several other phytoseiids and some predatory insects, and efforts are currently being made also with parasitic insects. (See Croft, 1976; Hoy, 1985a, 1986.)

In any program of genetic improvement, the attributes to be improved should be clearly defined. Adequate genetic variability, either natural or induced by radiation or by mutagenic chemicals, should be present in the stock to be selected, and effective rearing methods should be available. Release strategies for the improved strain may depend on the nature of the genetic mechanism involved. When the improved trait is governed by a single dominant gene, the new strain may be simply released into an established population and interbreeding would result in the injection, so to speak, of the favorable gene into it. If, on the other hand, the improved trait is polygenic, interbreeding may result in its being swamped in an established population. Such strains should, therefore, be released in habitats where the original stock does not exist, or in a manner that would ensure that stock's replacement by the new strain. Continued selective pressure (such as pesticidal applications) may be required in such cases. Of course, if the improved strain is also thelytokous, or otherwise reproductively isolated from the original stock, its release is equivalent to the introduction of an exotic species.

The potential for genetic improvement of natural enemies is almost unlimited. Selection for pesticide resistance is the most obvious step in this direction but we consider, for instance, the possibility of changing the host preference of an enemy especially worthy of investigation because one can choose a highly effective enemy to start with, so that the major desirable characteristics are already 'built-in'. Another important area is in mass-production programs, where inbreeding and genetic drift could lead to deleterious changes. With further research emphasis, genetic improvement of predators, parasites and pathogens may become one of the important approaches to applied biological control. Recent advances in genetic engineering techniques are especially intriguing in this respect (see Beckendorf & Hoy, 1985).

Chemical cues

As mentioned in Chapter 3, the searching behavior of natural enemies is mediated by various semiochemical cues, emanating from the prey itself (in which case they are termed kairomones), from the host plant (synomones), from other organisms associated with the prey, or from various interactions between them. Recent studies have indicated that some of the chemicals involved in such cues may be manipulated to alter the behavior of natural enemies and enhance their efficacy.

Certain kairomones and food substances may be used to increase the searching activity of a natural enemy, or to retain it in the site of release or in an area of low prey density. Host-marking pheromones may also be used to improve egg distribution and reduce superparasitism by parasites. Conceivably, such pheromones may be employed to protect primary parasites from attack by hyperparasites, or predators and the natural enemies of weeds from their own primary parasites. Certain kairomones may even aid in the collection of exotic natural enemies in foreign exploration. (See Gross, 1981; Vinson, 1977, 1986.)

Much further research is required to identify the sources and determine the chemical identity of candidate substances, to elucidate their role in natural enemy behavior, and to overcome potential problems such as possible overstimulation of the target species or subtle deleterious effects on the ecosystem, before this interesting approach can be practically employed in biological control.

8

Utilization by the public

Anyone who grows plants, and to a lesser extent those who raise animals, are automatic recipients of a great deal of natural biological control. Additionally, they can take certain measures to increase the degree of biological control they enjoy, if it is not sufficient to start with. Satisfactory natural biological control on ornamental and shade trees is the rule; exceptions are rare, as inspection of domestic gardens and parks will reveal. The exceptions, of course, are the things we hear about – the occasional serious pest species which can cause great damage if not controlled.

The dooryard horticulturist, home gardener or the small-scale grower who utilizes local markets, often has certain advantages over the large commercial farmer as far as the utilization of biological control and the minimization of pesticides is concerned. The former do not have to be concerned with 'cosmetic' appearance of their produce nor of its shipability, so they can overlook minor insect damage and thus suffer no industry-imposed cullage, while being able to harvest at the peak of flavor. The absence of non-essential cullage means a few fruit or vegetables can actually be lost entirely to insects, without necessarily suffering any appreciably greater losses per plant than the large-scale farmer who sprays regularly and so has large pest-control bills. In addition to vast monocultural acreages that tend to be faunistically impoverished, the large-scale commercial farmer often has more con-straints imposed upon him by outside sources such as government and industry, which may make it more difficult to utilize biological control to its maximum. He also has more at stake economically, and this may tend to keep him on the insecticide treadmill. These are differences in degree only, however. One should not infer he cannot utilize or increase biological control, but only that more expertise and sophisticated methodology may be required, as is discussed in Chapter 10.

On the other hand, the commercial farmer may do very well without resorting at all to insecticides, as the following excerpts from *Natural History Magazine* (Hall, 1972) indicate.*

> 'On . . . January 17, . . . I . . . read an Associated Press dispatch . . . crediting our new Secretary of Agriculture with stating that if pesticides were banned from farming, we could not feed all Americans. That dispatch followed the departmental line, as expressed by the administrator of the Research Service of the US Department of Agriculture: "As we cope with . . . a more populous world . . . one of the greatest needs is production of food for billions of people. At present such production requires the use of pesticides" (*Science*, June 19, 1970).
>
> The statement that pesticides are required to grow food crops is repeated so often by employees of the Department of Agriculture that many citizens believe it is true. Actually, it isn't. Individuals and local organizations of agriculturists in many parts of the United States have been, and are, demonstrating that the use of pesticides (herbicides and insecticides) is not necessary in food production.
>
> One example is a 107-acre farm that I manage in Marysville Township, Miami County, Kansas. Five acres are pasture, 22 are woodland along a creek, and 80 are cultivated. No pesticides have ever been used on growing crops on this farm. Corn, milo, soybeans, and wheat are the crops raised now. For any given year in the period from 1886 through 1964 there was little or no difference in the yields from farms with comparable ground throughout the township. However, in the seven years since then, herbicides were introduced on neighboring farms and the yield per acre on my 80 acres has been greater than on comparable land, especially when soybeans were planted where corn had been grown the year before. On the comparable land, residues of herbicides applied to inhibit the growth of broad-leaved weeds in corn fields accumulate in the soil, causing the soybeans, themselves broad-leaved plants, to be puny. Other kinds of herbicides applied in order to inhibit the growth of grasses in soybean fields accumulate in the soil and cause the corn plants, themselves grasses, to be puny, especially when corn is planted where soybeans had been grown the year before . . . Considering the common interest, it would seem that, in Miami County, for each 2500 acres of cultivated land there should be at least five instead

---

* From 'Down on the Farm' by E. Raymond Hall, *Natural History Magazine*, March 1972. Copyright © The American Museum of Natural History, 1972.

of four farmers and that they should control weeds by cultivation and should not use pesticides. This system produces more bushels per acre and more total bushels.

Furthermore, under this system the fish in the creek grow big and do not die prematurely because of pesticide residues, which have already stilled the spring voices of five birds in the woodland on and around the land that I manage.

The example outlined above is only one strand of the web that regulates the lives of man, other animals, and plants in the current green (agricultural) revolution, but illustrates why I am very tired of being informed that pesticides are required to grow enough food crops.'

If the reader is inclined to discount the observations and opinions of the author as coming from just an unscientific farmer, it should be mentioned that Dr Hall was Chairman of the Department of Zoology of the University of Kansas and a Vice-President of the American Association for the Advancement of Science. This is only one instance of hundreds or thousands of growers in the United States and other countries, who are producing crops of superior quantity and quality while using pesticides rarely, if ever.

The idea that the private individual can utilize and even increase biological control is not new. Almost a century ago, C. V. Riley (1893) wrote:

'There are . . . two methods by which these insect friends of the farmer can be effectually utilized or encouraged, as, for the most part, they perform their work unseen and unheeded by him . . . These methods consist in the intelligent protection of those species which already exist in a given locality, and the introduction of desirable species which do not already exist there.'

Regarding the former, i.e. conservation of natural enemies, Riley considered that the 'husbandman' would be restricted in what he could accomplish. Also, Riley did not yet recognize the potential possibility of mass production and periodic colonization of natural enemies. (Importation work with new natural enemies was then, as today, restricted to official agencies and so is outside the scope of private endeavors, aside from individuals utilizing the results.) However, he went on to say:

'That a knowledge of the characteristics of these natural enemies may, in some instances, be easily given to him [the public] and will, in such instances, prove of material value, will hardly be denied. The oft-quoted experience which Dr Asa Fitch recorded, of the man who complained that his rosebushes were more seriously affected with aphides than those of his neighbors, notwithstanding he conscientiously cleaned off all the old parent bugs (he having mistaken the beneficial ladybirds for the parent aphides) may be mentioned in this connection . . . But for the most part the nicer discriminations as to the beneficial species must be left to the trained entomologist.'

Riley was certainly correct for his time, and to a considerable extent remains so today. There are limitations on what the individual can do to utilize or augment biological control of certain pests, because many pest problems require considerable time, money and professional research for their solution. However, there is no question but that the interested home-owner, gardener, farmer, forester, poultry producer, etc., often can take advantage of today's increased knowledge and commercial availability of natural enemies to foster biological control of pests in the area of his concern.

Conservation

The greatest possibility for utilization of biological control by the public lies in the conservation of established natural enemies along the lines discussed in Chapter 7. We have already emphasized the great extent of biological control that occurs naturally, and the importance of not disrupting it. In many cases in the home garden, dooryard or commercial farm this will be all that is necessary for adequate pest control. Some good references for the layman on practical biological and ecological control include *Common Sense Pest Control* by Olkowski (1971), *Organic Gardening: Yesterday's and Tomorrow's Agriculture*, ed. Wolf (1977), *The Encyclopedia of Natural Insect and Disease Control*, ed. Yespen (1984), and numerous articles in journals such as *Organic Gardening*.

The activities of natural enemies can be conserved and fostered by individuals in various ways. The most important is by restricting the use of pesticides to the minimum that is absolutely necessary to prevent pest damage. Chemicals should never be used on a so-called 'insurance' or 'calendar' basis, without regard to whether the pest actually constitutes a hazard at the time or whether natural enemies are in the process of reducing or regulating the pest. When it appears on the basis of careful counts or observations that a pesticide may be necessary, only short-lived, non-persistent materials, highly selective for the target pest, should be used. Microbial insect pathogens such as preparations of *Bacillus thuringiensis* are especially suitable in this regard, but there are also relatively safe selective chemical pesticides. Among them, plant-derived materials such as rotenone, derris, ryania and nicotine are generally safe when properly used. Another possibility are the new insect growth regulators, discussed in Chapter 9. Unfortunately, all these selective pesticides, and commercially available natural enemies, presently account for only one percent of the world market of crop protection products (Jutsum, 1988).

Table 8.1 lists the microbial pesticides registered in the United States. Most of them are available commercially but some, like Elcar[R], are no longer in production. Several of these microbials, or similar ones, have been registered in other countries. The Canadian Forestry Service, for instance, in 1983 registered two NPV products, for use against the Douglas-fir tussock moth (Virtuss[R]) and the redheaded pine sawfly (Lecontvirus[R]). Nematodes do not have to be registered in the United States. Some microbial pesticides that are in commercial production in various countries are listed in Table 8.2. The most commonly available, *Bacillus thuringiensis* variety *kurstaki*, has been produced by at least six companies in the United States and six in other countries. Various other microbial pesticides are currently being developed experimentally and may soon be available. Some of the basics, applications and future potential of these selective pesticides are discussed by Falcon (1985) and Payne (1988), and in several of the publications listed in Chapter 2.

Good lists, sources and discussions of selective materials, as well as other alternative non-chemical measures, may be found in Briggs (1970), Olkowski (1971) and Yespen (1984). It should be recognized that any pesticide – no matter how non-toxic to humans or other vertebrates; no matter how short-lived or how selective – has the potential of causing upsets or resurgences in any given ecosystem or habitat. Each decision to use a chemical should reflect this possibility, but this is not to say that in any given instance a decision to spray or dust is not the best one under the circumstances. Shirley A. Briggs, Executive Director, Rachel Carson Council, Inc., 8940 Jones Mill Road, Chevy Chase, MD 20815, USA, maintains extensive files and has summary charts of pertinent information on 600 or so major pesticide ingredients, as well as various relevant publications issued by the Council. Information about organic gardening can also be obtained from the Institute for Alternative Agriculture, 9200 Edmonston Road, Suite 117, Greenbelt, MD 20770,

Table 8.1. *Microbial pesticides registered in the United States.**

Group	Microbial agent (trade name)	Registrant (Year)	Uses
Bacteria	*Bacillus popilliae*, *Bacillus lentimorbus* (Doom[R], Milky Spore[R])	Fairfax, Hyponex, Reuter (1948)	Japanese beetle larvae on turf
	Bacillus thuringiensis var. *aizawai* (Certan[R])	Sandoz (1981)	Wax moth larvae in honeycombs
	Bacillus thuringiensis var. *israelensis* (*Bti*) (Tenkar[R])	Sandoz (1981)	Larvae of mosquitoes and blackflies
	Bacillus thuringiensis var. *kurstaki* (*Btk*) (Thuricide[R], Dipel[R])	Sandoz, Abbott, others (1961)	Against lepidopterous caterpillars on many agricultural crops and sites, including forest, ornamental and shade trees. [Exempt from tolerance for all raw agricultural commodities]
Fungi	*Hirsutella thompsoni* (Mycar[R])	Abbott (1981)	Citrus rust mite
Protozoa	*Nosema locustae* (Noloc[R])	Sandoz (1980)	Rangeland grasshoppers
Viruses	Douglas-fir tussock moth NPV (TM BioControl 1[R])	USDA Forest Service (1976)	Douglas-fir tussock moth on Douglas fir
	Gypsy moth NPV (Gypcheck[R])	USDA Forest Service (1978)	Gypsy moth on forest, shade and ornamental trees
	Heliothis NPV (Elcar[R])	Sandoz (1975)	Bollworm and tobacco budworm on cotton. [Exempt from tolerance for all raw agricultural commodities]
	Neodiprion sertifer NPV (Neocheck[R])	USDA Forest Service (1983)	Pine sawfly

*Adapted from Falcon (1985). Copyright © Academic Press, Orlando. Reprinted by permission.

Table 8.2. *Some microbial pesticides in commercial production.*★

Group	Microbial agent	Product (Country)	Target
Bacteria	*Bacillus moritai*	Lavillus M (Japan)	Diptera
	Bacillus thuringiensis var. *kurstaki*	Plantibac (France) Biospor (Germany)	Lepidopterous caterpillars
	Bacillus thuringiensis var. *israelensis*	Bactimos (France)	Mosquitoes
Fungi	*Beauveria bassiana*	Boverin (USSR)	Colorado potato beetle
	Metarrhizium anisopliae	Metaquino (Brazil)	Spittle bugs
	Verticillium lecanii	Vertalec (UK) Mycotal (UK) Thriptal (UK)	Aphids Whiteflies Thrips
Viruses	*Lymantria dispar* NPV	Virin-ENSh (USSR)	Gypsy moth
	Mamestra brassicae NPV	Mamestrin (France)	Cabbage moth
	Neodiprion sertifer NPV	Virin-Diprion (USSR) Monisarmio-Virus (Finland) Virox (UK)	Pine sawfly
	Autographa californica NPV	VPN 80 (Guatemala)	Noctuids
	Spodoptera sunia NPV	VPN 82 (Guatemala)	*Spodoptera* spp.
Nematodes	*Heterorhabditis* spp.	various (Europe) Otinem (Australia)	Garden pests Black vine weevil
	Steinernema spp.	BioSafe (USA)	Lawn and garden pests
		Nemasys (UK)	Black vine weevil

★No intent to endorse certain products over others is intended.
Sources: Cunningham (1988), Khachatourians (1986), Payne (1988).

USA and the International Alliance for Sustainable Agriculture, 1701 University Avenue SE, Room 202, Minneapolis, MN 55414, USA. Similar organizations exist in many other countries. One example is the Informationcentre for Low External Input Agriculture (ILEIA), P. O. Box 64, 3830 AB Leusden, The Netherlands.

Pests are often mistakenly sprayed at a time when their natural enemies would have controlled them completely within a few days and before damage resulted. Such procedure invariably has a greater adverse effect on the parasites and predators than on the pest, with the consequence that the pest rapidly increases again and requires another treatment. The mere fact that 90 or 95 percent of the pest are killed – by whatever means – has a seriously upsetting effect on the attainment or maintenance of natural balance. Various so-called non-poisonous materials such as oil sprays, talc (used as a carrier for insecticides), nutritive zinc formulations and even road dust, have been shown to decimate natural enemies and cause upsets. Some highly toxic organophosphorus insecticides and others which have quite a short residual life may be looked upon by wildlife conservationists as relatively safe chemicals to use because they do not accumulate and build up in the food chain; nevertheless, various of these can cause catastrophic upsets of non-target insects because of their intense effect against insect parasites and predators, and such effects are compounded when these materials come to be applied frequently. How much have we gained when we eliminate DDT or a like material because of its long residual life, if we apply one or more highly toxic substitutes so frequently that they are, in effect, present constantly?

In home gardens and yards, simple water-spraying with the garden hose, hand-picking of larger insects and various other methods can often be used in place of insecticides in cases where natural enemies do not do a satisfactory job. Water-sprays have the added advantage of removing airborne dust, which is deleterious to many small natural enemies and may be the reason why biological control is not satisfactory. When hand-picking is used to remove pest eggs, larvae or pupae, these can be placed in suitable containers having a mesh screen cover which will permit parasites to emerge, escape and continue to propagate but prevent the pest from escaping. Another protective control device is the use of wide burlap or corrugated cardboard bands around the trunks of ornamental or fruit trees to trap caterpillars, pupae or adults of certain species that seek such concealing shelter. In 1971, an entomologist in Connecticut completely eliminated an infestation of caterpillars of the notorious gypsy moth from his yard trees by applying a wide strip of burlap around the trunks, overlapping the top, then later removing the strips and knocking the accumulated larvae into a container of kerosene to kill them. Nearby neighbors' trees were defoliated.

Aside from the use of upsetting chemicals, the presence of honeydew-seeking ants, where they occur at all commonly, possibly has the next greatest adverse effect on natural enemies. Ants of this type attend many homopterous pests such as aphids, soft-scales and mealybugs for their syrupy secretions, and at the same time kill or drive away natural enemies. Since they do not discriminate between species of natural enemies, ants can also cause increases in non-honeydew-producing associated pests. For satisfactory biological control, ant control is a must if they are at all abundant and are tending colonies of honeydew-producing insects. Often they can be excluded from trees or large shrubs by mechanical or sticky barriers, otherwise commercially available ant poisons can be used. However, it should be recognized that many ant species prey on pest insects and are beneficial. Only ants definitely running on plants and primarily obtaining honeydew secretions from aphids, mealybugs and the like, should be controlled.

One thing that is very important to natural enemies is that they have a favorable habitat. This can vary greatly with different species, but the greater the floral and faunal diversity that is present in a backyard or garden, the greater is the chance that a given natural enemy species will find a suitable source of alternative hosts or prey when the primary one becomes scarce; will find the necessary free water, nectar, pollen or honeydew essential for adult nutrition and reproduction; and will find suitable nesting sites or protective shelter. A diversity of pest species, even though in low numbers, helps to maintain a diversity of general predators or parasites, which assist greatly in maintaining balance if a more specific natural enemy becomes temporarily decimated for one reason or another. A chronological sequence of plants is highly important to the maintenance of natural balance. If a vegetable garden is abandoned during part of the growing season, serious upsets between predator–prey interactions will result. The natural disappearance of certain or all annual plants during the winter in temperate climates is not of so much consequence, because both pest insects and natural enemies have evolved and adapted to such natural changes.

Augmentation

Individuals can foster biological control by personally collecting and introducing natural enemies from other nearby places where they are abundant, or by purchase from commercial insectaries and dealers. However, the public should clearly realize that not every species of natural enemy, in fact usually not many different species, will do the particular job they want. Parasites and predators tend to be specialists to a greater or lesser degree, especially the best ones, which are invariably limited to a restricted prey or host list. Not just any natural enemy will do, and the problem for the non-specialist is to learn which will be suitable for a particular pest problem. Once this is determined, the enemies should be colonized in close proximity to the pests, under favorable weather conditions.

There are many popular and professional misconceptions, stemming from the lack of knowledge concerning this type of manipulation of natural enemies. As in the case of medicine, there are bona fide, scientifically-based cures, and there are quack cures. The size or ferocity of a natural enemy is no indication of its effectiveness. The spectacular praying mantis, which is commonly offered for sale, is a very general predator which is not likely to materially reduce the population of any particular target species, but rather will feed indiscriminately. The ladybeetle *Hippodomia convergens* has been collected by the million for years in the western United States from overwintering aggregations of adults and sold by the gallon from California to Texas or even farther afield. The hitch is that these beetles are in a state of ovarian diapause, and are not in a physiological condition to feed and reproduce until they have left their aggregation sites naturally, in response to changes in weather conditions. Hence, when they are collected and released in the fields artificially they fly away, seeking to aggregate again, rather than remaining to feed and reproduce. Later, often the same or other species will increase naturally in the field and the farmer, failing to recognize the difference, will assume he has been responsible for the results, whereas actually he has wasted his money. Today certain insectaries feed the adult *Hippodomia* beetles with a special diet after they are collected, so that their reproductive ability is awakened and they are ready to feed and lay eggs when they are released in the field. Only such preconditioned ladybeetles should be purchased.

The mass colonization of egg parasites of the genus *Trichogramma* is carried on worldwide against a variety of pests, often with spectacular success but many times with utter failure. One main reason for the failures has been that species of *Trichogramma* can be very specific to a particular pest or habitat, but in the past entomologists have failed to recognize the minute differences between certain species, so that sometimes the wrong ones have been produced and colonized. Thus, when purchasing such parasites, it should be known that the species in question prefers the pest one wishes to control.

There are also excellent, completely successful examples of how natural enemies can be conserved and augmented in rather small, restricted habitats. In Europe, outstanding pest management in glass-houses has been demonstrated after a great deal of basic biological research (see Hussey & Bravenboer, 1971; Hussey & Scopes, 1985; van Lenteren, 1988). It relies principally upon the mass production and periodic colonization of various natural enemies, the initial seeding of certain others as well as the pest itself at appropriate times in order to establish the necessary balance early, and the judicious use of selective chemicals, especially fungicides, when necessary.

The two-spotted spider mite (*Tetranychus urticae*) is controlled on glasshouse cucumbers by deliberately introducing about 10 mites onto every cucumber plant within a few days of planting. Subsequently, two predatory *Phytoseiulus persimilis* mites are introduced onto the same plants, and a continuing natural balance is established below the economic injury level. This program can be modified to fit particular circumstances. The greenhouse whitefly (*Trialeurodes vaporariorum*) is commonly controlled by rearing its parasite *Encarsia formosa* in an insectary, introducing whiteflies early at the rate of ten for every fifth cucumber plant, then placing ten parasites on the same plants at three bi-weekly intervals. Even lower numbers are required on tomatoes. However, the temperature must average over 18 °C; development of tomato varieties that produce well at lower glasshouse temperatures may create a problem, as *Encarsia* is less efficient under these conditions.

The cotton aphid (*Aphis gossypii*) on cucumber was successfully controlled by releasing one third-instar larva of the ladybeetle *Cycloneda sanguinea* for every 20 aphids, followed by the introduction of small cotton plant seedlings bearing about 50 aphids, half of which were parasitized by *Aphelinus flavipes*. This aphid can also be effectively controlled with preparations of the pathogenic fungus *Verticillium lecanii*.

Biological control of thrips is a recent development. The onion thrips (*Thrips tabaci*) can be a serious pest of glasshouse sweet peppers in the Netherlands. A predatory mite, *Amblyseius cucumeris*, is mass-reared on a flour mite and, because it can survive on alternative prey, is released at the rate of 50–60 mites on every seventh plant even before the thrips are present in the glasshouse. This results in effective biological control.

With glasshouse chrysanthemums, boxes of rooted cuttings may be artificially inoculated with both spider mites and *Phytoseiulus* prior to planting. The same boxes may also be 'seeded' with green peach aphids (*Myzus persicae*) that have been exposed to the parasite *Aphidius matricariae*. This ensures control of both pests until the flowers are harvested. The chrysanthemum leafminer (*Chromatomyia syngenesiae*) may be controlled by early introduction of the endoparasite *Dacnusa sibirica* at the rate of three adults per 1000 plants, followed by a release of the ectoparasite *Diglyphus isaea*. Caterpillars may be controlled by *Bacillus thuringiensis*.

Yield increases of glasshouse cucumbers of at least 20 percent, amounting to US $18 000 per hectare, have been reported where biological control was used. These yield increases are attributed partly to superior red spider mite control by the predatory mites, as compared to chemical control, and partly to the lack of phytotoxicity under a biological control program, whereas with the alternative chemical method frequent applications of petroleum oils and acaricides cause plant damage. In 1979, the cost of biological control of the greenhouse whitefly on tomatoes was only 9 cents per square meter, as compared to 25 cents for chemical control.

Although growers may be initially reluctant to adopt techniques that involve deliberately infesting their glasshouse crops with injurious pests, they have been turning to this method in increasing numbers. By 1985, biological control was applied on about 8000 ha of glasshouses in some 23 countries, mainly in Europe. In the Netherlands, where more than half of this activity takes place, the greenhouse whitefly is controlled by parasites on 50 percent of the total area of glasshouse tomatoes, spider mites are controlled by predators on 85 percent of the cucumbers, and thrips are controlled by predators on more than 50 percent of the sweet peppers. Several commercial companies produce the natural enemies and sell them to the growers, along with the required specialized guidance, which is included in the price. (See also Bal & van Lenteren, 1987.)

It will not always be easy for the public to obtain positive answers as to the best natural enemy for a particular problem, or where to obtain them. Some likely sources of such information include ecologically-oriented entomologists of State University Agricultural Experiment Stations, of other universities, of the United States or other national Departments of Agriculture, the state or provincial Departments of Argriculture, county agents or extension specialists. Once an individual knows the correct natural enemy or enemies for a particular pest problem he may either purchase it, if available, or personally obtain it from naturally occurring infestations of the same pest. Obviously, a certain amount of expertise is needed, but this is not beyond the scope of the interested, intelligent layman.

One of the best sources of information now available will be found in the International Organization for Biological Control (IOBC). One of the main objectives of this global organization is the dissemination of information and the promotion of biological control in all its aspects. Any individual interested in biological control can join the IOBC for a nominal annual fee and receive the journal *Entomophaga* and a newsletter, which are devoted to articles and information on biological control. Information regarding the organization may be obtained from the President, Mr J. R. Coulson, Beneficial Insects Introduction Laboratory, USDA, BARC East, Building 476, Beltsville, MD 20705, USA or from the Secretary General, Dr J. P. Aeschlimann, CSIRO Biological Control Unit, 335 avenue P. Parguel, 34100 Montpellier, France. Currently there are four regional sections, namely, the West Palearctic, East Palearctic, South and East Asian and Nearctic Regional Sections, and two others are being organized for Tropical Africa and Latin America. The West Palearctic Regional Section (WPRS) is the oldest and most active, maintaining more than 20 international working groups that collaborate in the study and implementation of various aspects of biological control. The names and addresses of the Presidents and Secretaries of the various Regional Sections are listed on the inside cover of *Entomophaga*. Franz (1988) has recently reviewed the history of IOBC.

Lists including some important sources of natural enemies that are available commercially in the United States and elsewhere have been presented by Ridgway & Vinson (1977*b*) and Hussey & Scopes (1985), appear periodically in *The IPM Practitioner* (a journal published by the Bio-Integral Resource Center, Berkeley, California), and also are available on request from the State of California Department of Food and Agriculture, Biological Control Services Program, 3288 Meadowview Road, Sacramento, CA 95832, USA. The Association of Applied Insect Ecologists, 1008 10th Street, Suite 549, Sacramento, CA 95814, USA, can furnish commercial sources of natural enemies and lists of consultants on biological and integrated control. Species that can be obtained commercially include green lacewings, ladybeetles, predatory mites, praying mantids, *Trichogramma* egg parasites and various other parasitic wasps. One of the largest and most reliable private insectaries producing a variety of insect natural enemies for sale to the public is Rincon-Vitova

Insectaries, Inc.,* PO Box 95, Oak View, CA 93022, USA. All of the natural enemies they produce have been demonstrated to be effective when used under the proper conditions. To aid the public in utilizing these beneficial insects to their greatest advantage, this firm issues information leaflets regarding each. The following are statements from their *Technical Bulletins* concerning green lacewings, which are proving to be one of the better kinds of natural enemies for periodic mass release in the garden, orchard and farm; and concerning parasites for biological-integrated control of pest flies. †

GREEN LACEWING

Versatile and voracious, the green lacewing is used as an efficient method of biological control

The Green Lacewing, *Chrysopa carnea* and *Chrysopa rufilabris* species, are noted for their truly voracious appetite. Attacking almost any soft-bodied insect that crosses its path, this beneficial insect is used in a wide variety of agricultural and horticultural situations making it an extremely versatile predator. The lacewing is used primarily for the control of aphids, mealybugs, thrips, mites, small larvae, caterpillar eggs, scales, leafhopper nymphs, etc. Lacewings are used in place of, or in alternation with some pesticides, and are considered a highly effective alternative for pest control.

Biology and Lifecycle. Lacewings are considered to be the most important beneficial insect from the insect order Neuroptera because of their usefulness to farmers, greenhouses, and backyard gardeners.

Because the Lacewing larva has such a ravenous appetite from birth, it is necessary for the female to lay each egg on top of a hair-like stalk in order to keep the young from eating each other at birth. This method of egg laying also helps prevent ants and other enemies attacking the Lacewings before they hatch. The female may lay up to 300 eggs over approximately three weeks. In 4 to 7 days, the eggs hatch into 'alligator-shaped' larvae that will begin [to feed upon a] wide variety of insect pests. They have been known to feed upon 60 aphids in one hour. Lacewings use their sharp pincer-like mandibles to pierce the bodies of their prey, and then suck out the body fluids. After feeding for approximately three weeks, the larva pupates by spinning a cocoon with silken thread. In about 5 to 7 days the adult emerges, cutting a hole in the

---

* No intent to endorse products of the firms mentioned over certain others is intended.
† Reprinted by permission of J. Everett Dietrick, Rincon-Vitova Insectaries, Inc.

cocoon top. As an adult, the Lacewing will mate and begin to lay eggs. This cycle is directly influenced by climatic conditions; in Summer a complete life cycle can occur within 28 to 30 days. Lacewings overwinter as pupae or adults. Lacewings should be released early each Spring to attain the large numbers necessary to quickly control pest species.

Lacewings should be released in the garden, grove or field where ample prey is available. There, they will lay eggs and provide several additional generations of predators. Lacewing larvae tend to be nocturnal in feeding habits, and usually stay out of strong direct sunlight that could dehydrate them. Ample humidity is important.

A few main nutrition sources should be available to adult Lacewings. These include flowering plants that provide pollen and nectar, honeydew that is produced by species of Homopteran insects (aphids, whiteflies, scales, mealybugs, etc.), and free water given off by evapotranspiration of plants and by dew.

How it works. Lacewing eggs that are shipped to the customer are hatching or close to hatch by the time the customer receives them. The eggs are generally mixed with a packing material, either rice hulls or ground corn grit, and moth eggs as a food source. The packing material serves two functions: 1. to provide separation for the hatching larvae to minimize cannibalism, 2. to increase the volume of the contents to facilitate application and distribution of the insects.

After receiving the eggs, the customer is advised to keep them between 80° and 90°F [26.7°–32.2°C] to ensure a timely hatch. Refrigerating the eggs to delay hatching is *not* recommended unless absolutely necessary; they should be ordered in quantities as needed, and released *as soon as possible* after the majority of eggs have hatched. When freshly laid, the eggs are pale green. When held at 80°F, they are grey in color by the third day and by the fourth or fifth day, the larvae hatch, and the hatched eggs are white. The tiny larvae can be seen crawling on the sides of the container. Soon after hatching, they crawl around the container searching for food and may not be easily visible. They must be released within 1 or 2 days of hatching to avoid cannibalism.

The larvae are quite hardy, and can travel as far as 80 feet [24 m] for their first larval meal. Moving through the foliage, they may travel some 6 or 7 miles [10–11 km] in their larval stage.

Method of application. Lacewings may be applied by hand to very small areas, and by hand-held blower on small acreages. Larger acreages are applied by ground rig/hopper system, and by helicopter or fixed-wing aircraft. Techniques and equipment recommendations are available.

The amount of Lacewings to apply depends on a number of factors including pest population, climatic conditions, anticipated application of pesticides, time of season, etc. The general release recommendations start at about 5 000 per acre [12 500 per hectare], and subsequent applications are used in most crop situations, with up to 20 000 Lacewings released per acre per season.

When to start. Pest management systems based on biological control must be implemented as early in the season as possible to insure good results. It is very difficult attempting to start a bio-control program after the pest has reached damaging levels.

In the interest of providing timely delivery of product in the quantity desired, we recommend that the customer contact us in December or January to place the order. This gives us time to help you design your pest management program and schedule production for delivery 3 or 4 months later.

Helpful Hints. 1. Call early in order to schedule well in advance for the season. 2. Make releases in the cooler times of day. This will allow any unhatched larvae a better chance of surviving the hotter, dehydrating temperatures. 3. Call the insectary if you have any questions about the insects or problems with your shipment.

FLY PARASITES

Integrated pest management of filth flies utilizing beneficial insects

This information was compiled to address pest fly management in commercial livestock facilities such as feedlots, dairies, race tracks, poultry ranches, horse ranches and hog operations as well as wastewater treatment facilities. The integrated pest management system of using beneficial insects is fast becoming a popular, cost-effective way to control pest flies.

It has long been known that there are many insects that are beneficial to man. Some of these insects can be used to control pest insects. Fly parasites are natural enemies of flies, and there are species present throughout the world, wherever flies breed. Fly parasites are beneficial insects that either consume or lay their eggs in the immature stages of flies (larva and pupa). They serve as an important preventative measure, since a female fly killed in the immature stage [is prevented] from laying up to 800 eggs as an adult.

How it works. The need for chemical sprays can be greatly reduced by adopting a program of Biological Fly Control. Such a program involves periodic releases of fly parasites, aimed at destroying flies at their sources, or breeding sites.

Fly parasites are very small, harmless insects that nature has programmed to attack and kill flies. They lay their eggs inside the immature stages of flies in the breeding site, and the fly becomes food for their young. Since they eat only flies, they will not bother anything else. And their need for flies, in order to reproduce, provides a strong and natural incentive to do all the work: search and destroy. They do not bite, sting, or swarm, and will not become a nuisance to horses. And because they are nocturnal, these small, gnat-sized wasps are rarely seen during daylight hours unless one searches carefully in areas where flies are breeding. They will attack common house flies, the biting stable fly, blow flies, garbage flies, little house flies, horn flies, flesh flies, and false stable flies. These are the most common types of filth-breeding, irritating flies around livestock and poultry.

Fly parasites are able to complete a generation every two to four weeks, giving rise to an increase in their own population as they reduce the number of flies. For several reasons, however, this population of fly parasites needs continuous reinforcement to maintain a high level of fly control. First of all, the flies have a distinct advantage in actual numbers produced. A single fly will lay up to 800 eggs, but a single parasite will attack less than 50 pupae. The life cycle of a fly is also much shorter than that of the parasite, and this contributes to the pest's advantage in reproductive capability. Also, the size and strength of the fly are important advantages that enable it to travel greater distances, and to some extent, resist the effects of pesticides more effectively. Therefore, to offset the fly's natural advantages, a high level of parasites must be maintained in the breeding site.

Method of control. Shipped to the customer in the form of parasitized fly pupae, the insects are easily handled as they are still developing within the fly pupa they are consuming. Once their growth cycle has been completed, they cut a hole in the pupal casing, and leave as adults to search out and parasitize more flies. Releases are usually done when the first of these adults begin to emerge and involves nothing more than sprinkling the pupal cases in and around an area of fly breeding. Once released, each insect will then kill 40–50 flies, and in a period of two to three weeks, another generation of beneficial insects is born. To enhance the effectiveness of our program, we include several different species of fly parasites in our shipments. Each of the species differ slightly in the host fly they prefer, the depth of search in manure strata, and preferred geographic/climatic conditions.

To achieve the greatest reduction of flies, one should employ a combination of measures that work hand-in-hand. This is known as Integrated Fly Control, which would include manure management, trapping, etc.

When to start. Fly control should be preventative. It is much easier to prevent a buildup of flies than it is to get rid of them. Starting a Biological Control Program as early in spring as possible is critical to the success of Integrated Fly Control.

We recommend releasing the required amount of fly parasites weekly or semi-monthly, depending on your level of fly infestation. For the best results, order your first shipment in early spring, before the flies are numerous.

If the fly population in your area increases rapidly at any time during the year, you should (3 weeks prior to the time) order a special shipment of 2 to 5 times your usual amount of fly parasites. A special shipment should also be ordered if you fail to start the program before the flies are numerous. These special shipments, along with the use of traps and baits, will reduce the fly population quickly.

Helpful hints. 1. Cultural control – good sanitation practices. No fly control program will be successful if manure is allowed to pile up around the premises. Also, elimination of as many wet areas as possible and the use of hydrated lime on the areas where your animals urinate frequently are also very helpful. 2. Avoid applying poisons to the breeding habitat – which does more harm than good, as fly parasites and other beneficial organisms are thus destroyed. 3. Complementary methods. The use of traps and poisonous baits in non-release areas can be useful. This way the poison can help knock down adult flies without disrupting the parasites in the breeding areas.'

The largest commercial producer of natural enemies in Europe, Koppert BV, Vellingweg 64, 2651 BE Berkel en Rodenrijs, The Netherlands, distributes similar information. The following are excerpts concerning pesticide-resistant predatory mites; and concerning whitefly parasites★.

Phytoseiulus persimilis, the efficient red spider control

The red spider mite (*Tetranychus urticae*) is one of the many pests in glasshouse horticulture. The mite is a virtually ever present potential danger to many vegetable and flower crops. It sucks the underside of the leaf and causes yellowing of the leaves.

★ Reprinted from a 1986 publication by permission of P. C. Koppert.

The red spider mite can be effectively controlled with the aid of the predator *Phytoseiulus persimilis*. Koppert have used it in glasshouse horticulture for many years with good results

The predatory mite, originating from the sub-tropics, is a lively, shiny mite coloured pink to red. Immediately noticeable are the long-legged body and the fast movements of these mites. The predator actively hunts for its prey in the crop and can travel quite appreciable distances. An adult may devour some 20 young or 5 adult spider mites a day. The female predators require this nourishment to produce eggs. A female can lay 3 to 4 eggs a day, i.e. an overall production during her life of 50 to 60 eggs. The eggs hatch after 2 to 3 days, the predator subsequently reaching adulthood through several larval stages. It therefore takes some 5 to 7 days overall for the adult mite to develop, almost twice as fast as the red spider mite.

The predator is totally dependent on the red spider and does not harm or interfere with any crop.

Application. *Phytoseiulus persimilis* can be used to control red spider mite in most crops under glass or plastic. The only conditions to be observed are the relative air humidity, which may not be too low, and the temperature, which must be regularly above 20 °C.

The predator is applied mainly to cucumber and pepper crops. On a small scale, it is also used on tomatoes, beans, gherkins, aubergines, melons, grapes, strawberries and various flower crops. In southern regions, the predator can also be used to control the red spider in fruit trees.

The predatory mite should be introduced as soon as the presence of red spider is detected. This often means starting in a particular spot and subsequently providing the entire establishment with predators. Generally, 8 to 10 predators have to be introduced per m².

The predator will keep the red spider under control throughout the growing season. It is hardly affected by controlling other diseases and pests. For example, organo-phosphorous compounds can be used besides the predatory mite, without harming it.

Product. Koppert supplies the predatory mite (SPIDEX) in handy shaker bottles. The predators can easily be spread rapidly and uniformly through the crop by means of these shaker bottles. Use of these bottles has important advantages over supply of predators on leaf material. Leaf material bears the disadvantage of being not only difficult in distribution but also in practicability in production and transport. Moreover, there is also the risk of other pests and diseases being transmitted by the leaves.

Encarsia formosa, the parasite with a long residual action

The greenhouse whitefly (*Trialeurodes vaporariorum*) is a very common pest in glasshouse-cultures. The adult whitefly looks like a small moth, 1.5 mm in length, and is easily seen at the top of the plant. It lives on the underside of the leaves and has a highly characteristic behaviour – when the plant is shaken a cloud of whitefly will fly up and soon after they will return to the underside of the leaves. The larvae are located lower on the plant, again on the underside of the leaves. Throughout their development they stay on the same leaf.

Both the adult whitefly and its larvae obtain their food from the plant. The larvae in particular excrete a great deal of honeydew. This makes the crop sticky and causes damage through inhibited growth and dirty fruits. The sooty moulds growing on the honeydew, in particular, seriously damage the plant.

Encarsia formosa is a small parasitic wasp about 0.6 mm in size, with a black thorax and a yellow abdomen. It lays its eggs in the whitefly's larvae. The egg develops within and at the expense of the whitefly larva. After some 10 days, the larva turns black. Parasitization of the whitefly will then be quite evident and instead of a whitefly, a parasite will emerge from the pupa. This then continues the work of its predecessor.

At a temperature of 20 °C, the parasites develop from egg to adult within 25 days. The parasite also feeds on the younger larvae of the whitefly, which then die.

Application. When applying *Encarsia formosa* against the greenhouse whitefly it is important that the system of repeat introductions should be commenced as quickly as possible, as soon as the first whiteflies are found. Each crop demands a different approach for successful control of whitefly. The number of introductions and intervals between them therefore differ from crop to crop. The parasite is used widely in tomato growing and to a lesser degree in cucumbers and aubergines. Application of *Encarsia formosa* is also feasible in many other protected crops provided that this fits within the integrated control scheme. The parasite is very susceptible to many chemical products. Caution must therefore be taken that the control of other pests and diseases is proceeded with the utmost of care. The yearly issued 'Koppert spray cards' provide you with full information on this point.

Product. *Encarsia formosa* (EN-STRIP) is supplied by Koppert on paper cards. Each of these cards contains the black pupae from which the parasites emerge soon after the introduction in the glasshouse. The cards can be hung quickly and easily on the plants. The advantage of the system is that it is easy to apply, the material can be safely transported and the product is delivered clean.

In connection with green lacewings, it should be noted that Dr Kenneth Hagen and colleagues at the Division of Biological Control, University of California, Berkeley, have found an effective supplementary food for these predators and others, Wheast^R, which is available commercially. Dr Hagen is a pioneer in the development of supplementary foods for entomophagous species. He has stressed the absolute necessity of such species either having natural food available in their habitat, or of man supplying it artificially. Experiments carried out with natural populations in the field, using sprays of Wheast^R plus sugar, honey, or molasses plus water, increased the abundance and oviposition of green lacewings, ladybeetles and syrphid flies and reduced the abundance of various pests on several crops, as compared to unsprayed plots. (See Hagen, 1986.)

A source of natural enemies, primarily for importation and establishment purposes, is the CAB International Institute of Biological Control (CIBC), which maintains highly professional insectaries in various parts of the world. Headquartered in England, they have permanent stations in Switzerland, India, Pakistan, Kenya and Trinidad, and a base in Malaysia, and furnish a wide variety of natural enemies on a contract basis to various governments, organizations or even individuals. CIBC also provides information and advice on biological control, publishes the abstract and news journal *Biocontrol News and Information*, holds training courses and undertakes research and implementation programs on biological control of major pests and weeds. Information may be obtained by writing to Dr D. J. Greathead, Director, CIBC, c/o Imperial College at Silwood Park, Ascot, Berkshire SL5 7PY, United Kingdom. Another possible source is the recently established Biological Control International, Inc., which specializes in foreign exploration, importation, surveys and evaluation of natural enemies. Information may be obtained from Mike Rose, Vice President and General Manager, BCI, P.O. Box 1569, College Station, TX 77841, USA.

Political action

Individuals, organizations, farming industries and other commercial enterprises interested in fostering biological control for their own use or as a means of reducing chemical pollution of the environment are greatly hampered by the general lack of information and emphasis on biological control furnished by federal, state, university and commercial agencies dealing with pest control. Political action is a potentially powerful means of increasing support for and the use of biological control. Various conservation, environmental protection and consumer organizations are already capable of influencing legal decisions, legislation and financing in the right direction. Interesting and informative accounts of such possibilities are given in the books *Since Silent Spring* (Graham, 1970, especially Chapter 21), *Unfit for Human Consumption* (Harmer, 1971, especially Chapters 16 and 17), *Politics and Pesticides* (Tallian, 1975, especially Chapter 23) and *The Pesticide Conspiracy* (van den Bosch, 1978). The environmentally oriented public can and should effectively press the issues of clean food *vs* poisoned food, the right of a farmer not to have his crops receive drift of toxic chemicals applied by another nearby farmer, the adverse effects of the use of non-selective and environment-polluting pesticides [as the Environmental Defense Fund (EDF) and other organizations have done successfully with DDT and are continuing to do with other dangerous pesticides], and the setting by the agricultural industry and government of ridiculously high standards for 'cosmetic' appearances of foodstuffs even though no actual damage or loss of crop occurs, so that the farmer is forced to apply un-needed insecticides or fungicides in order to produce a product without a single blemish. Taste and nutritional quality should be the goals, but they are often sacrificed in today's market.

In Europe, the West Palearctic Regional Section of IOBC has been pressing for legislation that would grant official recognition of an 'IPM label' for agricultural produce grown without resort to broad-spectrum chemical pesticides. This would be an important step toward reducing pesticide usage in the European Community.

It is generally estimated by ecologists that at least half of the insecticides now used are unnecessary. With proper emphasis on biological control there is no telling to what minimum the use of insecticides can be reduced. Such emphasis can be greatly increased by group pressure for expansion of research on and support of biological control by federal, state and local governments and universities.

9

Other biological methods

Other non-chemical approaches to pest management include four methods of proven practical usage in the field, as well as several still in the idea or pilot-testing stage. The four main methods now in practical use include: (1) the development of plants resistant to insects, plant diseases and nematodes, (2) cultural techniques or habitat modification designed to control pests, (3) the genetic technique of sterile-male releases in the field in order to mate with wild females and prevent progeny production, and (4) the use of insect pheromones to attract and trap or otherwise neutralize pests. The competitive displacement of pests by non-pests has emerged from the idea stage in at least one major project. Genetic techniques such as habit modification, use of lethal genes, etc. mainly await the first pilot tests.

There are other potentially useful control techniques which may be borderline between being a biological method or a chemical method (which indeed pheromones may be). These include the insect growth regulators – juvenile hormones or their analogs, chitin synthesis inhibitors, etc. – which may be used to prevent successful completion of the life cycle or otherwise disrupt physiological processes. These much publicized 'third-generation pesticides', however, are still mainly in the pilot testing stage, although some have been registered for use.

A promising technique, albeit of limited use, is the post-harvest rinsing of fruit to remove 'cosmetic' damage. Still other approaches that are non-biological but may be ecologically non-disruptive include use of electromagnetic energy, especially in light traps, which have shown promising results in some large-scale experiments with the tomato and tobacco hornworms (*Manduca* spp.) and other pests. Other methods, none of which have more than limited application, include mechanical exclusion, heat, cold, and light and sound to disrupt biological patterns; use of chemosterilants, which prevent reproduction but are not yet developed as a practical commercial method and there remains a serious problem of safety in field applications; use of repellents, which as yet have no application in agriculture but are used against certain pests of public health or veterinary importance; and antifeedants, which also have not as yet been developed to the stage of commercial usefulness, with one or two exceptions. Of course, another chemical technique that can be ecologically non-disruptive involves the use of selective insecticides. This actually is of much greater immediate practical importance than any of the other chemical or physical methods mentioned above, and will be discussed further in Chapter 10. The remainder of this chapter will be devoted mainly to the four biological methods mentioned in the first paragraph and an evaluation of their relative potential as compared to each other and to biological control. Some of the other methods will also be discussed. Useful general references covering these subjects are Burn, Coaker & Jepson (1987), Flint & van den Bosch (1981), Huffaker & Messenger (1976), Kilgore & Doutt (1967), Knipling (1979), Metcalf & Luckmann (1982), National Academy of Sciences (1969, 1975) and Pimentel (1981).

Plants resistant to pests

This method would appear to offer such an excellent means of controlling insect pests that it seems surprising that its application has been relatively limited in the overall scheme of pest management. The principle involved consists of artificially altering the plant species genetically, or selecting naturally available plant varieties, so that the plant is not as satisfactory a host to the pest as it was originally. The changes can be classed as physical or biochemical. The former include such characteristics as hardness or hairiness, the latter involve chemicals in the plant that might repel the insect or mask the plant as a host (*nonpreference*, or *antixenosis*), or have adverse effects on the development, survival or reproduction of the insect (*antibiosis*). Resistance may also be due to the ability of the plant to withstand the pest's attack (*tolerance*). These types may be interdependent and appear in various combinations. The resistance of wheat to the cereal leaf beetle (*Oulema melanopus*), for example, is due to the presence of pubescence on the leaves, which deters adults from ovipositing, as well as reducing egg hatch and larval feeding. Resistance to the wheat stem sawfly (*Cephus cinctus*) is largely due to solid stems as contrasted to hollow stems which permit greater egg and larval survival. Resistance to the Hessian fly (*Mayetiola destructor*) is principally antibiotic in nature as most larvae die when feeding on resistant plants. Resistance to the greenbug (*Schizaphis graminum*) is mainly due to the ability of the plant to tolerate aphid attacks. The principle applies as well to resistance of animals to various parasites. Levels of resistance of practical value to internal parasites have been shown in cattle, sheep, goats and chickens, but even less work has been done here than with plant resistance.

The earliest and perhaps most spectacular example of this method was the use of resistant rootstocks to control the grape phylloxera (*Daktulosphaira vitifolii*) in Europe in the 1870s. The development of plant varieties resistant to the Hessian fly, the European corn borer (*Ostrinia nubilalis*), the spotted alfalfa aphid (*Therioaphis maculata*) and the boll weevil (*Anthonomus grandis*) are outstanding examples in the United States (Beck & Maxwell, 1976). Others are the wheat stem sawfly, the greenbug and the chinch bug (*Blissus leucopterus*). In Hawaii, the development of sugar-cane varieties resistant to the beetle borer (*Rhabdoscelus obscurus*) was credited, along with introduced parasites, with successful control. There are various examples around the world, including ones that have been empirically adopted from naturally occurring resistant varieties. Success in breeding resistance to *Empoasca* leafhoppers into cotton has been achieved in Africa, Australia and India. More than 20 million ha of rice land in Asia are planted with varieties resistant to the brown planthopper (*Nilaparvata lugens*) (Oka, 1983). Some other plant varieties more or less resistant include: corn to the corn leaf aphid (*Rhopalosiphum maidis*), corn earworm (*Heliothis zea*) and corn rootworms (*Diabrotica* spp.); alfalfa to the pea aphid (*Acyrtosiphon pisum*); rice to the striped borer (*Chilo suppressalis*), the green leafhopper (*Nephotettix virescens*) and the rice delphacid (*Sogratodes orizicola*); beans to the Mexican bean beetle (*Epilachna varivestis*); and cabbage to the potato leafhopper (*Empoasca fabae*) and the striped flea beetle (*Phyllotreta striolata*).

The economic savings from certain projects have been enormous. The total cost of research to develop resistant varieties for the Hessian fly, the wheat stem sawfly, the European corn borer and the spotted alfalfa aphid was about US $9 300 000, resulting in annual savings of $308 million, or a return on each dollar invested of $300 over a ten-year period (Bottrell, 1979). About eight million ha of wheat are now being planted with some

40 Hessian-fly-resistant varieties in the United States. Painter (1968) concluded that $15 million a year in increased yield was being attained in Kansas alone from the use of Hessian-fly-resistant varieties, and Luginbill (1969) calculated the savings for the entire nation for the year 1964 to be $238 million. Reduction in damage by the wheat stem sawfly has resulted in similar great savings. No chemical control of this pest has been economically possible, but the use of the resistant varieties Rescue and Chinook permits the profitable growing of wheat on 720 000 ha in Canada and 600 000 ha in the United States. Sawfly populations were reduced to a minimum over a period of several years. Previously, losses in the Great Plains area of Canada and the Unites States ranged up to 75–85 percent. A $6 million loss was estimated for North Dakota alone in 1944. It has also been estimated that the use of resistant corn hybrids, particularly for the European corn borer (about 7.4 million hectares in 1975), has saved about $100 million per year in the United States alone.

It is impossible in such a short review to include details of the principles or the research involved in plant resistance to pests, nor can all sources be cited. The following will be found useful to the interested reader: Beck & Maxwell (1976), Harris (1980), Harris & Frederiksen (1984), Kennedy *et al.* (1987), Kogan (1982), Maxwell & Jennings (1980), National Academy of Sciences (1969), Painter (1951, 1958), Tingey (1981) and van Emden (1987).

The Kansas Agricultural Experiment Station, under the leadership of Reginald H. Painter, was the pioneer in the United States with a formal project since 1926. They have approved for distribution to farmers numerous crop varieties resistant to various insects, including varieties of sorghums, alfalfa, corn, barley and wheat (Painter, 1960). Other leading institutions include Cornell, Purdue, Texas A & M, Oklahoma State and North Carolina State Universities, the USDA, and such international centers as the International Rice Research Institute (IRRI, Los Baños, the Philippines), International Crop Research Institute for the Semi-arid Tropics (ICRISAT, Hyderabad, India), Centro Internacional de Agricultura Tropical (CIAT, Cali, Colombia) and Centro Internacional de Mejoramento de Maíz y Trigo (CIMMYT, El Batán, Mexico).

According to Painter (1960), the following are the important characteristics of insect resistance as a pest control measure:

'1. In contrast to chemical control, which is sudden and often rapidly decreasing in effect and often presents a toxic residue problem, resistance is cumulative and continuing and is without residue problems. The combined use of plant resistance and insecticides may therefore give greater than additive control and perhaps be longer lasting.

2. The breeding of insect-resistant varieties requires the continued cooperation of plant breeder and entomologist.

3. The production of a resistant crop plant variety requires ordinarily ten to fifteen years, although sometimes there are means of shortening this time.

4. The effect of a resistant variety has been to increase the yield or quality of the crop and often to decrease the insect population even in neighboring susceptible varieties.

5. The use of resistant varieties involves no direct cost to the user and perhaps adds to the value of an otherwise improved variety. Where the margin of profit per acre is small the use of resistant varieties reduces the cost of production and increases stabilization.'

A distinct advantage of using resistant plants is its compatibility with other methods of control. Insects feeding on resistant plants may be less vigorous and more easily killed by weather or more easily handled, if necessary, with reduced amounts of insecticides. Predation or parasitism has been shown to be greater on some resistant varieties, and as a rule natural enemies are given a relative advantage if the rate of increase of their prey is diminished, as it is even on moderately resistant plants. Thus, even a slight increment in resistance could lead to an increase from partial to complete biological control. Pests stressed by antibiosis may be more susceptible to certain natural enemies, but on the other hand antibiotic chemicals acquired from the plant may adversely affect the natural enemies, and certain morphological traits of resistant plants may deter natural enemies. (See Boethel & Eikenbary, 1986.)

Currently, plant resistance to over 100 species of pest insects is known in dozens of crops, and many instances are being studied in depth. Numerous resistant lines are being developed and incorporated into crop varieties. A few varieties have been developed that are resistant to more than one insect pest species and even to some pathogens. Needless to say, these will have greater practical utility and more ready acceptability to the grower. Additionally, and actually on a much grander scale, plant resistance has been developed to many plant pathogens and nematodes. This also is advantageous to biological control, because the chemicals otherwise used against those organisms can cause upsets.

If the method is so potentially effective, valuable, compatible with other pest control techniques and ecologically non-disruptive, as we concur it is, why hasn't it been utilized to a much greater extent? There are a complex of reasons.

The research involved usually requires a multidisciplinary approach which, although often relatively simple, is expensive. A large team effort may be required including entomologists, plant breeders, agronomists, geneticists, plant pathologists and economists. Large-scale governmental support may be necessary, and sometimes this should be co-ordinated into co-operative international programs. This, for instance, would be the only logical means of obtaining an adequate source of the enormous genetic variability available within a given crop. A correlative objective of such teams would be to check new plant varieties developed for other purposes for possible loss in resistance to pests. In cotton, for instance, considerable effort has been devoted to developing commercial varieties having low gossypol content because this chemical is toxic to non-ruminant animals and has disadvantages in processing cottonseed. However, low-gossypol cotton has proved to be very attractive and susceptible to several cotton insects, and in fact such cotton is attacked by insects not previously known as cotton pests.

Painter states that the production of a resistant crop plant variety requires ordinarily 10 to 15 years, although it may be shortened in certain cases. However, with tree crops the time required may be even longer. Obviously, with a team effort involved, this means a large outlay of money before results are obtained. This time factor also may not be attractive to scientists who are often under pressure to 'produce'.

There are also continuing biological and ecological screening problems, for example that the development of plant varieties resistant to the target pest does not lead to reduced resistance to some other pest or potential pest or otherwise change the variety, or the ecosystem, in some undesirable way. In cotton, the same genetic factor that is responsible for high resistance to the boll weevil was found to increase susceptibility to the tarnished plant bug (*Lygus lineolaris*), leafhoppers, thrips and certain other pests. Similarly, strawberry varieties that were grown in California because of their resistance to pathogens lost much of this advantage when they were severely damaged by the cyclamen mite (*Steneotarsonemus pallidus*), which had been only a minor pest on the earlier varieties. Possible adverse effects on natural enemies should also be taken into account. There are also cases, such as the pea weevil (*Bruchus pisorum*) in

Finland, where rather careful programs of selection carried out for at least five years with a large number (94) of plant strains resulted in no increase whatsoever in resistance, so results are not guaranteed. In fact, resistance to oligophagous insects is often unknown in either the plant species or genus. For instance, this is the case with the saddle gall midge (*Haplodiplosis equestris*), a pest of wheat and other small grains.

Technical problems, such as the fact that screening methods are insufficiently developed for large numbers of plants and that the genetic basis of resistance to insects is often complex, pose additional difficulties. For example, in some aphid species, resistance even differs between the alate and the apterous generations.

Finally, once a resistant variety has been developed, there remains the formidable problem of grower and consumer acceptance. If, for example, the pest can easily be controlled by insecticides, there may be considerable reluctance to plant a new variety. Habits are not easy to change. It is probably not a coincidence that the most outstanding results have been obtained against pests for which no satisfactory chemical control existed – the Hessian fly, the wheat stem sawfly and the grape phylloxera.

Even after a resistant variety has met with commercial acceptance, the development of new biotypes, or physiological races, of the pest may occur. The artificial development of resistant plants is a genetic and evolutionary procedure, and similar events can just as well occur naturally in the pest to overcome the plant resistance, especially if the resistance is governed by a single dominant gene. In the Hessian fly, at least 13 biotypes have been identified in the field and laboratory, and as many as nine may be present in some local populations (S. G. Wellso, personal communication, 1989). Selective breeding for plant resistance may therefore become a continuing process. Thus far, biotype development has not occurred nearly as rapidly or as generally with insect pests as it has with fungus diseases, such as wheat rust, which has several hundred biotypes. van Emden (1987) recorded only 14 pests which had developed biotypes to overcome plant resistance. Such 'virulent' biotypes usually do not appear generally over the range where the resistant varieties are grown, and they may be controlled by additional sources of resistance. In fact, resistant wheat varieties are now available for all known biotypes of Hessian fly. No evidence of the development of biotypes in the wheat stem sawfly with respect to the famous resistant variety Rescue has been observed. This is true also with corn resistant to the European corn borer, although some of the varieties have been grown continuously since 1949.

Leafhopper-resistant varieties of cotton developed in Africa have remained stable for more than 50 years, and the apple variety Winter Majetin, which was recorded as resistant to the woolly apple aphid (*Eriosoma lanigerum*) as early as 1831, still retains its resistance. Thus, biotypes represent a potential problem in any given case, but do not appear to be serious enough with insect pests to restrict increased research and development.

It can be concluded that this is an elegant method having potentially broad application for pest population regulation, in spite of some drawbacks. It is under-researched and undersupported. Large-scale co-operation and financing on a national and international basis is indicated. It has such obvious advantages it certainly should be exploited more thoroughly.

Cultural methods

This is one of the oldest methods of pest control, and was probably in successful empirical use before the underlying biological reasons were known. It is reported that, for centuries, Chinese farmers have used particular grain-planting dates to prevent damage by the rice borer, the wheat stem maggot and a pest of millet. The disposal of crop residues was recommended in Europe as early as 1897 in order to reduce overwintering populations of the European corn borer.

Cultural control involves changing agrotechnical and other practices in any of many diverse ways, in order to alter the habitat and make it less favorable for pest reproduction and survival. Effects can be direct on the pest or indirect, such as by favoring natural enemies or increasing plant tolerance, or they can include a combination. Habitat modification can involve manipulation of such parameters as planting or harvesting times, cultivation, plant spacing, irrigation, crop rotation, trap crops, habitat diversity, fertilizers, pruning, thinning, soil preparation, and sanitation. Manipulation of water levels is an example in aquatic ecosystems.

The research and application of cultural methods must be based on a biological and ecological foundation just as much as any other technique. A thorough knowledge of the pest's life history and habits is particularly essential, including precise information on its natural enemies. Similar knowledge of its plant or animal host and agrotechnical or husbandry practices may be just as necessary, so that any alterations made do not have undesirable effects. The discovery of the most vulnerable developmental stage or stages of the pest's life cycle is a prime requisite. Then the habitat, farming practices or others must be modified so as to strike at the pest in its most susceptible stage. At the same time an understanding of the ecosystem is necessary, because habitat modification deleterious to one pest could well favor one or more others.

Cultural control techniques were more commonly used before the advent of organic insecticides made such methods sometimes seem cumbersome and complicated. Now that very serious side-effects have arisen from the use of insecticides, there is a strong resurgence of interest in cultural methods. Many have been and still are quite effective. Today, as a rough estimate, there may be 100–200 instances of cultural control being used more or less effectively against a diversity of pests in various countries. However, there does not appear to be any general review of the number, type and effectiveness of cultural methods on a world basis, so it is difficult to compare the potentialities, costs and returns with other methods. There is no doubt that such methods may furnish the key to the achievement of overall satisfactory pest management in any given crop or area. In fact, studies of tobacco insects by R. L. Rabb and colleagues in North Carolina clearly suggest that since the suppression of hornworms by cultural methods, there are probably no real pests on that crop except occasional ones. Rabb's (1969) data on the tobacco hornworm (*Manduca sexta*) distinctly show the importance of planting dates, early stalk destruction and sucker control in reducing pest populations. The pink bollworm (*Pectinophora gossypiella*), a key cotton pest, is successfully controlled in central Texas almost entirely by cultural controls aimed at reducing overwintering populations. Various component techniques, including area-wide stalk destruction, legally set plowing-under times for the shredded crop residues, delayed uniform planting resulting in suicidal emergence of the adult moths before cotton fruit is available, preharvest defoliation, early harvesting, modern cotton-ginning operations, and sanitation around the gins, add up to provide satisfactory control. These practices are now mandated by law in Texas. (See Bottrell & Adkisson, 1977; Cate, 1985.)

Thorough discussions of this subject will be found in Coaker (1987), El Titi (1987), Flint & van den Bosch (1981), National Academy of Sciences (1969), Pimentel (1981) and Stern *et al.* (1976). Examples, derived from a variety of sources, follow. They have been listed under some of the general headings given in the second paragraph of this section.

Planting time

The crop can be grown when the pest is inactive, or so that the susceptible stage of the crop occurs when the pest is least abundant. The Hessian fly in the United States furnishes an outstanding example in use for many years. However, it only applies to the fall brood, not to the spring generation. The adults of the fall brood emerge and live for only 3–4 days. When the winter wheat crop is planted after this generation is over, few eggs will be deposited on the plants. Dates for delayed sowing of winter wheat have been established by entomologists in Hessian-fly-infested states. The method is strongly complementary to the use of resistant varieties, but also can be used alone. It has been estimated to have a potential saving, if rigorously followed on the entire wheat acreage, of $114 million a year against a total research cost of about $2 million and a yearly maintenance cost of about $20 000. This represents an exceptionally high return on the research investments, and costs the growers nothing. On the other hand, if the entire acreage were treated with chemicals, the cost would be about $80 million per year. This result alone should offer sufficient incentive for society and the responsible institutions to finance research on cultural methods on a much grander scale. Similarly, it has been estimated that delayed uniform planting of cotton to control the boll weevil (*Anthonomus grandis*) in the Rolling Plains region of Texas has had an average annual economic impact of $192 million in that region, with the impact on the entire State being more than $300 million (Frisbie & Adkisson, 1986).

Growing pickling cucumbers in the spring, before the pickleworm (*Diaphania nitidalis*) invades from the Gulf Coast states where it overwinters, provides control. The notorious pink bollworm (*Pectinophora gossypiella*) and boll weevil (*Anthonomus grandis*) of cotton can be held down by planting all cotton in an area within a short period of time, so as to prevent extension of the fruiting period which favors increases of these pests. Also, planting during the earliest favorable period for an area makes early stalk destruction possible and reduces overwintering populations. The early planting of corn in the northern United States may enable it to mature before the corn earworm (*Heliothis zea*) and the fall armyworm (*Spodoptera frugiperda*) can migrate north from the southern states, where they overwinter. The seedcorn maggot (*Delia platura*) can be controlled by timing planting so that the corn seed is not present when the females emerge and are capable of laying eggs. The sorghum midge (*Contarinia sorghicola*) is effectively controlled in the high plains of Texas if the crop is planted soon enough to bloom before the first week in August. In certain areas, the beet leafhopper (*Circulifer tenellus*) is likely to cause damage unless planting is delayed until after the spring leafhopper migration. In the Sudan, timing of planting and size of acreage of alternate hosts influence the extent of infestation by the *Heliothis* bollworm. In the USSR, late planting of summer wheat caused reduced infestations by a flea beetle and a thrips. Early sowing of wheat during May in India freed the crop from attacks by a gall moth, whereas sowings made in late June or early July were heavily infested. In Israel, postponing the planting of corn by several weeks has effectively solved the problem of maize rough dwarf virus, which is transmitted by leafhoppers but is inactivated in their bodies at high temperatures (Harpaz, 1972).

Cultivation

Methods of tilling the soil that coincide with susceptible life-history stages of pests have long been used. Not only is a thorough knowledge of the biology of the pest in question necessary, but the effects must be compatible with other faunal interactions and with good agronomic practice. As mentioned earlier, the whole ecosystem must be studied in relation to any given proposed habitat change. High mortality of overwintering pupae of the corn earworm is accomplished by autumn plowing. Deep spring plowing that buries all plant fragments destroys many larvae of the European corn borer, and the survivors crawl to the surface where they are exposed to weather and natural enemies and are further decimated. The wheat stem sawfly has been reduced by as much as 75 percent by turning-under of stubble. The grape berry moth (*Endopiza viteana*) can be greatly reduced by plowing so as to bury the overwintering cocoons under a layer of soil through which they are unable to emerge.

The best control for the pale western cutworm (*Agrotis orthogonia*) and some others in Canada is to summer-fallow fields and then *not* till them in August and September. The soil becomes crusted, and the moths will not lay eggs on it. On the other hand in Canada, clean summer fallowing every second or third year to eliminate plants on which wireworms feed, will reduce their populations to non-economic levels.

The cereal leaf miner (*Syringopais temperatella*) used to be the most important pest of wheat and barley in the eastern Mediterranean Basin. The newly-hatched larvae aestivate in the soil, and the introduction of deep plowing – a good agronomic practice for many other reasons – has completely solved the problem by exposing them to the hot dry weather. In the USSR, on the other hand, it was found that deep plowing destroys the egg parasites of a weevil pest, whereas surface disc tillage conserves the parasites and is more effective in reducing the pest.

Harvesting

Both time and manner of harvesting can be manipulated to reduce pest populations. Harvesting the crop just as soon as it is mature reduces damage to sweet-potatoes by the sweet-potato weevil (*Cylas formicarius elegantulus*), to Irish potatoes by the potato tuberworm (*Phthorimaea operculella*), to peas by the pea weevil (*Bruchus pisorum*) and to cabbage by the imported cabbageworm (*Pieris rapae*) and related species, as well as by the cabbage looper (*Trichoplusia ni*). Losses from the sugar-cane borer (*Diatraea saccharalis*) in heavily infested sugar-cane can be reduced if the cane is cut as early as possible. A selective cutting system works well in control of the western pine beetle (*Dendroctonus brevicomis*) in ponderosa pines in some areas. It involves the early detection, cutting and removal of those trees most likely to be attacked by the beetle.

In alfalfa, strip-harvesting on a chronological basis not only may give strikingly better pest management in that crop, but lygus bug infestations in adjacent crops may be reduced (van den Bosch & Stern, 1969). The alfalfa field is cut in alternate strips, so that two different ages of plants occur simultaneously in the same field and it becomes a much more stable environment than if an entire field were cut at one time and laid bare. One series of strips are cut when the others are about half-grown. The lygus bugs move into the younger strip when the older one is cut, instead of flying into adjacent crops as they would if the entire field were cut at once. The natural enemies also move with them and are conserved. Although the bugs lay eggs in the half-grown hay, the hay is harvested about 15 days later, hence many of the eggs and newly hatched nymphs are removed or destroyed. In Wisconsin, damage to alfalfa by the potato leafhopper (*Empoasca fabae*) could be reduced in most years by making the second cutting at the late-bud or very-early-bloom stage. The alfalfa weevil (*Hypera postica*) may be controlled by early cutting of the first and second crops. Similarly in Israel, cutting alfalfa for hay a few additional times, thereby postponing the seed harvest until later in the season, effectively prevents damage by the alfalfa seed chalcid (*Bruchophagus roddi*).

Irrigation and water management

Reservoirs, irrigation developments and natural aquatic habitats provide breeding sites for pest mosquitoes, biting midges, horse flies and others. Populations of these pests often can be minimized or eliminated by proper manipulation. Many techniques adapted to particular circumstances are well developed, including use of water-impervious canals to prevent seepage, use of drains to prevent ponding and to provide self-draining, the regulation of the amount of irrigation water used in fields so that none remains standing, management of reservoirs, etc. by completely removing vegetation from the summer fluctuation zone of the permanent pool, by artificially varying water levels to minimize conditions favorable to mosquito reproduction, and diverse others. Artificial impoundment is a practical means of minimizing the reproduction of salt marsh mosquitoes and biting midges in some marshes. For example, if the marsh is continually flooded by pumping or by stream diversion, the mosquitoes and midges will then breed only at the water's edge inside the impoundment, instead of throughout the marsh. Wave action and predation by fish at the edges may virtually eliminate mosquito breeding.

Irrigation or flooding can be manipulated to control various soil-inhabiting insects. In some countries sugar-cane borers in the seed cane are subjected to flooding as a means of control. Flooding of rice paddies for a minimum of five days after harvest substantially reduces larval and pupal populations of rice stem borers. The introduction of summer irrigation has effectively controlled the almond borer (*Capnodis carbonaria*), which used to be a serious pest of non-irrigated almonds in Israel. Manipulation of irrigation regimes has been shown in California to have a great impact on pest populations in cotton. However, irrigation may be a two-edged sword – increases in various pests have been recorded following irrigation or flooding. This again emphasizes the fact that no cultural technique should be considered only from the viewpoint of controlling a given pest. The whole agro-ecosystem must be taken into account.

Sanitation

This general technique has been successful against a number of pests. Destroying harvest residues or plant refuse may prevent damage to future crops. The method may cause direct mortality to vulnerable stages, or remove breeding or hibernating sites. As in other cultural techniques, a thorough knowledge of the ecology, biology, life cycle and habits of the pest must be obtained before the most scientifically sound methods can be developed.

The housefly is a good example of the need for sanitation. Satisfactory control by insecticides or any other means is virtually impossible if proper disposal or manipulation of animal dung, etc., is not carried out. Substantial reduction of cotton boll weevil populations has been obtained in Texas by early-autumn destruction of cotton plants, as soon as possible after harvesting but before the first killing frost. Such a procedure resulted in only 27 percent punctured squares by late July of the next season, as compared to 63 percent punctured squares where the plants were not destroyed. The former figure is economically acceptable. If the stubble of the harvested corn crop is plowed out of the soil in the autumn, low winter temperatures will kill the overwintering larvae of the southwestern corn borer (*Diatraea grandiosella*) and effect control. In areas of warmer winters, control can be obtained by plowing stubble under several inches of soil, so that the adults cannot emerge. The practice of cutting and burning cotton plants soon after harvest, as well as fall plowing, has been carried on for many years in order to destroy hibernating larvae of the pink bollworm. However today, area-wide stalk-shredding following harvest, along with plowing-under of the refuse, is looked upon as the best pink bollworm control technique. These combined operations can reduce late-autumn populations by at least 95 percent.

Control of the sweet-potato weevil is aided by destruction of sweet-potato crop residues by plowing during the winter to expose the roots and vines to freezing temperatures, by good sanitation of storage facilities and the elimination of infested potatoes prior to storage. Overwintering sugar-cane borers may be killed in large numbers by plowing-under the millable sugar-cane and high stubs left in fields after harvest. Similarly, destruction of tobacco stalks after harvest decimates tobacco hornworm larvae, tobbaco budworm (*Heliothis virescens*) and corn earworms, thus resulting in lower initial populations the following season. Removal and destruction of slash residues following logging operations is especially important in reducing the amount of breeding by several species of bark beetles and thus preventing outbreaks from developing. Similarly, destruction of dying trees and prunings may effectively prevent bark-beetle damage in orchards.

There are other applications too numerous to mention. Sanitation appears to offer wide potential in insect control and justifies more research emphasis, even though, alone, it does not always provide completely satisfactory control. It would appear to be particularly useful against most stalk-boring insects and rootworms of annual crops, where harvesting and plowing techniques may be manipulated to reduce pest populations.

Crop rotation

The principle involved here is to follow one crop by another botanically unrelated one, so that the pests of the first crop are unable to feed on the subsequent one. However, such techniques may or may not coincide with good agronomic practice, and each case must be decided on its own merits. Crop rotation tends to be most successful with pests having both a limited host range and powers of dispersal, as is the case with the cereal leaf miner in Israel. Larvae escaping deep plowing and emerging from aestivation into a non-cereal crop will not be able to survive, nor to migrate into a suitable crop.

Grass crops following legumes generally greatly reduce damage from white grubs. For example, the white-fringed beetle (*Graphognathus leucoloma*) reproduces freely on such legumes as peanuts, soybeans and velvetbeans, but adults feeding on grasses, including corn and small grains, lay few eggs and additionally the grasses are not damaged as much by a given number of larvae. Proper rotation gives adequate control; legumes followed by legumes results in rapid increase in this pest.

Space limitation precludes discussion of other proven techniques such as use of trap crops, habitat diversity, manipulation of fertilizer regimes, pruning, thinning and soil preparation. However, it is obvious that there are a great number of actual and potential pest control techniques available by manipulation of cultural practices. This is the only non-chemical method of pest control that approaches biological control in the number of successful applications.

There are some difficulties and disadvantages to the method. Conservatism and reluctance to change established ways may be formidable obstacles. The practical application of scientifically sound cultural techniques may remain difficult to put into practice because nearly complete adoption over a wide area may be necessary, especially where free-flying or migratory pests are involved. Also the measures are largely preventative and must be put into practice a considerable time previous to insect damage. This is a psychological disadvantage in getting users to adopt the method, because they prefer to see immediate results. There also is likely to be a reluctance to adopt a method that may not provide complete economic control. Then there is a formidable extension task in seeing to it that proper timing and application of the particular technique is carried out. From the entomologist's viewpoint, he must realize that any cultural control technique may have widespread interactions in the agro-ecosystem; so broad, yet basic, interdisciplinary long-term research is required. As in the case of insecticides, it must be determined that the treatment to reduce the target pest does not have other effects which are adverse. Also, as crop production techniques change or new varieties are introduced, new cultural controls may be necessary, so research must be considered a continuing proposition.

The advantages are that many cultural techniques can be integrated into normal practices with little or no additional expense to the grower. Some methods in use are so simple and cheap that growers do not even realize they are practicing pest control. Additionally, they do not result in pollution problems, nor are they subject to the development of resistance as are insecticides. It seems obvious that cultural methods must play an increasing role in integrated pest management.

Genetic methods

Genetic pest control connotes the manipulation and use of genetic material in a manner deleterious to pest insects. Except for the 'sterile-male technique' (also known as the 'sterile insect release method', or SIRM), these methods are still in the idea, laboratory or pilot-test stage. The potential applications are based on principles and techniques discovered and proven by geneticists working principally in laboratories.

Genetic methods can be divided into two categories: (1) autocidal methods, and (2) habit modification. In the former, lethal and sterility factors are incorporated into the field population in order to reduce its density. In the latter, the pest population is not necessarily reduced but its habits are changed, so that for example a mosquito species might be modified from a human-biter to an animal-biter. The principal obstacle to the realization of most possibilities lies not in the creation of suitable genetic traits in the laboratory, but rather in their incorporation into a field population in such a way that they will not be rejected and eliminated. This is theoretically feasible, and an interesting field experiment was conducted in Burma, where a population of the southern house mosquito (*Culex quinquefasciatus*) was locally eradicated by the release of cytoplasmically incompatible males. This is about all that can be claimed at present, except as discussed below. The reader is referred to articles and books by Boller (1987), Cavalloro (1983), Davidson (1974), Hoy & McKelvey (1979), International Atomic Energy Agency (1982), Knipling (1979), Waterhouse, LeChance & Whitten (1976), Whitten (1985) and Whitten & Foster (1975) which, along with the references cited therein, cover this complex field very adequately.

The control of insect populations by the release of sterile males has been demonstrated with at least ten insect species. This spectacular technique has been termed *autocidal control*, and involves using an insect species to bring about its own self-destruction. Very briefly, it is accomplished by irradiating laboratory-reared males of the species to an extent sufficient to disrupt the genetic function of the sperm nucleus but not appreciably to interfere with the normal ability of the male to mate or of the sperm to penetrate the eggs of the female. However, such 'fertilized' eggs fail to develop, so a wild female mated with a sterilized male produces no progeny. If sterile males are released in large enough numbers in relation to the wild population, they will mate with more females than will the wild males. The advantage is cumulative in each generation, hence eradication may be achieved within a few generations under ideal conditions. Unfortunately, ideal conditions, either biologically, ecologically or economically, do not occur very often. Thus the only really major success thus far has been with the screw-worm fly (*Cochliomyia hominivorax*), which was eradicated from the Island of Curaçao in 1954, from Florida in 1958–9, and from the entire southeastern United States during 1962–82. In spite of serious setbacks in the 1970s due to production of poor-quality flies, weekly releases of up to 180 million sterilized males have now successfully eradicated this major economic pest from the United States and most of Mexico. Efforts are currently being made to push the fly all the way back to the Isthmus of Panama, where a 'barrier zone' of sterilized males would prevent it from spreading northward. These outstanding projects have paid handsome dividends, and they illustrate the value that such unique approaches can have on occasion. The benefits have been estimated to average US $20 million per year in the southeastern states and about $100 million per year in the southwest.

Another huge eradication project, 'Programa Moscamed', has been directed against the Mediterranean fruit fly (*Ceratitis capitata*) in Central

America. This major pest was first recorded in that region in Costa Rica in 1955, and by 1976 had spread all over Guatemala. By 1977 it had invaded southern Mexico, where potential losses were estimated at nearly US $2 billion per year, mainly due to the loss of foreign markets. In an attempt to prevent its further spread over Mexico, and eventually into the United States, a co-operative project including quarantine inspections, massive trapping, bait sprays and sterile male releases was embarked upon by the governments of Mexico, Guatemala and the United States. A large rearing facility was established in Metapa, Mexico, which by 1981 was producing 500 million flies per week at an annual operating cost of $8 million, and 1000 sterile males were released per hectare per week over the entire eradication zone. Bait spraying was completely suspended in Mexico in 1980, and the Mediterranean fruit fly was declared eradicated from that country by the end of 1982. Efforts are continuing to eradicate the pest from Guatemala and the rest of Central America. (See Hendrichs *et al.*, 1983.)

Promising results have been obtained against the Mediterranean fruit fly in Procida Island (Italy), the Canary Islands, Hawaii and Nicaragua; the Mexican fruit fly (*Anastrepha ludens*) in Mexico; the Queensland fruit fly (*Dacus tryoni*) in Australia and the European cherry fruit fly (*Rhagoletis cerasi*) in Switzerland. A low infestation of the Oriental fruit fly (*Dacus dorsalis*) was eradicated from Guam in 1963, but tests on three other islands were not completely successful. The melon fly (*Dacus cucurbitae*) was eradicated from Rota Island (Marianas) in 1963 using a combination of a bait spray and sterile-insect releases, and from Kume Island (Okinawa, Japan) in 1976. However, Kume was reinvaded as soon as the sterile-male releases were discontinued. By 1987, the melon fly was eradicated from several islands in the Okinawa and Miyako Archipelagos. The southern house mosquito was eradicated from Seahorse Key (Florida) in 1969, the stable fly (*Stomoxys calcitrans*) was reduced by more than 99.9 percent on St Croix (Virgin Islands) in 1976, and the tsetse fly (*Glossina palpalis*) was locally eradicated in Upper Volta in 1979 and in Nigeria in 1985.

The European cockchafer (*Melolontha melolontha*) was eradicated from a 30-hectare plot in Switzerland in 1962. Sterile-male releases have been credited with preventing the spread of the pink bollworm (*Pectinophora gossypiella*) into California's San Joaquin Valley. The boll weevil (*Anthonomus grandis*) is reported to have been eradicated from a small isolated field in Louisiana by the release of chemically sterilized males. Chemosterilants have been under intensive investigation but are not as yet practical. They offer the advantage of treating the wild population directly and thus avoiding the mass-rearing necessary with gamma radiation.

Sterile-male releases are applied commercially as a control measure against the onion fly (*Delia antiqua*) in the Netherlands. In addition, tests with the olive fly (*Dacus oleae*) and grape moth (*Eupoecilia ambiguella*) in Europe, the codling moth (*Cydia pomonella*) in the United States, Canada and Switzerland, the malaria mosquito (*Anopheles ludens*) in El Salvador and several other pests have shown sufficiently effective results to indicate this to be a promising method, either for control or perhaps for eradication under ideal circumstances. However, practical application of the technique in these cases has not been achieved even, for instance, after 30 years of research in Canada and the United States on the codling moth. The conclusion reached in that project was that although control of the codling moth was technically feasible, it would cost about twice as much as conventional chemical control. Various other pilot tests have failed to produce appreciable results.

From these results, representing more than three decades of intensive and quite expensive research, it is apparent that the sterility principle will not be practical for controlling or eradicating many insect species, but it may be spectacularly successful against some major insect pests. In fact, E. F. Knipling, the guiding force behind the method, has stated that the sterile insect technique will not be practical for controlling or eliminating established populations of most of our destructive insect species. The method will not be feasible for those that have a wide host range, are sporadic in appearance, or which do not cause high economic losses. Additionally, the released insects must be distributed evenly throughout the range of the wild population, and such field populations must be low enough to achieve an effective ratio between released and natural populations. Laboratory and other aspects also have to be favorable; mass-rearing must be cheap, sterility must be produced without other adverse effects so that sterilized males are competitive with wild males, and the females preferably should mate only once during their lifetime.

The advantages of the method are that if eradication can be achieved the problem is solved permanently, barring reintroductions, and the method is highly selective for the target pest and ecologically non-disruptive, at least where radiation-sterilized insects are concerned. Great savings can be obtained against serious major pests of expensive commodities, as illustrated by the screw-worm.

Use of pheromones

Various powerful attractants have been used in pest management for many years, either in monitoring systems designed to assess the population density of a pest and the need for its control, or in actual control programs. In Israel, for instance, the populations of the Mediterranean fruit fly on citrus are monitored with a network of trap-jars containing trimedlure, a powerful synthetic attractant for male flies, and control is effected when necessary by aerial applications of poisoned bait, containing protein hydrolysates as an attractant for female flies (see Chapter 10).

On a few occasions, attractants have been employed successfully for pest eradication. An example of the possibilities was furnished by the eradication of the Oriental fruit fly on Rota Island (Marianas), where concurrently with the sterile-male technique the synthetic attractant, methyl eugenol, was used to lure the male flies to a poison-coated surface. A similar but more important achievement was the eradication of the Mediterranean fruit fly from about 400 000 ha in Florida in 1956–7 with poisoned protein hydrolysate baits. The cost was about $11 million. On the whole, however, the use of such attractants has been rather limited.

A major breakthrough has been achieved during the last two decades or so with the chemical identification and synthesis of numerous insect pheromones. These are natural behavioral chemicals that mediate the interactions among conspecific organisms. There are sex pheromones, which attract sexes to each other and induce copulation; aggregation pheromones, which attract members of the species to a newly found host plant or food source; trail and alarm pheromones, and various others. They are often effective in incredibly small concentrations, and are becoming increasingly available for use in pest management programs. Inasmuch as pheromones are now commonly synthesized, it becomes a moot question whether we are dealing with a chemical or a biological method, as is similarly the case with some pathogens. Comprehensive reviews of this fast-developing area can be found in Birch & Haynes (1982), Hummel & Miller (1984), Jutsum & Gordon (1989), Kydonieus & Beroza (1982), Mitchell (1981), Nordlund, Jones & Lewis (1981), and Shorey & McKelvey (1977).

By far the most prevalent use of pheromones is in monitoring pest populations, either for early detection of new invasions in sensitive locations such as around airports or near international borders, or for determination of the density of established populations. Sex pheromones usually are used to attract males, and this method has enabled the development of accurate economic thresholds and exact timing of control measures, and has often resulted in considerable reduction of pesticidal applications. The codling moth (*Cydia pomonella*) is a good example. Catches of five moths per trap per week have been determined as a threshold in many areas, and although problems still exist regarding trap density and design, reliance on pheromone traps has reduced the number of spray applications by 50 to 75 percent. An extensive network of pheromone traps is used in the United States to monitor the spread of the gypsy moth and the pink bollworm, and similar programs are now in operation against numerous pests all over the world.

Both sex and aggregation pheromones have been used in actual control by mass trapping. The recent development of an effective control program for the citrus flower moth (*Prays citri*) on lemons in Israel is an example. From an earlier method of population monitoring by traps baited with virgin females, a rather sophisticated system has been perfected, whereby sticky traps baited with a synthetic pheromone are used for mass trapping of male moths. This 'male annihilation' results in the females remaining unfertilized, and effective control is achieved without resort to chemical pesticides and at about the same price as chemical control. Promising results have been reported against the olive fly (*Dacus oleae*) in Greece and against forest bark beetles in the United States and Europe.

Even if pheromones or other attractants are used in combination with insecticides in poisoned baits, rather than in sticky traps, this is still much safer and less disruptive than conventional full-coverage chemical control, because the poisons are used in small amounts and are usually localized (e.g. in traps or spot sprays) and the baits are highly species-specific. However, the possibility that natural enemies may be attracted to the pheromones of their hosts should be taken into account.

Another approach is to mask natural pheromone cues by volatilizing synthetic sex pheromones to saturate an entire area, so that the pest insects are unable to locate their mates. This 'mating disruption' technique has been demonstrated experimentally with the cabbage looper (*Trichoplusia ni*) in California, where a 99 percent reduction of mating was obtained even when the evaporative sources were as much as 300 m apart. Major successes have been achieved against the pink bollworm over large acreages of cotton in California, Egypt and Pakistan, where control by mating disruption has been equivalent to that obtained by chemical pesticides. Tests with other cotton, vegetable, orchard and forest pests have also been encouraging.

The main advantage of pheromones lies in their being specific to the target species, and it is this high selectivity that makes their use so desirable. In a sense, this too could be termed autocidal control. Although the use of pheromones does not provide the permanent control that classical biological control does, it holds a decided advantage over chemical control in having no effect on non-target organisms, including honey bees and most natural enemies. Pheromones do not induce pest upsets or resurgences, and leave no toxic residues. Their use should therefore be encouraged.

Competitive displacement of pests by non-pests

Competitive displacement between ecological homologs is now generally accepted as an ecological principle. Briefly stated, it says that different species having identical ecological niches (i.e. ecological homologs) cannot coexist for long in the same habitat. Ecological niche in this context means the role played by an animal in the habitat, based on its *precise* food, spatial and habitudinal requirements. In other words, and even more briefly, two species of insects having *exactly* the same food or other essential requirements cannot coexist in the same habitat for any extended period of time. They do not have to fight actively for the food or be aggressive towards one another, but the food must be exactly, not superficially, the same. Feeding on the same plant, or even on the same leaf, is not necessarily enough. One example of ecological homologs would be where two species need to feed on the upper midrib of the same leaves in the same area of the same plant species. However, they could coexist if one fed and lived on the upper leaf surface whereas the other was restricted to the lower leaf surface. In this latter case they would not be homologs. In cases of true homologs, one will always eliminate the other, barring immigration, alternative hosts and other possible modifying factors. There are now many recorded cases of competitive displacement in the field, and a large number have been demonstrated experimentally in the laboratory (see DeBach, 1966).

The fact that such displacement (i.e. eradication or extinction) of one organism by another has occurred naturally, eventually led to conjectures that the phenomenon might be manipulated to man's advantage in pest control or eradication (see DeBach, 1964*b*). However, from the precise nature of the interactions it is evident that the principle will not be of broad application to all, or even most, pest species. It should, however, have as broad, or even broader, a potential as the sterile-male technique. In an ideal case, competitive displacement should be attained with a minimum of cost and effort, and eradication or permanent high-level control could result. The technique merely involves the discovery of a harmless homolog of a pest and its introduction and colonization. If the introduced homolog has a competitive advantage, as ultimately expressed by the greatest F_1 progeny production, it will eventually eliminate the other. The principle operates very much like natural selection and survival of the fittest.

In applying the principle of competitive displacement for control or eradication of a pest, obviously one cannot risk trading one pest for another, which might be as bad or possibly even worse. Especially suitable cases need to be chosen, and extremely careful selection of the candidate species would have to be based upon sound biological and ecological studies. The most likely applications involve use of non-pests to control or eradicate pests, and this would seem to generally rule out trials with phytophagous insects. However, it must be borne in mind that critical competition can occur between any active stage of the life cycle, i.e. any larval or nymphal stage, or the adult stage. Thus it is conceivable that the adults of a phytophagous pest, which require a specific nectar from a wild host plant, might be displaced by an introduced phytophagous species with the same requirements as an adult, but whose immature stages feed only on weeds. Other potentially suitable targets among phytophagous insects include insect vectors of plant pathogens. Often such vectors occur in low numbers and are only damaging because they transmit a plant disease, which, however, is very serious. Such vectors could be replaced by an ecological homolog that was a non-vector and it could exist on the plant even in considerably higher numbers without doing the same amount of damage, perhaps even none. An even better possibility exists where vectors overwinter on an obligate alternative wild host plant, as the beet leafhopper (*Circulifer tenellus*) largely does in California on Russian thistle. The introduction of another leafhopper restricted to Russian thistle, which could out-compete the beet leafhopper on that plant, might well solve this major problem. Some other possibilities are mentioned in the previous citation.

The likelihood remains that most potential applications involve public health, livestock, or nuisance pests such as mosquitoes, biting flies and midges, and others, where the larval food is of no economic importance or actually is unwanted. Thus, competition in water between mosquito larvae of different species could lead to displacement of one species by another with no detrimental effects to man or the environment. There is no reason why a species with adults that are non-human biting, or non-vectors of disease, could not be introduced and displace a serious mosquito through the medium of larval competition. The same idea applies to biting flies, gnats, house flies and others. E. F. Legner and E. C. Bay have suggested (personal communication) that the *Hippelates* eye-flies that are such a nuisance in many parts of North America, include species in the Caribbean area that are not attracted to man. Should their larvae, which develop in the soil or in decaying organic material, be ecological homologs, this problem could possibly be solved.

Some tests have indicated good possibilities. The house fly (*Musca domestica*) has various competitors for dung or other types of larval media. There is no problem here with introducing any coprophagous competitor, which has neither nuisance nor disease transmission potential. Pilot tests in California have shown that the black soldier fly (*Hermetia illucens*), which has little nuisance value, can reduce or eliminate the house fly under certain conditions, apparently by liquefying the medium and rendering it unsuitable for house fly pupation (see Sheppard, 1983).

In the 1920s, several species of dung beetles were successfully introduced into Hawaii to compete with the pestiferous horn fly (*Haematobia irritans*) and have reportedly caused a marked reduction in the fly populations. In experiments, dung pads exposed to horn flies but protected against beetles produced hundreds of flies each, whereas pads not protected against beetles yielded a few stunted flies at the most. Similar projects are currently under way in Texas and California. In Australia, a large-scale project for competitive displacement of dung-breeding flies has been carried out by the CSIRO since the 1960s. There, the bush fly (*Musca vetustissima*) and the blood-feeding buffalo fly (*Haematobia irritans exigua*) breed in huge numbers in bovine dung pads, which are not decomposed by local dung beetles. Several species of Old-

World scarab beetles were imported and established in Australia since 1967, and their numbers have increased spectacularly over large areas. Special quarantine techniques had to be developed, to ensure that the introduced beetles would not be accompanied by any cattle diseases that do not occur in Australia. Rapid elimination of dung pads has been evident in many pastures, and this in itself has been of considerable economic benefit because accumulated dung pads had been estimated to reduce the effective area of pastures in Australia by as much as 2 400 000 ha each year and this effect is partly cumulative, as it may last for several years. As far as the 'biological control of dung' is concerned, this project has already been a remarkable success. Additional coprophagous species (as well as some parasites and predators) are being sought, especially in southern Africa, and it is hoped that their establishment will eventually solve the fly problem. (See Doube, 1986; Waterhouse, 1974.)

After World War II, attempts with chemicals to eradicate a serious mosquito vector of malaria, *Anopheles labranchiae*, in Sardinia, led after two or three years to another, previously rare species, *A. hispaniola*, becoming dominant, while *A. labranchiae* became scarce. *A. hispaniola* is not a vector of malaria. This apparently occurred because of differences in habits of both adults and larvae of the two species, which favored relative survival of *A. hispaniola* under the chemical treatment imposed. Had such treatment been continued longer, extermination of *A. labranchiae* could have occurred, not from the chemical treatments alone but from the competitive advantage conferred on *A. hispaniola* by these measures. Thus, the process of competitive displacement can be swung in favor of the species we wish to be the winner, by using treatments selectively adverse to the other. High-level ecological research was behind the work described, and will be essential to any project of this nature.

This method definitely has good possibilities that are largely being overlooked when one considers that it probably applies most readily to insects of public health, nuisance and veterinary importance, and that such insects have not as yet yielded as readily to traditional biological control methods. These insects, such as malaria mosquitoes, also continue to constitute one of the great sources of environmental pollution with 'hard' insecticides. Competitive displacement offers one of the few techniques where eradication can be expected or achieved. In most cases its application could be extremely cheap, because the natural reproduction of the introduced competitor would do the job. However, only highly competent and careful ecologically oriented scientists should carry out such projects.

Other methods

Certain endocrine processes, including the regulation of molting, metamorphosis, reproduction and diapause, are unique to arthropods and do not occur in vertebrates. Insect growth regulators (IGRs), which are aimed at disrupting these processes, such as the synthetic juvenoids (juvenile hormone analogs) or juvenile hormone antagonists, are indeed highly selective to insects and mites. Another group are the benzoylphenyl ureas, which inhibit chitin synthesis, another process peculiar to arthropods. Some of the IGRs, celebrated because of their safety to humans and other vertebrates, have been registered for use against insect pests. However, some IGRs are quite harmful to certain arthropod parasites and predators, and they are not immune to the development of resistance among arthropod pests. Certainly, IGRs do not appear to be the panacea for insect pest problems that they were formerly expected to be. (See Retnakaran, Granett & Ennis, 1985.)

Physical and mechanical control techniques may range from the extraction of wood borers with a hooked wire, or the use of a sticky barrier to prevent caterpillars from climbing upon a tree, to highly sophisticated – but still largely experimental – uses of various forms of electromagnetic energy. Sometimes a very simple method may be highly effective. A fine example in this category is the use of refrigeration to prevent the passage of the Mediterranean fruit fly (*Ceratitis capitata*) with exported citrus fruit. By simply shipping the fruit under carefully controlled temperatures in refrigerated boats, every single immature fly will be killed during the voyage without resort to toxic chemicals. Another simple technique, the removal of scale insects and sooty mold from citrus fruit by high-pressure rinsing in the packing house, appears to have the potential of greatly reducing certain pest control practices in citriculture. If this new technique indeed proves as effective as early trials indicate, it may eventually permit the raising of economic thresholds in the citrus grove to much higher levels. Rather than control light scale-insect infestations that cause 'cosmetic' damage to the fruit, the grower would then be concerned only with infestations that threaten the tree itself, and this would drastically reduce the need for pesticidal treatments. (See Bedford, 1990.)

Comments and conclusions

In this chapter we have tried objectively to discuss and evaluate the potential of biological methods of pest control other than biological control, especially in agriculture. Aside from biological control, there are really only four methods of proven practical importance in the field: cultural control, use of resistant plants, use of the sterile-male technique and chemical attractants, including pheromones. The practical use of the sterile-male technique remains so limited after three decades of intensive research and trials, as far as the number of pest species involved is concerned, that it cannot be expected to play other than an occasional, very specialized role in pest management. Other genetic methods remain in the idea or pilot-test stage. Another 'biological' method mentioned, insect pheromones and their synthetic doubles, artificial chemical attractants, show good practical results in several cases and additional practical indications of promise, especially in pest monitoring, but again their use in actual control will probably be limited. The use of competitive displacement of a pest by a non-pest, or some similar phenomenon, appears to have promising practical applications, especially in public health and veterinary entomology, but it has not achieved a single complete success to date. Various other techniques – biological, chemical and physical – have received considerable publicity, but they remain largely unproven, or of limited use at best, for practical pest management in agro-ecosystems. They include juvenile hormones, chemosterilants, repellents, antifeedants and others.

Thus, if we are to avoid the unilateral use of insecticides in crop ecosystems and rely instead on ecologically sound insect pest management programs, we are currently predominantly dependent upon three principal biological techniques or methods and one chemical technique, namely: biological control, cultural control, the use of insect-resistant host plants and the use of selective chemical insecticides (or the selective use of insecticides).

From the historical record as well as projected promise, and from comparisons of advantages, disadvantages, costs *vs* returns, ease of application and other considerations, it is clear that the first priority on a general basis should go to research on biological control, and especially on importation of new natural enemies. Cultural control methods are indicated to deserve a strong second place in emphasis and the development of pest-resistant host plants should also receive strong support. In spite of the relatively few pest species that have been controlled by this method, it appears to have much greater potential than has been realized. The sterile-male technique should be pilot-tested wherever there seems to be a reasonable chance of success against an otherwise refractory major economic pest. Pioneering research should, of course, continue along the lines mentioned previously, as well as others, but not at the expense of more proven or more likely methods. These, of course, are general priorities, and with some certain specific pests could well be reversed but adequate preliminary ecological studies would have to determine this. However, unless biological control research had been *thoroughly* carried out previously, there would seem to be no reason to change priorities. The word 'thoroughly' is emphasized because, for instance, it will be recalled that we consider that the possibilities for biological control of the gypsy moth have not been thoroughly followed through in spite of an enormous amount of work expended on it years ago. Various other previous biological control projects also fit in this same category. A very good case in point involves the codling moth (*Cydia pomonella*), the most destructive pest of apples and other pome fruits. As mentioned earlier, research and pilot-testing directed toward the sterile-male technique has been going on in the northwestern United States and in Canada for about 30 years, beginning in 1956. Extensive experimentation, involving very large amounts of money for 20 years or so, eventually led to the conclusion that the method was not acceptable from an economic standpoint. Nevertheless, the project has not been discontinued and is still going on. Meanwhile, there have been no plans to import new natural enemies of the codling moth, even though there are over one hundred recorded enemy species, and only two species have thus far been imported into North America from the native home in the Old World. A bona fide biological control importation project emphasizing this pest has never existed, yet it could be done for a small fraction of the cost of the sterile-male project. This is a classic example of mis-direction of research priorities.

10

Escape from the pesticide dilemma

A very substantial reduction in the application of insecticides, going as far even as their complete withdrawal in some agro-ecosystems, can best be attained by the application of basic ecological principles to pest insect problems. Pest problems are bio-ecological problems, not chemical ones. The empirical and unilateral use of chemicals to attempt to hammer pests into submission by repeated costly blows is, as we have seen in Chapter 1, increasingly failing to provide a solution. Pest resurgence and upsets in natural balance, combined with the ever-increasing development of resistance by pests to the poisons used against them – all problems brought on by the pesticides themselves – clearly show that a rapid and drastic change is necessary in order to achieve control of pests in an ecologically and economically satisfactory manner.

Even with the continual development of new and presumably better pesticides, combined with their greatly expanded usage, chemicals are not controlling pests in general as well as previously. Percentage losses due to insects and plant diseases increased since the 1940s according to USDA records (Pimentel *et al.*, 1981; see Chapter 1). For example, a 3.5 percent loss of corn from insects was reported for the period 1942–51, as compared to a 12.0 percent loss for the period of 1951-60 and 8.1–12.9 percent for 1977.

In the United States, the production of synthetic organic pesticides increased 3000 times in less than 30 years – from an estimated 210 000 kg in 1951 to about 635 million kg in 1977 (Bottrell, 1979). Worldwide, pesticide imports rose 2.6-fold between 1972 and 1984, to $5.3 billion (Postel, 1987). World sales of pesticides in 1985 totaled $15.9 billion, and until the year 2000 were expected to increase by 2–3 percent per year (Jutsum, 1988). In California agriculture alone, the number of hectare-treatments rose from less than 3 million in 1960 to over 6 million in 1971. This increase continues, in spite of growing public awareness of the hazards involved. Better pest control is not resulting from these massive increases. The need for change becomes even more urgent when the serious side-problems generated by the use of pesticides are considered, such as hazards to human health, killing of fish and wildlife, adverse soil effects, hidden social costs and others. These so-called 'external costs', as well as many of the other economic considerations to be analyzed in a pest management program, are multi-faceted and too involved to consider here. A good overview of them has been presented by Pimentel & Perkins (1980).

An excellent summary of the pesticide problem as illustrated in cotton, the importance of natural enemies, and the need for an alternative ecological approach has been presented by Professors Ray F. Smith & Harold T. Reynolds (1972), two of the world's leading exponents of integrated pest management, as follows:*

> 'The actions of parasites, predators, and pathogens are important causes of pest mortality in many cotton agro-ecosystems. In other areas, especially where heavy use of pesticides has eliminated the natural controls, such biotic mortality may be minimal. The importance of parasites and predators in cotton agro-ecosystems has been most clearly demonstrated by the unleashing of secondary pests through the use of broad-spectrum organic insecticides. Also, in many instances, there have been tremendous resurgences of 'primary target' pest species following the use of pesticides. These pest resurgences and secondary outbreaks are largely the result of the elimination of parasites and predators. There are many examples of such pesticide-induced outbreaks including those of bollworm, tobacco budworm, cotton aphids, and spider mites in the Cotton South of the United States. Jassid control with DDT, Sevin, or Zectran results in increased incidence of whitefly in the Sudan . . . It should be emphasized that it would not long be possible to produce most crops, including cotton, economically or at all without the regulatory impact of beneficial species upon the pest complex . . .
>
> The use of chemical pesticides without regard to the complexities of

---

* From *The Careless Technology: Ecology and International Development*, edited by M. Taghi Farvar and John P. Milton, copyright © 1969, 1972 by the Conservation Foundation and the Center for the Biology of Natural Systems, Washington University. Reprinted by permission of Doubleday and Company, Inc. and Tom Stacey, Ltd.

cotton agro-ecosystems, especially the fundamental aspects of the population dynamics of pest species, has been the basic cause of the exacerbation of cotton pest problems over the past ten to fifteen years. A review of these problems in many cotton-growing regions of the world shows a common pattern. Typically, the initial introduction of the widespread use of modern pesticides to cotton agro-ecosystems results in significantly increased yields of seed and lint. But then more frequent applications are needed, the dosages must be increased to achieve control and the treatment season is extended. It is noted that the pest populations soon resurge rapidly to new, higher levels after treatment. The pest populations gradually become so tolerant of the pesticides that the latter become useless. Other insecticides are substituted and the pest populations become tolerant to them too, but this happens more rapidly than with the chemicals that were first used. At the same time, pests that had previously never, or only occasionally, caused damage become serious and regular ravagers of the cotton fields. This combination of pesticide resistance, pest resurgence, and unleashed secondary pests causes greatly increased production costs and often brings on an economic disaster . . .

The integrated control approach attempts to avoid the pitfalls of pest control . . . There is now ample evidence that the integrated control approach with its strong ecological inputs has proved to be a better approach to solve a wide variety of pest problems than has a total reliance on pesticides . . . Integrated control attempts to use a combination of suitable control techniques, in as compatible a manner as possible, to maintain the pest populations below defined economic injury levels. Integrated control derives its uniqueness of approach from its emphasis on the fullest practical utilization of the existing mortality and suppressive factors in the environment.

The cotton pest control situation . . . is not unique. Similar crises are upon us or are soon to be faced in the control of mosquitoes, fruit pests, spider mites on many crops, house flies, cabbage loopers on vegetables, ticks on cattle, and in many other situations where pesticides have been used intensively. In many of these situations, time is running out. The world population is facing a most critical food shortage. This is a crisis shared by crop protection specialists all over the world. We believe that with an imaginative approach to the management of agro-ecosystems this challenge can be met, that pests can be controlled without continued exacerbation of the magnitude of the problem . . .'

The change to sound ecological and biological principles in pest control has come to be referred to variously as integrated control or integrated pest management (IPM). There are somewhat different ideas as to what these terms mean, as well as different definitions. One thing is certain, however; integrated control does not merely refer to some more or less random mixture of different control methods. It must be based primarily upon careful biological and ecological studies of the pest complex and the natural control factors involved in a particular crop habitat. In a broad sense, integrated control means the integration of chemical control *with* biological control and other non-chemical approaches, not the opposite. Professor Ray F. Smith has characterized integrated control thus:

> 'It is clear that not all the important crop pests have adequate biotic controls as they are now handled in modern agro-ecosystems. However, even with these pests we may have biotic agents which give partial controls. The task of the plant protection specialists is to conserve and protect these hidden natural controls and partial control agents and mold them with other controls (resistant plant varieties, chemical controls, cultural controls) into a sound economic protection system. This is another way of describing the integrated control approach.'

Flint & van den Bosch (1981) have characterized IPM as follows:

> 'Integrated pest management (IPM) is an ecologically based pest control strategy that relies heavily on natural mortality factors such as natural enemies and weather and seeks out control tactics that disrupt these factors as little as possible. IPM uses pesticides, but only after systematic monitoring of pest populations and natural control factors indicates a need. Ideally, an integrated pest management program considers all available pest control actions, including no action, and evaluates the potential interaction among various control tactics, cultural practices, weather, other pests, and the crop to be protected.'

There is now a huge amount of literature dealing with the subject of integrated pest management, but much of it is either general or theoretical. Some good articles and books of broad interest include Adkisson *et al.* (1982), Apple & Smith (1976), Burn, Coaker & Jepson (1987), Delucchi (1987), Doutt & Smith (1971), Flint & van den Bosch (1981), Frisbie & Adkisson (1986), Huffaker (1980), Metcalf & Luckmann (1982), Pimentel (1981) and Wood (1971). A wealth of important articles too numerous to list, pertaining to pest management and biological control, will be found in the *Proceedings of the Tall Timbers Conference on Ecological Animal Control by Habitat Management.* Outstanding specialists from various countries are invited to present papers at these conferences. Eight volumes have so far been issued by the Tall Timbers Research Station, Tallahassee, Florida.

The objectives of pest management in a crop ecosystem are to optimize yield and quality while minimizing pesticidal pollution, other adverse environmental effects and cost. The opinion is held by many ecologically oriented entomologists that a good portion of these objectives could be met almost immediately, because probably at least 50 percent of pesticide chemicals are now used unnecessarily. A good example of what can be done in this regard was illustrated by the management of sugar-cane borer populations (*Diatraea saccharalis*), the key sugar-cane pest, in Louisiana (see Hensley, 1971). The maximum number of annual insecticide applications was reduced from twelve to three – a 75 percent decrease – in one decade, from 1960 to 1970. This somewhat primitive system of pest population management (in the author's own words) took full advantage of the suppressive effects of a large complex of arthropod predators, varietal resistance and adverse weather conditions. These were supplemented by the judicious use of insecticides, only when necessary. The reduced chemical application system was developed by research which led to the adoption of the following principal practices: (1) discontinuance of insecticidal control of the first generation after research had shown that larvae of this generation did not destroy enough young tillers to reduce yield and that insecticidal control of the first generation did not provide sufficient suppressive effect on the second generation to justify cost; (2) discontinuance of fixed weekly application schedules after a reliable economic injury threshold had been established which recommended treatment or retreatment of second and third generation infestations only when 5 percent of the stalks were infested with small larvae; (3) improvement in field survey methods for detecting potentially damaging infestations, by replacing a system of observing plants for superficial feeding signs of larvae that may have later been eliminated by predators or other ecological mortality agents with a system based on the presence of live larvae in leaf sheaths; (4) use in control programs of highly effective synthetic organic insecticides that permit long intervals between applications and that cause minimum of damage to natural enemy populations; and (5) more emphasis on host plant resistance as a means of controlling the sugar-cane borer and reducing insecticide use.

Other early examples, demonstrating the feasibility of integrated pest management, were presented by Adkisson (1971), Smith (1972) and many others. It was shown, for instance, that insecticide use in cotton in the United States could be reduced by perhaps 50 percent without reduction of yields, and that the amounts of insecticides used on grain sorghum for control of the greenbug (*Schizaphis graminum*) were two to five times greater than actually needed to control the pest. This, and the increasing realization that unilateral chemical control will not solve pest problems, eventually led to the endorsement of IPM by various federal agencies in the United States in the 1970s. Major efforts to establish a holistic, integrative approach to pest control began with the intitiation of the 'Huffaker Project', a nation-wide program engaging some 300 scientists from 18 universities, directed by Professor Carl B. Huffaker and funded jointly by the National Science Foundation, the Environmental Protection Agency (EPA) and the United States Department of Agriculture (USDA) during 1972–8. The goal was to promote the development of integrated pest management programs in major agro-ecosystems, including soybeans, cotton, alfalfa, pome and stone fruits, citrus and coniferous forests. This was accompanied by the development, with federal funding, of numerous pilot IPM programs on various crops by the Cooperative Extension Service in virtually all states. Then, in 1979, the Consortium for Integrated Pest Management (CIPM), directed by Professors Perry L. Adkisson and Raymond E. Frisbie and funded initially by the EPA and later by the USDA, developed a multi-disciplinary, 17-university program focusing on the multiple pest complexes – arthropods, weeds, diseases and nematodes – attacking each of four major crops in the United States: alfalfa, apple, cotton and soybeans. Other crops, such as grain sorghum and peanuts, also received intensive research and implementation efforts. More that 250 scientists participated in this program, which was terminated in 1986. (See Frisbie & Adkisson, 1986; Huffaker, 1980.)

Unfortunately, as we have pointed out in Chapter 1, despite the wide recognition of the central role of biological control in IPM, research and utilization of natural enemies were in fact rather neglected within these multi-million-dollar frameworks. Nevertheless, these much-publicized programs were rather instrumental in promoting the development of monitoring systems and economic thresholds, and in demonstrating the ability to eliminate unnecessary pesticidal treatments. As a result, pesticide usage on certain crops in the United States has decreased markedly. On a major pesticide receiver such as cotton, the use of insecticides decreased from 6.14 kg of active ingredients per hectare in

1976 to 1.68 kg per hectare in 1982, a dramatic decline of 72 percent, and the proportion of hectares treated with insecticides decreased from 60 to 36 percent during that period. Thus, cotton in 1982 accounted for only 24 percent of the insecticides applied to major field crops in the United States, as compared to 49 percent in 1976 (Frisbie & Adkisson, 1986).

One case history, of many, will suffice to illustrate this remarkable change. In Arkansas, where the bollworm is the main pest of cotton, it was routinely controlled by ten chemical applications in 1977 and fourteen in 1978. IPM tests during the same years showed that by monitoring bollworms in all stages and applying pesticides only when actually required, treatments could be reduced to one chemical and one microbial application in 1977, and to two chemical and two microbial applications in 1978. The amount of conventional insecticides was reduced by 87 percent under the IPM program, and the cost of this program was $20 per hectare as compared to $70 per hectare for the conventional program. No wonder that growers all over the State soon adopted the new program (Carl B. Huffaker, personal communication).

On grain sorghum, too, the amount of insecticides used per hectare decreased by 40 percent from 1971 to 1982. An even more impressive decline of 80 percent was recorded on peanuts: from 4.48 kg per hectare in 1971 to 0.86 kg in 1982.

These encouraging results indicate what can be achieved by emphasis on development of integrated pest management programs, yet this formidable job has hardly begun. Pesticide usage on the two most extensive crops in the United States, corn and soybeans, as well as on many others has not decreased at all yet, and the situation in some other parts of the world is probably even worse. Imagine how much more could have been done in IPM to date had stronger efforts been made to improve and utilize biological control on an active basis!

On the international scene, the Consortium for International Crop Protection (CICP) was formed in 1978 by a group of 12 United States universities, plus the University of Puerto Rico and the USDA, for the principal purpose of promoting integrated pest management programs in developing nations. Directed initially by Professor Ray F. Smith and later by Professor Allen L. Steinhauer, it has been funded by the United States Agency for International Development. The Food and Agriculture Organization of the United Nations (FAO) also has extensive programs in many developing countries.

The role of biological control in integrated pest management

We have already stressed (Chapters 3 and 7) the great extent of biological control that occurs naturally in all agro-ecosystems. This natural control furnishes by far the greatest part of the foundation upon which ecological pest management must be built. Without it, we would be unable to produce crops of any magnitude or quality; with it, we can concentrate on the few problem species that for one reason or another are not held under natural biological control.

Biological control should be regarded as the backbone of any IPM program. As pointed out in Chapters 7 and 8, biological control can be fostered and increased in several ways. The optimization of *established* natural enemies through conservation practices offers the most immediate and direct means of decreasing insecticidal pollution and reducing pest control costs. It constitutes the main aspect of integrated pest management, since about 99 percent of all potential pests are already under biological control. The remaining one or so percent of the species that constitute actual pests may need to be controlled chemically at the moment, but with proper research emphasis may still be subjected to biological control through importation of new natural enemies, or augmentation of established ones by means of mass production and periodic colonization or other techniques of biological control (see Chapter 7). When biological control cannot be achieved by these techniques, the alternative non-chemical methods discussed in Chapter 9 should be rigorously investigated. When serious major pests are involved, probably several of the most promising approaches can and should be investigated simultaneously, because a rapid solution to the costly annual losses is necessary. Systems analysis is being developed as a potent tool in the optimization of pest management. Good discussions and illustrations of the possibilities involved are given by Getz & Gutierrez (1982) and Huffaker (1980).

Selective chemicals and selective means of application can furnish an immediate first step toward integrated control, by substituting them for the hard chemicals that kill a broad spectrum of natural enemies (see also Chapter 8), or by timing applications or reducing dosages or altering formulations so that more natural enemies are spared. Another selective technique is to treat only a portion of the crop (strip, spot or skip-treatments) at a particular time, so as to leave reservoirs of natural enemies (see Hull & Beers, 1985). Manipulation of the crop by strip-cutting or other cultural practices may further reduce the use of pesticides. An elegant example of this approach was detailed by Stern (1969). He and colleagues found that early-season chemical treatment for lygus bugs in cotton eliminates many natural enemies and often triggers outbreaks of other pests such as bollworm, cabbage looper and spider mites. These in turn require special treatment. However, early-season lygus bug treaments can largely be avoided. First, movement of the lygus bug adults into cotton from alfalfa (where they are a minor problem) can be restricted by strip-cutting of alfalfa or by growing trap-strips of alfalfa in the cotton fields. Second, much higher populations of lygus bugs can be tolerated in cotton without economic loss than was previously considered to be possible. The application of this knowledge has greatly reduced the use of insecticides in the first half of the growing season and therefore minimized the occurrence of upsets and resurgences. (See the section before last in this chapter for further details on this program.)

It should be realized that to seek the highest possible percentage kill of a given pest with a chemical generally is ecologically disruptive. The degree of mortality produced should be designed primarily only to prevent the peak populations that will cause economic loss. Thus, 80 percent mortality, rather than 98 percent, may be sufficient for the immediate chemical control needed, and it will cause fewer upsets or resurgences and additionally will reduce the rate of development of resistance to pesticides. This is illustrated by an integrated mite management program, developed recently on almonds in California. This program relies on the presence of pesticide-resistant phytoseiid predators in the orchard, or inoculative releases of a laboratory-selected resistant strain when necessary, combined with applications of small amounts of acaricides at one-fifth to one-tenth the standard rates. If label rates of acaricides are used, the near-complete elimination of spider mites causes starvation or dispersal of the predators, resulting in subsequent resurgence of the pest mites. The low rates, on the other hand, do not essentially eliminate the prey but still result in effective control. Other pests are controlled with pesticides selective to the predators. The program requires weekly monitoring of pest and natural enemy populations. Savings to growers have been estimated at $60 per hectare for the first year if an inoculative release is required, and $110 per hectare annually thereafter. By 1985, when 25 percent of growers with spider mite problems had adopted the program, the benefit-cost ratio to the

California almond industry was calculated at 15:1 (see Headley & Hoy, 1986; Hoy, 1985*b*).

What is the real need for chemical control?

We have just estimated and shown examples that strongly indicate that approximately 50 percent or more of insecticides currently used in agriculture are unnecessary. Generally speaking, however, the people who use them consider that insecticides are basically required for crop protection, hence they must be shown that their belief is incorrect if they are to change. Thus the real, as opposed to the apparent, need for chemical control must be objectively investigated in all crop ecosystems by means of bio-ecological study and experimentation as well as by economic analysis, and questions such as the following answered.

Have chemicals used previously decimated natural enemies, causing upsets and resurgences resulting in new or worsened man-made pests? Has resistance to the pesticides developed, so that the chemicals used no longer kill the target pest, only other organisms? Is chemical overkill practiced, i.e. are greater dosages and more frequent applications used than are necessary? Are erroneous or misleading crop-loss claims used to justify the need for chemicals? Are asserted gains in production from chemical use accurate, or could other factors such as fertilizers, etc., be responsible? Are artificial crop quality standards of a cosmetic or other nature imposed on the farmer, which force him to use costly chemical treatments and thereby trade inconsequential blemishes or minor damage for toxic residues in the produce? Are the economic thresholds and economic injury levels (which are indications of the need for treatment or the level at which economic loss starts) scientifically accurate, merely rough estimates or only educated guesses? For example, sprays for the eye-spotted bud moth larvae (*Spilonota ocellana*) in Canada used to be applied arbitrarily. It is now known that five bud moth larvae or less per 100 apple leaf clusters do not constitute an economically important population. Hence, sprays are now omitted at these densities, because resulting damage is below the 5 percent level generally tolerated by the industry. Relatively few economic thresholds have been determined by economic entomologists using appropriate experimental field study, and even fewer have been adequately determined in chemically untreated check plots as compared to chemically treated ones. The fact that the economic threshold is not a constant for a given species is often overlooked. It will be completely different in a crop that has been frequently sprayed, as compared to the same crop which has been unsprayed or very little sprayed for a period of time. The decimation of natural enemies causes this. In frequently sprayed crops, economic thresholds are reached at much lower pest population levels, because when natural enemies are eliminated the pest population resurges much more rapidly. In California, the recommended threshold for spider mite control on almonds is 22.0 percent infested leaves in orchards lacking an adequate predator population, as compared to 43.6 percent infested leaves in orchards where predators are present (Hoy, 1985*b*). More such values are needed. Of course, when a pest is under complete biological control by natural enemies, the economic threshold becomes meaningless because economic injury levels are never reached unless the balance is upset. An international conference on prediction, forecasts and assessments of economic thresholds was held in Vienna, Austria, in 1969. Many of the papers presented related to means of determining ahead of time whether or not a given insect will become a problem in a particular season. This obviously is basic to our ability to reduce the use of pesticides (EPPO Report, 1970; see also Pedigo, Hutchins & Higley, 1986, and Stern, 1973).

When the preceding questions, and doubtless others, have been appropriately answered, it will become apparent how much chemical treatment is being carried out unnecessarily. It will also indicate the directions in which further research is needed to solve the remaining problems. In our experience and from numerous literature reports it seems definite that in any major crop the answers to the preceding questions will be affirmative for several to many pests, as well as non-target organisms. Many years of field research have shown that in California alone, many thousands of hectares of citrus annually receive from one to several unnecessary pesticide applications.

We cannot emphasize too strongly that the answers to most of the questions posed can only be obtained by conducting simultaneous experimental comparisons between adequately designed test plots receiving minimal or no chemical pesticides, as compared to similar plots receiving the usual commercial quota of pesticide treatments. (See Chapters 3 and 7.)

Supervised control

Even after the answers are known regarding the degree of unnecessary use of pesticides and their adverse effects, and after research has shown to what extent effective natural enemies of the various pests are present, it is not necessarily an automatic or easy matter to switch from a heavy chemical to an integrated control program. First, recall the habit-forming, narcotic-like effect of insecticides (see Chapter 1) and the concomitant, very real and often difficult problem of withdrawal from chemicals. The change to ecological pest management is a practical matter to the farmer, and he is not prepared to suffer any interim financial losses during the changeover. Even his imagined possibility of loss may make him quite reluctant to change his traditional program. The integrated control pioneers, A. D. Pickett and N. A. Patterson, wrote in 1953:

> 'There are great difficulties involved in combining biological and chemical control under field conditions. It is similar in many ways to the relationship between preventive and curative medicine. As Wiggles-worth has stated "The public loves the hospital, the doctor and the bottle of physic; while the advances in preventive medicine which have transformed our lives are scarcely noticed. So too it creates a greater impression on the mind to destroy an infestation of insects that can be seen, than by some simple change in practice to prevent any infestation from developing." Had the staff of our laboratory developed an insecticide that would control the oystershell scale as effectively as the mere changes in program that allowed nature to take its course, they would have been assured of a place in entomological history.'

In many, possibly most, crop ecosystems where chemical pest control has been relied on heavily, spraying and dusting cannot merely be stopped with the assurance that biological control will take over. Natural enemies may have to be reintroduced or otherwise manipulated, or other non-chemical measures taken, based on what research has shown to be feasible. The grower generally cannot be expected to have the detailed knowledge of pest and natural enemy biology and ecology essential to implement such a program. He needs a specialist trained in ecological pest management, i.e. an applied ecologist who can check and evaluate population developments and interactions in the field. Today's large scale agriculture almost makes it mandatory that pest management should be under the direction of a trained and licensed supervising applied ecologist who is responsible for all decisions relating to the matter. Such a specialist can be compared to a general medical practitioner, veterinarian or pharmacist in his use of complicated materials and techniques. There is a shortage of such trained personnel now, but more and more are entering the field. Various universities in the United States are now emphasizing graduate training in pest management, and the National Science

Foundation and the Ford Foundation have supported certain programs with financial grants. The Association of Applied Insect Ecologists is an organization of pest management specialists (see their address on page 315). Information can also be obtained from the Association of Registered Professional Entomologists (ARPE), Entomological Society of America, 9301 Annapolis Road, Lanham, MD 20706, USA.

The current shortage of applied insect ecologists must be rapidly overcome if ecological pest management is to be expanded at a rate commensurate with the need. California probably has more practicing applied insect ecologists than any area of comparable size in the world, yet their number is pitifully meager compared to those employed in the unilateral use of pesticides. In the early 1970s, the income of this infant industry for consulting pest management services amounted to only about one percent of the total product value of agricultural pest-control-related industries, whereas pesticide salesmen (some 6000 in the State alone at that time) and commercial pesticide applicators took in about 99 percent of the total spent on agricultural pest control. Although the number of salesmen has recently declined to about 4000, and the share of IPM consultants has risen dramatically to about 20 percent, there is still room for great improvement. (See also Lambur, Kazmierczak & Rajotte, 1989.)

The extension entomologist is a necessary link between the researcher, who finds out what it is possible to do, and the supervising entomologist, applied ecologist or farmer, who carries out the program. The role of the extension entomologist in pest management is well summarized by Palti & Auscher (1986).

The need for effective supervision and extension is nowhere more urgent than in developing countries, where trained ecologists are few and crop protection practices are often unsatisfactory. Such countries offer a great opportunity for the development of ecological pest management systems, because in many cases the use of chemical pesticides has been very limited in them and is economically unfeasible (see Brader, 1979). In response to this need, the World Bank has now supported so-called 'Training and Visit' agricultural extension systems in almost 40 countries, mainly in Asia, Central and South America, and Africa.

Examples of integrated pest management

Emphasis on the need for integrated pest management is so great today that some think it is a new idea; however, strong general *support* for integrated control is really what is new. Actually, it began developing more or less empirically many years ago, when combinations of biological control, cultural control, chemical control and sometimes plant resistance were used to control pests on a single crop. Practical applications in California, employing both insectary-reared natural enemies and chemicals, were carried out in citrus groves at least as early as the 1920s. C. V. Riley even envisioned integrated control in 1893 when he wrote:

> '. . . our artificial insecticide methods have little or no effect upon the multiplication of an injurious species, except for the particular occasion which calls them forth, and occasions often arise when it were wiser to refrain from the use of such insecticides and to leave the field to the parasitic and predaceous forms.
>
> It is generally when a particular injurious insect has reached the zenith of its increase and has accomplished its greatest harm that the farmer is led to bestir himself to suppress it, and yet it is equally true that it is just at this time that nature is about to relieve him in striking the balance by checks which are violent and effective in proportion to the exceptional increase of and consequent exceptional injury done by the injurious species. Now the insecticide method of routing the last, under such circumstances, too often involves, also, the destruction of the parasitic and predaceous species, and does more harm than good . . . it not only emphasizes the importance of preventive measures [cultural control], which we are all agreed to urge for other cogent reasons, and which do not to the same extent destroy the parasites, but it affords another explanation of the reason why the fight with insecticides must be kept up year after year, and has little cumulative value.'

Actually the terms integrated control or integrated pest management are somewhat inappropriate, because they imply that two or more distinct methods of pest control must be combined in an optimal manner to achieve the desired results, whereas in some instances, at least, biological control alone will suffice. In Hawaii, for example, complete biological control of all sugar-cane pests was obtained, and in California avocados enjoy essentially complete natural biological control. Nearly all major pests of coconut in Fiji were controlled biologically by the late 1930s. There are other examples (one was cited in Chapter 8) and larger proportions of more crops will be brought under complete or virtually complete biological control in the future. Perhaps ecological pest management would be a better term, because it is less restrictive and the term itself has broader implications. Be this as it may, pesticides are now being used more or less heavily in most commercially grown crops around the world, so the term integrated pest management will certainly continue into the foreseeable future to apply to the switch to ecological pest management in these crops. These terms are used more or less interchangeably in this Chapter.

All of the examples of integrated pest management that follow represent sequels to cases of non-target insect upsets, pest resurgences and/or resistance to insecticides, which were so serious that an alternative to the unilateral use of pesticides was urgently mandatory. The upset phase of most of the following examples has been discussed in Chapter 1.

Apples in Canada and the United States

CANADA. In many areas, apple growers are burdened by 15 or more pesticide applications each year. Most need relief from the increasingly intensive program. A good example of what can be done, and one of the pioneering and best known cases of integrated control, was developed in Nova Scotia by A. D. Pickett and colleagues, beginning just after World War II. A review of this work is given by MacPhee & MacLellan (1972). Apple production has been a well-established agricultural industry there for more than a century. The pest control program developed in Nova Scotia, as it did in the United States, with nearly total dependence upon the unilateral use of chemical pesticides. As time went on, new pest problems developed and some old ones became more serious. The European red mite (*Panonychus ulmi*) and the gray-banded leafroller (*Argyrotaenia mariana*), not important enough to be mentioned in earlier literature, were major pests by the early 1940s. The oystershell scale (*Lepidosaphes ulmi*) became an extremely destructive pest for the first time in the 1930s. The codling moth (*Cydia pomonella*) and the eye-spotted bud moth (*Spilonota ocellana*) became noticeably more serious.

These upsets and resurgences, combined with a loss of traditional overseas markets during the war and the consequent need to reduce costs, emphasized the need for a new approach. Ecological studies of the impact of the pesticides used showed that the traditionally used sulfur fungicides killed the parasite and a predaceous mite of the oystershell scale, but were harmless to the scale. The substitution of copper-based fungicides and Ferbam for sulfur proved to be harmless to the natural enemies, and natural biological control again resulted. Other selective fungicides have since been incorporated into the program. Native predators were found to be capable of controlling the European red mite if not interfered with by chemicals. Use of selective chemicals for other pests resulted in no further need for miticides.

The program, as it was successfully worked out by 1955, had narrowed the Nova Scotian growers' list of spray chemicals to glyodin, captan, bordeaux and mercury eradicants as fungicides for plant disease control; and nictone, lead arsenate and ryania as insecticides. All of these, as used, are relatively selective in sparing natural enemies. The virtual elimination of sulfur, dormant oils, dinitrocresol, chlorinated hydrocarbons, carbamates and organophosphorus insecticides from the program greatly reduced some of the more important apple pests.

Plant-feeding mites and scale-insects, as mentioned previously, became conspicuous by their absence. The eye-spotted bud moth, a serious pest for many years, was reduced to its lowest level in 30 years by 1955. At that time, by adopting modified spray programs, many Nova Scotian growers were averaging only two sprays per year. About 90 percent of them had adopted the program by the late 1950s and continued for many years. No acaricides were used, but adequate spraying for the apple scab disease was always followed. The conservation of the natural enemy fauna by this 'harmonized control' program substantially reduced production costs, increased production of good fruit, improved the growth of the trees, and reduced hazards to growers and consumers and of course to the environment.

It should be noted that this program, like several others, was developed because the growers had their backs to the wall. Had traditional pesticides continued to provide control without serious side-problems and increasing costs, very likely no change would have been sought. The integrated control program developed involved principally conservation of natural enemies through substitution of more selective pesticides. Stress on importation and augmentation of natural enemies might further appreciably enhance the program. Although many Nova Scotian growers eventually returned to the use of broad-spectrum pesticides, major components of the program are still practiced today, and its success has had a profound effect on the development of IPM in North America (see Whalon & Croft, 1984).

In Quebec, work on 'modified' and 'commercial' spray programs in apple orchards, designed to reveal gross effects of each program on predator-pest relationships, was initiated in 1951. The results have been reported by LeRoux (1960). A survey was made of the various spray programs in use in the major apple-growing regions of the Province, and their effects on the beneficial predator fauna were determined from predator counts made in orchards receiving each program. Orchards were categorized on the basis of the predominant spray materials in use as: (*a*) fungicide-DDT-parathion, (*b*) fungicide-lead arsenate, (*c*) fungicide alone, and (*d*) unsprayed or neglected. Results showed that such highly toxic pesticides as DDT and parathion, irrespective of the fungicide with which they were combined, markedly reduced the predator populations as compared to treatments with fungicides and arsenicals or with fungicides alone. There were still greater differences between insecticide-sprayed orchards and those that received no sprays. In the latter, however, much of the foliage was destroyed by apple scab. In orchards where DDT had been used for three or more consecutive years, pests of economic importance such as the European red mite and the red-banded leafroller (*Argyrotaenia velutinana*) were on the ascendency. Of all the programs surveyed, the fungicide-arsenical program destroyed the least number of predators while giving sufficient pest control to keep crop damages at a minimum.

The 'modified' program tested was of the fungicide-arsenical type, and its basic ingredients were the fungicide glyodin and the insecticide lead arsenate. Glyodin was applied each season; the insecticide was applied only as needed. Additional pesticides were used only when special control measures were required. The 'commercial' program with which it was compared was of the fungicide-DDT-parathion type, and the main materials used were the fungicides Magnetic 70 and glyodin, and the insecticides DDT, parathion, malathion and lead arsenate. The number of fungicide and insecticide applications made was the same as that made by the average grower, based on his own assessment of apple scab and arthropod pest situation in his orchard. Each program was defined as follows: *modified spray program* – using corrective sprays (insecticides, acaricides) only to control insect and mite pests, and preventative sprays (e.g. fungicides) mostly to control apple scab. Only spray chemicals known to be relatively innocuous to natural enemies of apple pests were selected; *commercial spray program* – using preventative sprays only (fungicides, insecticides, acaricides to control scab, insects and mites). Little or no emphasis placed on the selection of spray chemicals innocuous to natural enemies of apple pests. The program was in effect the one used by Quebec orchardists, guided by an annual spray calendar.

Results reveal that for a five-year period the 'modified' spray program gave as clean a crop as that obtained with the 'commercial' program, and for approximately half the cost. To carry out the 'modified' spray program, an average of 9.2 fungicidal and 2.2 insecticidal sprays were applied at a cost of $0.65 per tree. The average in apple yield per tree was

6.4 bushels. The program resulted in a build-up from low to high densities of beneficial forms and the maintenance of medium to low densities of destructive forms. The overall control of apple pests for the five years was good, considering that on average 90 percent of the crop was saleable. It would appear that much of the effectiveness of the 'modified' program was due to increased activity of arthropod predator and parasite complexes, and bird predation, which contributed an added degree of control of insect pests. To carry out the 'commercial' spray program for the same period, an average of 12.4 fungicidal and 6.0 insecticidal sprays were applied at an average cost of $1.14 per tree. The average yield in apples per tree was 5.2 bushels. The program resulted in the practical elimination of all natural control agents, with the maintenance of medium to high densities of destructive forms. The overall control of apple pests for the five years was good, with 90 to 95 percent of the crop being saleable. The dollar returns for the crop were substantially greater for the 'modified' or integrated control plots than for the 'commercial' plots. The total returns per 1000 trees for the five-year period were $10 500 more for the 'modified' plots than for the 'commercial' plots.

According to LeRoux, the results indicate that a more thorough and intensive approach is needed, if the mechanisms of orchard epidemiology are to be more clearly understood and methods of manipulating the orchard environment towards more efficient pest control are to be devised. Favoring effective populations of natural enemies of orchard pests is an ecological problem and to elucidate the problem more adequately, an attempt should be made first to study the natural limitations of orchard pest populations and their inter-relationships in the apple orchard habitat. Significant developments in these respects have been made in recent years in the United States and Canada (see Croft & Hoyt, 1983).

UNITED STATES. Apple IPM programs in the United States have focused on the prevention of secondary pest outbreaks. When DDT replaced lead arsenate for the control of codling moth and other pests after World War II, red spider mites soon increased and became major pests of apple, and rapid development of resistance to acaricides further exacerbated the problem. The first integrated mite control program was implemented in the early 1960s by S. C. Hoyt and colleagues in Washington State. Initial phases of the program emphasized the conservation of early-season populations of the apple rust mite (*Aculus schlechtendali*), which served as alternate prey for the predatory phytoseiid mite, *Metaseiulus occidentalis*. The rust mite itself rarely caused economic damage, but its presence enabled the predatory mite to effectively control the injurious McDaniel spider mite (*Tetranychus mcdanieli*) and European red mite. Then, when resistance to organophosphorus pesticides was discovered in *M. occidentalis*, the resistant mites became the backbone of the program. Their successful utilization, resulting in effective biological control of red spider mites and nearly complete elimination of summer acaricide sprays, has saved Washington State apple growers an estimated $5 000 000 per year.

Further major developments have included the establishment of pheromone-based monitoring systems for the codling moth and various other lepidopterous pests, refinement of other monitoring methods and economic thresholds, adoption of computerized phenological models for forecasting pest outbreaks, and improvement of cultural controls. Similar programs have been implemented in Michigan, Pennsylvania and most other apple-growing regions. By 1982, IPM technology was used on about 25 percent of the total apple acreage in North America, and the entire industry was at least indirectly affected. (See Asquith *et al.*, 1980; Croft *et al.*, 1984; Whalon & Croft, 1984.)

Citrus in California, South Africa and Israel

CALIFORNIA. The Fillmore Citrus Protective District in Ventura County, California, comprising some 3500 ha, is a unique co-operative pest control association that was under the supervision of a trained entomologist, Howard Lorbeer, for 47 years. One of the authors (PD) co-operated closely with Mr Lorbeer for many years. The district maintains its own insectaries to produce natural enemies as needed, and uses only a minimum of selective insecticides. They have nearly perfected integrated control, and as a result their pest control costs are the lowest of any district in California and their fruit quality and quantity among the highest. During 1971–80, the mean annual cost of pest control was $72 per hectare in the Fillmore district, as compared to $362 per hectare of Valencia oranges in Ventura County (Graebner, Moreno & Baritelle, 1984).

This has not always been the case. Prior to about 1940, most of the citrus orchards in the district were fumigated annually with hydrogen cyanide gas or sprayed with oil to control the Mediterranean black scale (*Saissetia oleae*), which was the major pest of citrus in southern California at that time. Other treatments were applied for a variety of less critical pests. The black scale problem was largely solved following the successful importation of a new parasite, *Metaphycus helvolus*, from South Africa in 1937 by the University of California at Riverside. This parasite generally produced satisfactory biological control, and has been supplemented by periodic release of insectary-reared parasites in some Fillmore district groves. For the ten-year period 1960–70, an average of less than five percent of the groves were sprayed each year to control black scale. This alone amounts to an annual saving of $100 per hectare, or more than $300 000 per year for the district.

The Fillmore Citrus Protective District originally was organized in 1922 to assist in the chemical eradication of the California red scale (*Aonidiella aurantii*), which had just been found for the first time in several orchards in the district. That was its sole purpose for the first four years, but gradually a program of pest control based primarily upon biological control has been evolved, and insecticides now are used mainly to quickly bring an infestation which threatens economic loss back to the level where effective biological control can be re-established. During the period of this development up to about 1960, the California red scale, a serious pest present elsewhere in southern California, was successfully prevented from establishing a foothold. However, by 1960, more and more small foci of infestation were found, so that it became economically unrealistic to continue red scale eradiction efforts. Fortunately, a promising new red scale parasite, *Aphytis melinus*, had been imported from India and Pakistan by the University of California at Riverside and established elsewhere in southern California in 1957. It was colonized widely in the Fillmore district immediately upon cessation of eradication efforts in 1961, and has given excellent biological control of the red scale ever since. Supplementary releases of insectary-reared *Aphytis* have been made as needed, totaling nearly 4.5 billion parasites since 1961. Some 372 million were reared and colonized during the 1986 and 1987 seasons (Rose, 1990). Otherwise, the integrated control program might have been largely nullified by the necessity to use major insecticidal applications against this key pest. As it turned out only a very few hectares – about one percent – in the district have had to be sprayed each year for the California red scale.

All told in the district, 12 insects that are serious major pests in other districts or other countries are being completely controlled by their natural enemies. These include the Mediterranean black scale, California red scale, yellow scale (*Aonidiella citrina*), brown soft scale (*Coccus hesperidum*), dictyospermum scale (*Chrysomphalus dictyospermi*), cottony-cushion scale (*Icerya purchasi*), woolly whitefly (*Aleurothrixus floccosus*), bayberry whitefly (*Parabemisia myricae*), and four species of mealybugs. Four of these were major pests at one time or another in the Fillmore district, but now are either minor or innocuous ones due to importation of new natural enemies. The two whiteflies were controlled by imported natural enemies even before they could become serious pests. Three other pests partially controlled by native natural enemies are the citrus red mite (*Panonychus citri*), spirea aphid (*Aphis spiraecola*) and navel orangeworm (*Amyelois transitella*). More acreage is treated for these latter than for anything else in the district, but these are generally selective minor treatments and the average for everything remains less than one spray per acre per year.

In addition to their emphasis on biological control (including control of ants which interfere with natural enemies), much of the success of the Fillmore District in achieving such outstanding integrated control must be ascribed to their long-standing policy of discouraging the use of highly toxic, broad-spectrum insecticides, particularly DDT, parathion, malathion and carbaryl.

SOUTH AFRICA. A successful integrated control program for citrus pests has been developed in South Africa (see Bedford & Deacon, 1984; Bedford, Vercueil & Deacon, 1985). Their key pest is the South African citrus thrips (*Scirtothrips aurantii*), which causes cosmetic damage to the fruit. Ironically, as pointed out by Bedford (1976), 'These external blemishes do not affect the internal quality of the fruit detrimentally; in fact, it is claimed that thrips marked oranges are sweeter. The market requirements are however that we must supply an attractive clean skin, which is thrown away. Keeping this skin clean is proving to be a costly business and causes many pest repercussions.' Although some predators have been recorded, the thrips is not under biological control in South Africa.

When an integrated control project is commenced, one should obviously attempt to bring as many pests as possible under complete biological control, especially if suitable enemies are available in other countries. For this purpose, the Department of Agriculture introduced the parasites and predators of four major introduced pests – the Florida red scale (*Chrysomphalus aonidum*), the purple scale (*Lepidosaphes beckii*), the California red scale (*Aonidiella aurantii*) and the citrus blackfly (*Aleurocanthus woglumi*). As a result, the Florida red scale has been reduced to an incredibly low level of subeconomic infestation by *Aphytis holoxanthus*, the purple scale has been brought under complete biological control by *Aphytis lepidosaphes*, and the citrus blackfly was effectively controlled by *Eretmocerus serius* along the coast even before it had a chance to spread into other citrus areas. The California red scale, once considered a key pest, has been brought under complete control by natural enemies in practically the entire Transvaal, except for the hotter areas of the Lowveld, as long as selective pesticides are used in the groves. Many other potentially injurious pests are maintained at subeconomic levels by native or introduced natural enemies, unless the balance is disrupted by application of broad-spectrum pesticides. (See Bedford & Cilliers, 1987.)

Up until 1950, HCN fumigation was usually used against scale insects. Although it was detrimental to natural enemies, it did not leave long-lasting residues on the trees. When it was replaced by parathion about 1950, severe upsets of the citrus red mite (*Panonychus citri*), the brown soft scale (*Coccus hesperidum*) and others became common, and these were dealt with by additional sprays. The situation was exacerbated by the introduction of temephos, which was highly effective against the thrips but caused upsets of other pests (see Chapter 1). The full chemical spray program that was developed might include a winter spray of parathion; another parathion spray at petal fall, followed by two low-dosage sprays of Abate; a spray of methidathion plus narrow-range mineral oil, or mercaptothion; and then a fungicidal spray with mancozeb or benomyl for the control of black spot disease. Variations might occur but the effect, including pest upsets and resurgences, was much the same. The situation became almost impossible when, in 1975, resistance to organophosphorus insecticides was discovered in California red scale populations, and up to five sprays a year were required for that pest alone.

The first step in developing an integrated pest management program is to replace broad-spectrum pesticides with more selective ones. This cannot be accomplished unless the effects of all pesticides on natural enemies are known. For this purpose, large-scale experiments with more than 50 pesticides and combinations of pesticides were carried out in citrus groves, where each pesticide was sprayed repeatedly at the recommended dosage over large blocks of trees. Only those that did not produce any pest upsets after several years of extensive use were considered suitable for the program.

The transition from conventional chemical control to an integrated program that is largely based on biological control may be difficult, because natural enemies are initially very scarce or absent in the groves. The California red scale in particular may flare up if pesticidal treatments are ceased altogether. To overcome this 'red scale barrier', groves in transition are monitored carefully, and corrective sprays are applied in them when necessary during the first one or two seasons. High-pressure rinsing of citrus fruit in the packing house, which reduces culling for scale insects to a minimum, also helps in the transition period.

The integrated pest management program in the Transvaal is based on biological control of the California red scale and most other pests, and on application of non-disruptive pesticides when necessary. These include chlorobenzilate or other selective acaricides for control of the citrus bud mite (*Eriophyes sheldoni*), endosulfan if the African bollworm (*Heliothis armigera*) is present, soil treatments with systemic insecticides against the citrus psylla (*Trioza erytreae*), and two or three sprays of tartar emetic plus sugar against the citrus thrips. The latter may be replaced by two sprays of triazophos or isofenphos, which are also effective against the citrus psylla and have not caused pest upsets. In the Lowveld, two sprays – benomyl plus narrow-range oil, followed by mancozeb – are used to control both black spot and the citrus rust mite (*Phyllocoptruta oleivora*). To minimize interference with natural enemy activity, nutritive sprays are applied early in the season, the trees are skirted and weeds are controlled to prevent ants from reaching the foliage, and injurious ants are controlled when necessary by spot treatments and sticky barriers.

Relaxation of marketing standards regarding thrips-scarred fruit would certainly go a long way toward improving the program. Of course, the best solution would be to obtain natural enemies that would bring this and some of the remaining key pests under biological control. But even as it is, the integrated control program has reduced the cost of pesticides for participating citrus growers by 50 percent or more. Nevertheless, acceptance was at first slow. In the words of Bedford & Cilliers (1987):

> 'Many growers were reluctant to accept the recommended integrated programmes until they were forced to do so when resistant red scale appeared in the Letaba-Letsitele area and spread to other parts of the Lowveld. These growers then had no alternative but to apply integrated

programmes as quickly as possible and to encourage the biological control of red scale.

Integrated programmes could have been applied much sooner, before resistant red scale had taken its toll, but there was considerable resistance from growers and entomologists who still put all their faith in chemical control. It took time and effort to bring about acceptance of the modern concept of IPM.'

ISRAEL. The integrated pest management program on citrus in Israel has been developed on a nation-wide basis and has affected practically the entire citrus industry (*ca* 35 000 ha) of the country. Aside from developing the necessary foundation of bio-ecological research to make the program technologically feasible, a major reason for this unique achievement lies in the organization of the citrus industry under the Citrus Marketing Board of Israel. The Board not only constitutes the sole marketing authority, but finances pest control research and maintains a large insectary – the Israel Cohen Institute of Biological Control, at Rehovot – for natural enemy importation and production. Perhaps most importantly, it is authorized to regulate and carry out certain pest control operations on a county-wide scale. Another important feature has been a highly effective extension service, provided until 1970 by the Board and since then by the Ministry of Agriculture.

Early developments included complete biological control of the cottony-cushion scale by the vedalia ladybeetle, introduced in 1912, and of the citriculus mealybug by a parasite introduced in 1940 (see Chapter 6). By 1955, the main pests of citrus in Israel were (1) the Florida red scale (*Chrysomphalus aonidum*), which required one or two sprays of mineral oil each year, (2) the Mediterranean fruit fly (*Ceratitis capitata*), which was controlled with full-coverage sprays of chlorinated hydrocarbons, and (3) the citrus rust mite (*Phyllocoptruta oleivora*), which was controlled with non-selective sulfur preparations.

Development of an integrated pest management program was made possible by the complete biological control of the Florida red scale in the late 1950s (see Chapter 6). It was soon realized that in order to conserve the natural enemies and maintain this outstanding success, all other pest control practices on citrus had to be modified. Hence, the treatments against the Mediterranean fruit fly were replaced by strip sprays of a poison bait, containing protein hydrolysates as a powerful attractant for female flies and malathion as a poison. The bait is sprayed from the air at a very low rate, down to one liter per hectare. The highly mobile flies find the scattered spots of bait, but natural enemies of non-target species are largely spared. The bait spray is applied only as needed, based on an effective lure-trap technique of sampling male flies. Concurrently, all pesticide formulations recommended for use against citrus pests were tested for their effects on natural enemies. As a result, selective acaricides were adopted against the rust mite, and this not only aided in obtaining better biological control of scale insects and mealybugs, but also enabled natural enemies of the mite to increase to the extent that the rust mite ceased to be a pest of major economic importance. Also, to prevent the adverse effects of pesticidal drift from adjacent fields, appropriate legislation has been enacted prohibiting aerial spraying of cotton and other crops with non-selective pesticides within 200 m of a citrus grove. Also, numerous natural enemies were imported and established against various citrus pests during that period.

These developments laid the foundations for integrated pest management. By 1966, the Florida red scale had virtually disappeared from citrus groves, the rust mite had declined, the medfly was effectively controlled by bait sprays, and several other pests were under good biological control. However, a complex agro-ecosystem is a dynamic, ever-changing system, and new measures had to be taken from time to time as new threats appeared. Thus, when the Mediterranean black scale (*Saissetia oleae*) gradually increased in importance and became a serious pest in the 1970s, it was brought under effective biological control by introduced parasites. Then, when the citrus whitefly (*Dialeurodes citri*) and the bayberry whitefly (*Parabemisia myricae*) invaded Israel, effective parasites were imported against them in the early 1980s and complete biological control was again achieved. Further developments included the use of pheromone traps to control the citrus flower moth (*Prays citri*) (see Chapter 9) and other pests, introduction of selective insect growth regulators against scale insects, adjustment of droplet size in bait sprays to minimize possible detrimental effects on natural enemies, post-harvest rinsing of fruit to remove scale insects and sooty mold, and refinement of the supervised control system, with a network of field scouts taking bi-weekly samples on designated trees. As a result of these concerted efforts, the use of pesticides on citrus has been drastically reduced. Development of resistance to organophosphorus and carbamate insecticides in scale insect populations also forced growers, who had increasingly turned to them in the 1970s, to return to selective mineral oil sprays. By 1988, some 30 percent of the citrus acreage had not been treated chemically at all against pests (except for the medfly bait sprays), and only about 10 percent of the total area received broad-spectrum pesticides. (See Rosen, 1980; Rössler & Rosen, 1990.)

Tropical perennial crops in Malaysia

Following the application of broad-spectrum, long-lasting insecticides, especially DDT, dieldrin and aldrin, serious pest upsets occurred on oil palms (Wood, 1971) and cocoa (Conway, 1972). On oil palms in West Malaysia, the occasional use of the insecticides mentioned for minor outbreaks of bagworms, nettle caterpillars and other insects, led to the ever-increasing need for additional treatments for several years, until the bagworms and nettle caterpillars were no longer occasional pests but constant ones. The more insecticide was used, the worse the problem became. Furthermore, even with these very strong but non-selective insecticides, sometimes a 40 percent crop loss occurred and it took several years for the trees to recover from defoliation. It was initially hypothesized, and later on detailed field studies showed, that natural enemies of these pests no longer existed in effective numbers in the treated groves and that the worst outbreaks occurred after the initial use of DDT, endrin and dieldrin. The recognition of these facts made the need for selective insecticides and selective application methods obvious. Experiments then showed that trichlorphon or lead arsenate gave sufficient control of the pest populations, but were relatively selective and allowed the natural enemies to survive. Further field studies showed the ideal time to spray; this was when young bagworms were emerging and before maximum parasite activity developed. Recent advances have included improved census methods, aerial application, trunk injection of systemic insecticides and introduction of even more selective IGRs and microbial insecticides. This has resulted in an effective integrated control program, where the pests are normally under biological control but occasional outbreaks are successfully controlled with a selective insecticide. (See Wood, 1987, 1988.)

Similar spectacular results occurred with cocoa in Sabah. The new cocoa plantations, planted in primary forest clearings with some trees left for shade, were attacked by native borers which sometimes caused dieback or even death of cocoa trees. This led the growers to adopt dieldrin sprays, which initially gave control. However, subsequent to the use of dieldrin, leaf-eating caterpillars became a problem and this led to more spraying with a wide variety of non-selective insecticides. As a result, several more pests in various taxonomic groups appeared in outbreaks of increasing intensity, until finally very severe defoliation by several species of bagworms occurred. Conway (1972), realizing that these outbreaks were probably upsets resulting from decimation of natural enemies, recommended that spraying be stopped. When this was done, some of the pests began to disappear more or less rapidly, and this was associated with increased parasite activity. However, bagworms still constituted a pest, so they were controlled with selective sprays of trichlorphon. Concurrently, studies revealed that a particular common native shade tree species was host to the original bark borer pest but was relatively uninjured by it. Removal of this tree solved the bark borer problem which had led to the whole chain of events.

The cocoa pod borer, a cocoa-infesting form of an indigenous moth (*Conopomorpha cramerella*), invaded Sabah in 1980 and Penisular Malaysia in 1986, causing heavy losses. Chemical control is rather ineffective and often results in upsets of various other pests, but regular complete harvesting, removing all pods with any sign of ripeness, effectively controls the pest. Other selective methods have been tried, including mass releases of *Trichogrammatoidea* egg parasites and utilization of pheromones, but cultural control has given the best results. A search for exotic natural enemies is currently in progress, and this may eventually solve the problem. Another key pest is *Helopeltis*, a plant-sucking mirid bug. Low dosage *gamma*-BHC (lindane) sprays, carefully timed according to population level, provide adequate control and have not caused any major upsets. A promising alternative approach is to establish and maintain black ants (*Dolichoderus bituberculatus*), which can effectively eliminate the bug. (See Ooi *et al.*, 1987; Wood & Chung, 1989.)

Cotton in Peru and the United States

Doutt & Smith (1971) describe how the chemical cart ran away with the biological horse. They were speaking of the closed, circular self-perpetuating system, employing a completely unilateral method of chemical pest control, that did not even bother to ask the question of whether a treatment was really necessary or could be justified economically. This they termed *the pesticide syndrome*.

In cotton, they discuss how the pattern of crop protection has tended to pass through a series of five phases, as follows: (1) *Subsistence phase* – primitive, non-irrigated local agriculture, low yields, and no organized program of crop protection. The crop protection that occurs results from natural control, inherent plant resistance, hand picking, cultural practices and rarely used chemical treatments. (2) *Exploitation phase* – irrigated agriculture on new land, crop protection measures introduced to protect the large and more valuable crops. These are often necessitated by the changed ecology. Such measures are nearly always solely dependent on chemical pesticides. They are used intensively, and often on fixed arbitrary schedules. At first these techniques are successful, resulting in high yields of seed and lint. (3) *Crisis phase* – develops after a variable number of years of heavy use of pesticides. Frequency of application of pesticides increases to maintain control; pest populations become so resistant to particular pesticides that they are useless. Another pesticide is substituted but the same events occur, only more rapidly. New pests arise and old ones resurge rapidly following pesticide treatment due to upsets of natural enemies. All this occasions a greatly increased application of pesticides and causes greatly increased production costs. (4) *Disaster phase* – because of high pesticide costs and failing control, cotton can no longer be grown profitably. Marginal farmers are the first to cease production. Eventually, cotton production becomes unprofitable in the entire area. This has occurred or is occurring in several countries. (5) *Integrated control* or *Recovery phase* – a crop protection system is devised, comprising an ecological approach to pest management with strong reliance on natural enemies in combination with other appropriate techniques, including cultural control, resistant host plants and selective pesticides.

PERU. The pattern outlined above is well illustrated by the famous case history of cotton in the Cañete Valley of Peru. The initial phases will be omitted here. Some of the later phases were described in Chapter 1 and will be covered only briefly. Beginning in the early 1950s, the new organic insecticides (especially DDT, BHC and toxaphene) began to be used and some culture methods (new cotton strains and irrigation) changed. Yields increased by over 50 percent in four to five years. The idea developed that the more pesticides used the better. They came to be applied like a blanket over the entire Valley. Beneficial parasites and predators were decimated. The number of treatments increased year after year and started earlier each year. By late 1952, resistance had developed to BHC by aphids. By mid-1954, toxaphene failed to control the leafworm. In 1955–6, boll weevil reached high levels early in the season, then a new moth pest, *Argyrotaenia*, appeared. Next, the bollworm (*Heliothis*) exploded and showed a high resistance to DDT. Organophosphorus compounds were substituted for the chlorinated hydrocarbons. The period between treatments was progressively shortened from 8–15 days down to three days. In the interim, a whole complex of previously innocuous insects rose to serious pest status. Not surprisingly, the 1955–6 season was an economic disaster for the cotton growers of the Cañete Valley. Millions of bales were lost. In spite of the tremendous amounts of insecticides used in futile pest control attempts, yields dropped to less than half of what they were only two or three years before.

This catastrophe forced on the Cañete growers the realization that they could not rely on pesticides alone. They appealed to their Experiment Station for help, and this led to the development of a successful integrated control method. Changes in pesticide practices were made, cultural control methods improved, perennial cotton (which enabled carryover of certain pests) prohibited, and natural enemies were reintroduced from other areas and fostered in various ways. Use of synthetic organic insecticides was generally prohibited, and there was a return to the more selective arsenicals and to nicotine sulfate. These changes and others were made legal regulations of the Ministry of Agriculture, with the approval of the growers. As a result, there was a rapid and striking reduction in the severity of cotton pests. The entire group of formerly harmless species, which cropped up in the organic insecticide period, reverted to their innocuous status. The intensity of key pest problems also decreased, so that there was an overall reduction in direct pest control costs, along with better control. By the next year, the yield increased markedly and since has averaged out at the highest yields in the history of the Valley.

Similar stories could be told for other cotton areas in other countries. Large areas in Mexico around Torreon (Coahuila), Culiacan (Sinaloa) and Hermosillo (Sonora), where treatments had become excessive yet were still failing, have developed programs leading to greatly reduced use of pesticides. Each of these areas has its own large insectary that produces millions of *Trichogramma* egg parasites for field release against bollworms.

UNITED STATES. Integrated pest management programs have been vigorously developed in various cotton-producing areas of the United States. Pest problems differ from area to area and many of them remain to be solved, but some of the programs have already been remarkably successful, and the prognosis for ultimate development of a high degree of sophisticated pest management in cotton on a nation-wide basis appears very good.

In Texas, the pink bollworm (*Pectinophora gossypiella*) is effectively controlled by cultural methods (see Chapter 9). Insecticides are now seldom required for this key pest. Another major pest, the boll weevil (*Anthonomus grandis*), is usually controlled with pesticides. However, the introduction of new short-season varieties of cotton, in which the bolls mature before the main population buildup of the weevil, has been the key to the development of a highly successful IPM program. Thus, one to three pesticidal applications are used to control the overwintering weevils early in the season, before they cause any crop damage. Low dosages of relatively selective chemicals are used, such as azinphosmethyl, which are effective against the boll weevil but cause only minimal harm to its natural enemies, and they are applied only when necessary, based on careful monitoring. This allows natural enemies to control *Heliothis* spp. effectively, so that such secondary pests can be largely ignored. Additionally, the short-season cottons require less fertilizer and water than other varieties. All in all, the IPM system has been widely adopted and has resulted in a 75 percent reduction in the use of insecticides. Net profits have been estimated as $350–475 per hectare higher than under the conventional system. In the Coastal Bend area of southern Texas, this program has resulted in an annual benefit estimated at $29 million.

In the Central Valley of California, the key pest of cotton is the lygus bug (*Lygus hesperus*). Other potential pests such as the cotton bollworm (*Heliothis zea*), the beet armyworm (*Spodoptera exigua*) and the cabbage looper (*Trichoplusia ni*) are secondary and do not become serious unless their natural enemies are destroyed. The main goal of the IPM program was to reduce the use of pesticides. This was made possible by detailed studies and modeling that indicated that a fixed economic threshold for the lygus bug was erroneous, because the bug does not cause any damage after the peak period of boll squaring. Thus, a dynamic economic threshold was developed, and the system was based on careful monitoring of plant development as well as pest and natural enemy populations, and on some of the cultural practices mentioned earlier in this chapter. Highly disruptive chemicals have been eliminated, and applications are avoided during the period that is most critical to the natural enemies of the secondary pests. In fact, certain pesticides such as methyl parathion were found not only to induce pest upsets, but also to reduce yields due to adverse physiological effects on the cotton plant. As a result, pesticide usage has been reduced considerably and the cotton bollworm, the beet armyworm and the cabbage looper are no longer pests of economic importance. (See Flint & van den Bosch, 1981; Phillips, Gutierrez & Adkisson, 1980; Reynolds *et al.*, 1982.)

Rice in Southeast Asia

Grown by tens of millions of small farmers, rice is by far the most important crop in tropical Asia. Major pests include stem borers, planthoppers and leafhoppers, as well as various diseases and weeds. Even before the introduction of 'green revolution' varieties, farmers became increasingly dependent on pesticides. In the Philippines, in the early 1950s only 3 percent of irrigated rice farmers used insecticides, as compared to 60 percent in 1965, when over half the irrigated rice land of the country was treated free of charge by government technicians. In an effort to improve rice production, governments have advised farmers to use more insecticides, have subsidized them and included them in loan packages. By the mid-1980s, over 95 percent of irrigated rice farmers in the Philippines used insecticides on every crop and believed them to be 'progressive', modern, effective and profitable. Yet, careful studies have repeatedly shown that insecticide use was not correlated with insect damage, that fields sprayed 9 to 11 times often did not give higher yields than unsprayed fields, and that reliance on economic thresholds was more profitable than preventative, calendar-based treatments. Worst of all, heavy pesticide use induced widespread upsets of the brown planthopper (*Nilaparvata lugens*), which has become the most destructive pest of tropical rice (see Chapter 1).

An Inter-Country Integrated Pest Control Program was established in 1980 by FAO in Bangladesh, India, Sri Lanka, Thailand, Malaysia, the Philippines and Indonesia. Based on conservation of natural enemies and utilization of pest-resistant rice varieties, this program aims to train farmers to monitor their fields once a week, to distinguish between pests and natural enemies and between insect damage and disease symptoms, and to use thresholds for minimal, need-based pesticide application. Teaching these skills to millions of farmers is a tremendous challenge to extension systems. In the Philippines, thousands of farmers and extension workers have undergone such training, and those that have adopted integrated control have already shown marked reduction in pesticide use, with increased yields, lower costs and higher profits than those resorting to conventional practices. Several Southeast Asian governments have declared IPM in rice to be national policy, and in 1986 President Suharto of Indonesia banned 57 kinds of pesticides from any use on rice. (See Kenmore *et al.*, 1987; Zelazny, Chiarappa & Kenmore, 1985.)

Conclusions

It is clear from the examples discussed, and from many others in the literature, that we are in either the crisis or disaster phases connected with unilateral pest control with chemicals on many major agricultural crops in one place or another. These failures of the chemical method are mainly due to adverse effects on natural enemies, compounded by increasing development of resistance by pests to pesticides. The achievement of complete control by natural enemies is the ideal goal and generally the most practical single approach to pest management. However, even though its use can be very greatly expanded, this must be based on increased bio-ecological research that will take time. Also, there will be refractory pests not amenable to control by natural enemies, and these will require other control strategies. Meanwhile, the immediate application of current ecological knowledge and technology to the pesticide dilemma can rapidly lead to general reductions of about 50 percent in the use of pesticides, with consequent greatly reduced environmental pollution and production costs, and often with increases in yield and quality. An immediate change to more selective chemicals and to selective means of application and timing will go a long way towards this initial achievement, but concurrent manipulation of certain natural enemies may be necessary or desirable. Adequately trained supervising entomologists will generally be necessary to implement such a program.

Further scientific sophistication in pest management, stressing conservation, importation and augmentation of natural enemies, will solve additional key pest problems and will further significantly reduce the need for chemical pesticides. In some cases the need will be eliminated completely. Refinement of cultural controls (including mechanical controls) and development of resistant host plants furnish additional potent general tools for use in integrated pest management. The other non-chemical methods of pest control discussed in Chapter 9 may furnish the key to optimal pest management in certain instances, but are not yet of very general applicability. Each ecosystem is different, and objective decisions for the ultimate in *optimal* pest management in a given instance must be based on thorough bio-ecological research on the organisms involved, and the integration of this knowledge with economic, social and other realities. Systems analysis and modeling should prove to be a potent tool in this regard.

BIBLIOGRAPHY

Adkisson, P. L. (1971). Objective use of insecticides in agriculture. in *Agricultural Chemicals – Harmony or Discord*, ed. J. E. Swift, pp. 43–51. University of California Division of Agricultural Sciences.

Adkisson, P. L., G. A. Niles, J. K. Walker, L. S. Bird & H. B. Scott. (1982). Controlling cotton's insect pests: a new system. *Science*, **216**: 19–22.

Altieri, M. A. & D. K. Letourneau. (1982). Vegetation management and biological control in agroecosystems. *Crop Protec.* 4: 405–30.

Andres, L. A., C. J. Davis, P. Harris & A. J. Wapshere. (1976). Biological control of weeds. In: *Theory and Practice of Biological Control*, ed. C. B. Huffaker & P. S. Messenger, pp. 481–99. New York: Academic Press.

Andres, L. A. & R. D. Goeden. (1971). The biological control of weeds by introduced natural enemies. In: *Biological Control*, ed. C. B. Huffaker, pp. 143–64. New York: Plenum Press.

Apple, J. L. & R. F. Smith (Eds). (1976). *Integrated Pest Management*. New York: Plenum Press.

Askew, R. R. (1971). *Parasitic Insects*. London: Heinemann Educational Books.

Asquith, D., B. A. Croft, S. C. Hoyt, E. H. Glass & R. E. Rice. (1980). The systems approach and general accomplishments toward better insect control in pome and stone fruits. In: *New Technology of Pest Control*, ed. C. B. Huffaker, pp. 249–317. New York: John Wiley & Sons.

Axtell, R. C. (1981). Use of predators and parasites in filth fly IPM programs in poultry housing. In: *Status of Biological Control of Filth Flies*, ed. R. S. Patterson, pp. 26–43. New Orleans: Science and Education Administration, US Department of Agriculture.

Bal, A. & J. C. van Lenteren. (1987). Integrated pest management in the Netherlands: practice, policy and opportunities for the future. *Med. Fac. Landbouww. Rijksuniv. Gent*, **52**: 385–93.

Baltazar, C. R. (1981). Biological control attempts in the Philippines. *Philipp. Entom.*, 4: 505–23.

Bartlett, B. R. (1978). Pseudococcidae. In: *Introduced Parasites and Predators of Arthropod Pests and Weeds: a World Review*, ed. C. P. Clausen, pp. 137–70. Agric. Handbook 480. Washington, DC: US Department of Agriculture.

Bartlett, B. R. & R. van den Bosch. (1964). Foreign exploration for beneficial organisms. In: *Biological Control of Insect Pests and Weeds*, ed. P. DeBach, pp. 283–304. New York: Reinhold.

Beck, S. D. & F. G. Maxwell. (1976). Use of plant resistance. In: *Theory and Practice of Biological Control*, ed. C. B. Huffaker & P. S. Messenger, pp. 615–36. New York: Academic Press.

Beckendorf, S. K. & M. A. Hoy. (1985). Genetic improvement of arthropod natural enemies through selection, hybridization or genetic engineering techniques. In: *Biological Control in Agricultural IPM Systems*, ed. M. A. Hoy & D. C. Herzog, pp. 167–87. Orlando: Academic Press.

Bedford, E. C. G. (1976). Citrus pest management in South Africa. *Proc. Tall Timbers Conf. Ecol. Anim. Contr. Habit. Mgmt*, 6: 19–42.

Bedford, E. C. G. (1990). Mechanical control: high-pressure rinsing of citrus fruit. In: *Armored Scale Insects, Their Biology, Natural Enemies and Control*, ed. D. Rosen, pp. 507–14. World Crop Pests, Vol. 4B. Amsterdam: Elsevier Science Publishers.

Bedford, E. C. G. & C. J. C. Cilliers. (1987). Highly successful biological control of four important citrus pests. In: *Research Highlights 1987. Plant Production*, ed. L. L. Lötter, pp. 96–9. Pretoria: Department of Agriculture & Water Supply.

Bedford, E. C. G. & V. Deacon. (1984). An integrated pest control program for citrus in the Transvaal Lowveld. *Farming in South Africa*, Citrus Leaflet **H.47.1/1984**.

Bedford, E. C. G., S. W. Vercueil & V. Deacon. (1985). A general guide to the integrated control of citrus pests with red scale under biological control. *Farming in South Africa*, Citrus Leaflet **H.47/1985**.

Bedford, G. O. (1980). Biology, ecology, and control of palm rhinoceros beetles. *Ann. Rev. Entomol.* **35**: 309–39.

Bedford, G. O. (1986). Biological control of the rhinoceros beetle (*Oryctes rhinoceros*) in the South Pacific by baculovirus. *Agric. Ecosyst. Environ.* **15**: 141–7.

Beglyarov, G. A. & A. I. Smetnik. (1977). Seasonal colonization of entomphages in the USSR. In: *Biological Control by Augmentation of Natural Enemies*, ed. R. L. Ridgway & S. B. Vinson, pp. 283–328. New York: Plenum Press.

Beirne, B. P. (1985). Avoidable obstacles to colonization in classical biological control of insects. *Canad. J. Zool.* **63**: 743–47.

Biliotti, E. (1977). Augmentation of natural enemies in western Europe. In: *Biological Control by Augmentation of Natural Enemies*, ed. R. L. Ridgway & S. B. Vinson, pp. 341–7. New York: Plenum Press.

Birch, M. C. & K. F. Haynes. (1982). *Insect Pheromones*. Studies in Biology No. 147. London: Edward Arnold.

Bodenheimer, F. S. (1931). Zur Fruhgeschichte der Enforschung des Insektenparasitismus. (To the early history of the study of insect parasitism). *Arch. Geschich. Math. Naturwiss. Tech.* **13**: 402–16.

Boethel, D. J. & R. D. Eikenbary. (1986). *Interactions of Plant Resistance and Parasitoids and Predators of Insects*. Chichester: Ellis Horwood.

Boller, E. F. (1987). Genetic control. In: *Integrated Pest Management*, ed. A. J. Burn, T. H. Coaker & P. C. Jepson, pp. 162–87. London: Academic Press.

Bottrell, D. G. (1979). *Integrated Pest Management*. President's Council for Environmental Quality. Washington, DC: US Government Printing Office.

Bottrell, D. G. & P. L. Adkisson. (1977). Cotton insect pest management. *Ann. Rev. Entomol.* **22**: 451–81.

Boyce, A. M. (1987). *Odyssey of an Entomologist*. Riverside, CA: UC Riverside Foundation.

Brader, L. (1979). Integrated pest control in the developing world. *Ann. Rev. Entomol.* **24**: 225–54.

Briggs, S. A. (1970). Safer pesticides for home and garden. In: *Since Silent Spring*,

Appendix I, by F. Graham, Jr., pp. 275–97. Boston: Houghton Mifflin.

Brown, A. W. A. (1978). *Ecology of Pesticides*. New York: John Wiley & Sons.

Burges, H. D. (Ed.). (1981). *Microbial Control of Pests and Plant Diseases 1970–1980*. New York: Academic Press.

Burges, H. D. & N. W. Hussey (Eds). (1971). *Microbial Control of Insects and Mites*. New York: Academic Press.

Burleigh, J. G., J. H. Young & R. D. Morrison. (1973). Strip-cropping's effect on beneficial insects and spiders associated with cotton in Oklahoma. *Environ. Entomol.* **2**: 281–85.

Burn, A. J., T. H. Coaker & P. C. Jepson (Eds). (1987). *Integrated Pest Management*. London: Academic Press.

Caltagirone, L. E. & R. L. Doutt. (1989). The history of the vedalia beetle importation to California and its impact on the development of biological control. *Ann. Rev. Entomol.* **34**: 1–16.

Caltagirone, L. E., K. P. Shea & G. L. Finney. (1964). Parasites to aid control of navel orangeworm. *Calif. Agric.* **18** (1): 10–12.

Canning, E. U. (1981). Insect control with Protozoa. In: *Biological Control in Crop Production*, ed. G. C. Papavizas, pp. 201–16. Beltsville Symposia in Agricultural Research 5. Totowa: Allanheld, Osmun.

Cantwell, G. E. (Ed.). (1974). *Insect Diseases*, 2 vols. New York: Marcel Dekker.

Carl, K. P. (1982). Biological control of native pests by introduced natural enemies. *Biocontrol News and Information* **3**: 191–200.

Carson, R. (1962). *Silent Spring*. Boston: Houghton Mifflin.

Cate, J. R. (1985). Cotton: status and current limitations to biological control in Texas and Arkansas. In: *Biological Control in Agricultural IPM Systems*, ed. M. A. Hoy & D. C. Herzog, pp. 537–56. Orlando: Academic Press.

Cavalloro, R. (Ed.). (1983). *Fruit Flies of Economic Importance*. Rotterdam: A. A. Balkema.

Charudattan, R. (1985). The use of natural and genetically altered strains of pathogens for weed control. In: *Biological Control in Agricultural IPM Systems*, ed. M. A. Hoy & D. C. Herzog, pp. 347–72. Orlando: Academic Press.

Clausen, C. P. (1962). *Entomophagous Insects*. New York: Hafner Publishing Co.

Clausen, C. P. (1976). Phoresy among entomophagous insects. *Ann. Rev. Entomol.* **21**: 343–68.

Clausen, C. P. (Ed.). (1978). *Introduced Parasites and Predators of Arthropod Pests and Weeds: a World Review*. Agric. Handbook **480**. Washington, DC; US Department of Agriculture.

Clausen, C. P. & P. A. Berry. (1932). The citrus blackfly in Asia, and the importation of its natural enemies into Tropical America. *US Dept. Agric. Tech. Bull.* **320**.

Clausen, C. P., D. W. Clancy & Q. C. Chock. (1965). Biological control of the oriental fruit fly (*Dacus dorsalis* Hendel) and other fruit flies in Hawaii. *US Dept. Agric. Techn. Bull.* **1322**.

Clausen, C. P., J. L. King & C. Teranishi. (1927). The parasites of *Popillia japonica* in Japan and Chosen (Korea) and their introduction into the United States. *US Dept. Agric. Bull.* **1429**.

Coaker, T. H. (1987). Cultural methods: the crop. In: *Integrated Pest Management*, ed. A. J. Burn, T. H. Coaker & P. C. Jepson, pp. 69–88. London: Academic Press.

Cock, M. J. W. (1985). The use of parasitoids for augmentative biological control of pests in the People's Republic of China. *Biocontrol News and Information* **6**: 213–23.

Cohen, I. (1975). *Citrus Fruit Technology*. Tel Aviv: Revivim. (In Hebrew.)

Compere, H. (1961). The red scale and its insect enemies. *Hilgardia*, **31**: 173–278.

Compere, H. (1969). The role of systematics in biological control: a backward look. *Israel J. Entomol.* **4**: 5–10.

Compere, H. & H. S. Smith. (1932). The control of the citrophilus mealybug, *Pseudococcus gahani*, by Australian parasites. *Hilgardia*, **6**: 585–618.

Conway, G. R. (1969). A consequence of insecticides – pests follow the chemicals in the cocoa of Malaysia. *Nat. Hist.* **78** (2): 46–51.

Conway, G. R. (1972). Ecological aspects of pest control in Malaysia. In: *The Careless Technology: Ecology and International Development*, ed. M. T. Farvar & J. P. Milton, pp. 467–88. Garden City, New York: The Natural History Press.

Cook, R. J. & K. F. Baker. (1983). *The Nature and Practice of Biological Control of Plant Pathogens.* St. Paul: American Phytopathological Society.

Corbet, P. S. (1980). Biology of Odonata. *Ann. Rev. Entomol.* **25**: 189–217.

Coulson, J. R., R. W. Fuester, P. W. Schaefer, L. R. Ertle, J. S. Kelleher & L. D. Rhoads. (1986). Exploration for and importation of natural enemies of the gypsy moth, *Lymantria dispar* (L.) (Lepidoptera: Lymantriidae), in North America: an update. *Proc. Entomol. Soc. Wash.* **88**: 461–75.

Coulson, J. R., W. Klaasen, R. J. Cook, E. G. King, H. C. Chiang, K. S. Hagen & W. G. Yendol. (1982). Notes on biological control of pests in China, 1979. In: *Biological Control of Pests in China*, pp. 1–192. Washington, DC: US Department of Agriculture.

Croft, B. A. (1976). Establishing insecticide-resistant phytoseiid mite predators in deciduous tree fruit orchards. *Entomophaga*, **21**: 383–99.

Croft, B. A. (1990). *Arthropod Biological Control Agents and Pesticides.* Somerset: John Wiley & Sons.

Croft, B. A., P. L. Adkisson, R. W. Sutherst & G. A. Simmons. (1984). Applications of ecology for better pest control. In: *Ecological Entomology*, ed. C. B. Huffaker & R. L. Rabb, pp. 763–95. New York: John Wiley & Sons.

Croft, B. A. & A. W. A. Brown. (1975). Responses of arthropod natural enemies to insecticides. *Ann. Rev. Entomol.* **20**: 285–334.

Croft, B. A. & S. C. Hoyt (Eds). (1983). *Integrated Management of Insect Pests of Pome and Stone Fruits.* New York: John Wiley & Sons.

Croft, B. A. & J. A. McMurtry. (1972). Minimum releases of *Typhlodromus occidentalis* to control *Tetranychus mcdanieli* on apple. *J. Econ. Entomol.* **65**: 188–91.

Cumber, R. H. (1964). The egg-parasite complex (Scelionidae: Hymenoptera) of shield bugs (Pentatomidae, Acanthosomidae: Heteroptera) in New Zealand. *New Zealand J. Sci.* **7**: 536–54.

Cunningham, J. C. (1988). Baculoviruses: their status compared to *Bacillus thuringiensis* as microbial insecticides. *Outlook Agric.* **17**: 10–17.

Davidson, G. (1974). *Genetic Control of Insect Pests.* London: Academic Press.

Davis, C. J. (1967). Progress in the biological control of the southern green stink bug *Nezara viridula* variety *smaragdula* (Fabricius) in Hawaii (Heteroptera: Pentatomidae). *Mushi*, **39** (Suppl.): 9–16.

Davis, J. J. (1916). *Aphidoletes meridionalis*, an important dipterous enemy of aphids. *J. Agric. Res.* **6**: 883–8, pl. CX.

Deacon, J. W. (1983). *Microbial Control of Plant Pests and Diseases.* Aspects of Microbiology 7. Washington, DC: American Society of Microbiology.

Dean, H. A., J. V. French & D. Meyerdirk. (1983). Development of integrated pest management in Texas citrus. *Texas Agric. Exp. Sta.* **B–1434**.

Dean, H. A., M. F. Schuster, J. C. Boling & P. T. Riherd. (1979). Complete biological control of *Antonina graminis* in Texas with *Neodusmetia sangwani* (a classical example). *Bull. Entomol. Soc. Amer.* **25**: 262–67.

DeBach, P. (1958*a*). Selective breeding to improve adaptations of parasitic insects. *Proc. 10th Int. Congr. Entomol.* (Montreal, 1956), 4: 759–68.

DeBach, P. (1958*b*). The role of weather and entomophagous species in the natural control of insect populations. *J. Econ. Entomol.* 51: 474–84.

DeBach, P. (1960). The importance of taxonomy to biological control as illustrated by the cryptic history of *Aphytis holoxanthus* n. sp. (Hymenoptera, Aphelinidae), a parasite of *Chrysomphalus aonidum* and *Aphytis coheni* n. sp., a parasite of *Aonidiella aurantii*. *Ann. Entomol. Soc. Amer.* 53: 701–5.

DeBach, P. (Ed.). (1964*a*). *Biological Control of Insect Pests and Weeds.* New York: Reinhold Publishing Corp.

DeBach, P. (1964*b*). Some ecological aspects of insect eradication. *Bull. Entomol. Soc. Amer.* 10: 221–4.

DeBach, P. (1966). The competitive displacement and coexistence principles. *Ann. Rev. Entomol.* 11: 183–212.

DeBach, P. (1969). Uniparental, sibling and semi-species in relation to taxonomy and biological control. *Israel J. Entomol.* 4: 11–28.

DeBach, P. (1971). Fortuitous biological control from ecesis of natural enemies. In: *Entomological Essays to Commemorate the Retirement of Prof. K. Yasumatsu,* pp. 293–307. Tokyo: Hokuryukan Publishing Co.

DeBach, P. (1972). The use of imported natural enemies in insect pest management ecology. *Proc. Tall Timbers Conf. Ecol. Anim. Cont. Habit. Mgmt,* 3: 211–33.

DeBach, P. & L. C. Argyriou. (1967). The colonization and success in Greece of some imported *Aphytis* spp. (Hym. Aphelinidae) parasitic on citrus scale insects (Hom. Diaspididae). *Entomophaga,* 12: 325–42.

DeBach, P. & B. R. Bartlett. (1964). Methods of colonization, recovery and evaluation. In: *Biological Control of Insect Pests and Weeds,* ed. P. DeBach, pp. 402–26. New York: Reinhold.

DeBach, P., E. J. Dietrick & C. A. Fleschner. (1949). A new technique for evaluating the efficiency of entomophagous insects in the field. *J. Econ. Entomol.* 42: 546–7.

DeBach, P., E. J. Dietrick, C. A. Fleschner & T. W. Fisher. (1950). Periodic colonization of *Aphytis* for control of the California red scale. Preliminary tests, 1949. *J. Econ. Entomol.* 43: 783–802.

DeBach, P. & K. S. Hagen. (1964). Manipulation of entomophagous species. In: *Biological Control of Insect Pests and Weeds,* ed. P. DeBach, pp. 429–58. New York: Reinhold.

DeBach, P. & C. B. Huffaker. (1971). Experimental techniques for evaluation of the effectiveness of natural enemies. In: *Biological Control,* ed. C. B. Huffaker, pp. 113–40. New York: Plenum Press.

DeBach, P., C. B. Huffaker & A. W. MacPhee. (1976). Evaluation of the impact of natural enemies. In: *Theory and Practice of Biological Control,* ed. C. B. Huffaker & P. S. Messenger, pp. 255–85. New York: Academic Press.

DeBach, P. & J. H. Landi. (1961). The introduced purple scale parasite, *Aphytis lepidosaphes* Compere, and a method of integrating chemical with biological control. *Hilgardia,* 31: 459–97.

DeBach, P., J. H. Landi & E. B. White. (1955). Biological control of red scale. *Calif. Citrogr.* 40: 254, 271–2, 274–5.

DeBach, P. & M. Rose. (1976). Biological control of woolly whitefly. *Calif. Agric.* 30 (5): 4–7.

DeBach, P. & M. Rose. (1977). Environmental upsets caused by chemical eradication. *Calif. Agric.* 31 (7): 8–10.

DeBach, P. & D. Rosen. (1976). Armoured scale insects. In: *Studies in Biological*

Control, ed. V. L. Delucchi, pp. 139–78. Cambridge: Cambridge University Press.

DeBach, P. & D. Rosen. (1982). *Aphytis yanonensis* n. sp. (Hymenoptera, Aphelinidae), a parasite of *Unaspis yanonensis* (Kuwana) (Homoptera, Diaspididae). *Kontyû*, **50**: 626–34.

DeBach, P., D. Rosen & C. E. Kennett. (1971). Biological control of coccids by introduced natural enemies. In: *Biological Control*, ed. C. B. Huffaker, pp. 165–94. New York: Plenum Press.

DeBach, P. & R. A. Sundby. (1963). Competitive displacement between ecological homologues. *Hilgardia*, **34**: 105–66.

Delucchi, V. (Ed.). (1987). *Integrated Pest Management: Quo Vadis?* Geneva: Parasitis.

Delucchi, V., D. Rosen & E. I. Schlinger. (1976). Relationship of systematics to biological control. In: *Theory and Practice of Biological Control*, ed. C. B. Huffaker & P. S. Messenger, pp. 81–91. New York: Academic Press.

Doane, C. C. & M. L. McManus (Eds). (1981). The gypsy moth: research toward integrated pest management. *US Dept. Agric. Tech. Bull.* **1584**.

Dodd, A. P. (1940). *The Biological Campaign Against Prickly Pear*. Brisbane, Australia: Commonwealth Prickly Pear Board.

Doten, S. B. (1911). Concerning the relation of food to reproductive activity and longevity in certain hymenopterous parasites. *Univ. Nevada Agric. Expt. Sta. Tech. Bull.* **78**.

Doube, B. M. (1986). Biological control of the buffalo fly in Australia: the potential of the southern Africa dung fauna. In: *Biological Control of Muscoid Flies*, ed. R. S. Patterson & D. A. Rutz, pp. 16–34. *Entomol. Soc. Amer. Misc. Publ.* **61**.

Doutt, R. L. (1947). Polyembryony in *Copidosoma koehleri* Blanchard. *Amer. Nat.* **81**: 435–53.

Doutt, R. L. (1964*a*). Ecological considerations in chemical control. Implications to nontarget invertebrates. *Bull. Entomol. Soc. Amer.* **10**: 83–8.

Doutt, R. L. (1964*b*). The historical development of biological control. In: *Biological Control of Insect Pests and Weeds*, ed. P. DeBach, pp. 21–42. New York: Reinhold.

Doutt, R. L. & J. Nakata. (1973). The *Rubus* leafhopper and its egg parasitoid: an endemic biotic system useful in grape-pest management. *Environ. Entomol.* **3**: 381–6.

Doutt, R. L. & R. F. Smith. (1971). The pesticide syndrome – diagnosis and suggested prophylaxis. In: *Biological Control*, ed. C. B. Huffaker, pp. 3–15. New York: Plenum Press.

Dowell, R. V., R. H. Cherry, G. E. Fitzpatrick, J. A. Reinert & J. L. Knapp. (1981). Biology, plant-insect relations, and control of the citrus blackfly. *Univ. Fla. Agric. Exp. Sta. Tech. Bull.* **818**.

Drea, J. J. & R. M. Hendrickson. (1986). Analysis of a successful classical biological control project: the alfalfa blotch leafminer (Diptera: Agromyzidae) in the northeastern United States. *Environ. Entomol.* **15**: 448–55.

Dysart, R. J., H. L. Maltby & M. H. Brunson. (1973). Larval parasites of *Oulema melanopus* in Europe and their colonization in the United States. *Entomophaga*, **18**: 133–67.

Ehler, L. E. & P. C. Endicott. (1984). Effect of malathion-bait sprays on biological control of insect pests of olive, citrus and walnut. *Hilgardia*, **52** (5): 1–47.

El Titi, A. (1987). Environmental manipulation detrimental to pests. In: *Integrated Pest Management: Quo Vadis?*, ed. V. Delucchi, pp. 105–21. Geneva: Parasitis.

Elton, C. S. (1958). *The Ecology of Invasions by Animals and Plants*. London: Methuen.

Embree, D. G. (1971). The biological control of the winter moth in eastern Canada by introduced parasites. In: *Biological Control*, ed. C. B. Huffaker, pp. 217–26. New York: Plenum Press.

Embree, D. G. & I. S. Otvos. (1984). *Operophtera brumata* (L.), winter moth (Lepidoptera: Geometridae). In: *Biological Control Programmes against Insects and Weeds in Canada 1969–1980*, ed. J. S. Kelleher & M. A. Hulme, pp. 353–7. Farnham Royal: Commonwealth Agricultural Bureaux.

Entwistle, P. F. & H. F. Evans. (1985). Viral control. In: *Comprehensive Insect Physiology, Biochemistry and Pharmacology*, ed. G. A. Kerkut & L. I. Gilbert, vol. 12, *Insect Control*, pp. 347–412. Oxford: Pergamon Press.

EPPO Report. (1970). *Report of the International Conference on Methods for Forecasting, Warning, Pest Assessment and Dectection of Infection*. Paris: European and Mediterranean Plant Protection Organization.

Essig, E. O. (1931). *A History of Entomology*. New York: Macmillan.

Falcon, L. A. (1985). Development and use of microbial insecticides. In: *Biological Control in Agricultural IPM Systems*, ed. M. A. Hoy & D. C. Herzog, pp. 229–42. Orlando: Academic Press.

Ferron, P. (1978). Biological control of insect pests by entomogenous fungi. *Ann. Rev. Entomol.* **23**: 409–42.

Ferron, P. (1985). Fungal control. In: *Comprehensive Insect Physiology, Biochemistry and Pharmacology*, ed. G. A. Kerkut & L. I. Gilbert, vol. 12, *Insect Control*, pp. 313–46. Oxford: Pergamon Press.

Flaherty, D. L. & C. B. Huffaker. (1970). Biological control of Pacific mites and Willamette mites in San Joaquin Valley vineyards. I. Role of *Metaseiulus occidentalis*. *Hilgardia*, **40**: 267–308.

Flint, M. L. & R. van den Bosch. (1981). *Introduction to Integrated Pest Management*. New York: Plenum Press.

Force, D. (1974). Ecology of insect host-parasitoid communities. *Science*, **184**: 624–32.

Franz, J. M. (Ed.). (1986). *Biological Plant and Health Protection*. Stuttgart: Gustav Fischer Verlag.

Franz, J. M. (1988). Highlights in the development of the International Organization for Biological Control of Noxious Animals and Plants. *Entomophaga*, **33**: 131–4.

Frisbie, R. E. & P. L. Adkisson (Eds). (1986). *Integrated Pest Management on Major Agricultural Systems*. Texas Agricultural Experiment Station, **MP–1616**.

Fulton, B. B. (1933). Notes on *Habrocytus cerealellae*, parasite of the angoumois grain moth. *Ann. Entomol. Soc. Amer.* **26**: 536–53.

Furuhashi, K. & M. Nishino. (1983). Biological control of arrowhead scale, *Unaspis yanonensis*, by parasitic wasps introduced from the People's Republic of China. *Entomophaga*, **28**: 277–86.

Fuxa, J. R. & Y. Tanada (Eds). (1987). *Epizootiology of Insect Diseases*. New York: John Wiley & Sons.

Georghiou, G. P. & T. Saito (Eds). (1983). *Pest Resistance to Pesticides*. New York: Plenum Press.

Gerson, U. & E. Cohen. (1989). Resurgences of spider mites (Acari: Tetranychidae) induced by synthetic pyrethroids. *Exp. Appl. Acarol.* **6**: 29–46.

Gerson, U., B. M. OConnor & M. Houck. (1990). Acari. In: *Armored Scale Insects, Their Biology, Natural Enemies and Control*, ed. D. Rosen, pp. 77–98. World Crop Pests, Vol. 4B. Amsterdam: Elsevier Science Publishers.

Getz, W. M. & A. P. Guiterrez. (1982). A perspective of systems analysis in crop production and insect pest management. *Ann. Rev. Entomol.* **27**: 447–66.

Goedaert, J. (1662). *Metamorphosis et historia naturales insectorum I*, pp. 175–8. Medisburg: Jacobum Fierensium.

Goeden, R. D. (1978). Biological control of weeds. In: *Introduced Parasites and Predators of Arthropod Pests and Weeds: a World Review*, ed. C. P. Clausen, pp. 357–414. Agric. Handbook No. 480. Washington, DC: US Department of Agriculture.

Goeden, R. D. & L. T. Kok. (1986). Comments on a proposed 'new' approach for selecting agents for biological control of weeds. *Canad. Entomol.* **118**: 51–8.

Graham, F., Jr. (1970). *Since Silent Spring*. Boston: Houghton Mifflin.

Graham, F. Jr., (1985). *The Dragon Hunters*. New York: E. P. Dutton.

Graebner, L., D. S. Moreno & J. L. Baritelle. (1984). The Fillmore Citrus Protective District: a success story in integrated pest management. *Bull. Entomol. Soc. Amer.* **30**: 27–33.

Greany, P. D., S. B. Vinson & W. J. Lewis. (1984). Insect parasitoids: finding new opportunities for biological control. *BioScience,* **34**: 690–96.

Greathead, D. J. (1971). A review of biological control in the Ethopian region. *CIBC Tech. Commun.* **5**.

Griswold, G. H. (1926). Notes on some feeding habits of two chalcid parasites. *Ann. Entomol. Soc. Amer.* **19**: 331–4.

Gross, H. R., Jr. (1981). Employment of kairomones in the management of parasitoids. In: *Semiochemicals, Their Role in Pest Control*, ed. D. A. Nordlund, R. L. Jones & W. J. Lewis, pp. 137–50. New York: John Wiley & Sons.

Gulmahamad, H. & P. DeBach. (1978). Biological control of the San Jose scale *Quadraspidiotus perniciosus* (Comstock) (Homoptera: Diaspididae) in southern California. *Hilgardia,* **46**: 205–38.

Hafez, M. & R. L. Doutt. (1954). Biological evidence of sibling species in *Aphytis maculicornis* (Masi) (Hymenoptera, Aphelinidae). *Canadian Entomol.* **86**: 90–6.

Hagen, K. S. (1986). Ecosystem analysis: plant cultivars (HPR), entomophagous species and food supplements. *Interactions of Plant Resistance and Parasitoids and Predators of Insects*, ed. D. J. Boethel & R. D. Eikenbary, pp. 151–97. Chichester: Ellis Horwood.

Hagen, K. S., R. van den Bosch & D. L. Dahlsten. (1971). The importance of naturally-occurring biological control in the western United States. In: *Biological Control*, ed. C. B. Huffaker, pp. 253–93. New York: Plenum Press.

Hakkonen, H. & D. Pimentel. (1984). New approach for selecting biological control agents. *Canad. Entomol.* **116**: 1109–21.

Hall, E. R. (1972). Down on the farm. *Nat. Hist.* **81** (3): 8, 12.

Hallenbeck, W. H. & K. M. Cunningham-Burns. (1985). *Pesticides and Human Health*. New York: Springer-Verlag.

Harmer, R. M. (1971). *Unfit for Human Consumption*. New Jersey: Prentice-Hall, Englewood Cliffs.

Harpaz, I. (1972). *Maize Rough Dwarf*. Jerusalem: Israel Universities Press.

Harris, M. K. (Ed.). (1980). *Biology and Breeding for Resistance to Arthropods and Pathogens of Cultivated Plants*. Texas Agricultural Experiment Station, Misc. Publ. **1451**.

Harris, M. K. & R. A. Frederiksen. (1984). Concepts and methods regarding host plant resistance to arthropods and pathogens. *Ann. Rev. Phytopathol.* **22**: 247–72.

Hassan, S. A. (1986). Side effects of pesticides to entomophagous arthropods. In: *Biological Plant and Health Protection*, ed. J. M. Franz, pp. 89–94. Stuttgart: Gustav Fischer Verlag.

Hassell, M. P. (1980). Foraging strategies, population models and biological control: a case study. *J. Anim. Ecol.* **49**: 603–28.

Hassell, M. P. & J. K. Waage. (1984). Host-parasitoid population interactions. *Ann. Rev. Entomol.* **29**: 89–114.

Haynes, D. L. & S. H. Gage. (1981). The cereal leaf beetle in North America. *Ann. Rev. Entomol.* **26**: 259–87.

Headley, J. C. & M. A. Hoy. (1986). The economics of integrated mite management in almonds. *Calif. Agric.* **40** (1–2): 28–30.

Heinrichs, E. A., G. B. Aquino, S. Chelliah, S. L. Valencia & W. H. Reissig. (1982). Resurgence of *Nilaparvata lugens* (Stål) populations as influenced by method and timing of insecticide applications in lowland rice. *Environ. Entomol.* **11**: 78–84.

Heinrichs, E. A. & O. Mochida. (1984). From secondary to major pest status: the case of insecticide-induced rice brown planthopper, *Nilaparvata lugens*, resurgence. *Protec. Ecol.* **7**: 201–18.

Helle, W. & M. W. Sabelis (Eds). (1985). *Spider Mites, Their Biology, Natural Enemies and Control.* World Crop Pests, Vol. 1B. Amsterdam: Elsevier Science Publishers.

Hendrichs, J., G. Ortiz, P. Liedo and A. Schwarz. (1983). Six years of successful medfly program in Mexico and Guatemala. In: *Fruit Flies of Economic Importance*, ed. R. Cavalloro, pp. 353–65. Rotterdam: A. A. Balkema.

Hendrickson, R. M. & J. A. Plummer. (1983). Biological control of alfalfa blotch leafminer (Diptera: Agromyzidae) in Delaware. *J. Econ. Entomol.* **76**: 757–61.

Henry, J. E. (1981). Natural and applied control of insects by Protozoa. *Ann. Rev. Entomol.* **26**: 49–73.

Hensley, S. D. (1971). Management of sugarcane borer populations in Louisiana, a decade of change. *Entomophaga*, **16**: 133–46.

Hodek, I. (1967). Bionomics and ecology of predaceous Coccinellidae. *Ann. Rev. Entomol.* **12**: 79–104.

Hodek, I. (1973). *Biology of Coccinellidae.* The Hague: W. Junk.

Hollingworth, R. M. (1987). Vulnerability of pests: study and exploitation for safer chemical control. In: *Pesticides: Minimizing the Risks*, ed. N. N. Ragsdale & R. J. Kuhr, pp. 54–76. Washington, DC: American Chemical Society.

Holloway, J. K. (1964). Projects in biological control of weeds. In: *Biological Control of Insect Pests and Weeds*, ed. P. DeBach, pp. 650–70. New York: Reinhold.

Horn, D. J. (1988). *Ecological Approach to Pest Management.* New York: Guilford Press.

Hörstadius, S. (1974). Linnaeus, animals and man. *Biol. J. Linn. Soc.*, **6**: 269–75.

Howard, L. O. & W. F. Fiske. (1911). The importation into the United States of the parasites of the gypsy moth and the brown-tail moth. *US Dept. Agr. Bur. Entomol. Bull.* **91**.

Hoy, M. A. (1976*a*). Establishment of gipsy moth parasites in North America: an evaluation of possible reasons for establishment or non-establishment. In: *Perspectives in Forest Entomology*, ed. J. F. Anderson & H. K. Kaya, pp. 215–32. New York: Academic Press.

Hoy, M. A. (1976*b*). Genetic improvement of insects: fact or fantasy. *Environ. Entomol.* **5**: 833–9.

Hoy, M. A. (1985*a*). Recent advances in genetics and genetic improvement of the Phytoseiidae. *Ann. Rev. Entomol.* **30**: 345–70.

Hoy, M. A. (1985*b*). Almonds (California). In: *Spider Mites, Their Biology, Natural Enemies and Control*, ed. W. Helle & M. W. Sabelis, pp. 299–310. World Crop Pests Vol. 1B. Amsterdam: Elsevier Science Publishers.

Hoy, M. A. (1985*c*). Improving establishment of arthropod natural enemies. In: *Biological Control in Agricultural IPM Systems*, ed. M. A. Hoy & D. C. Herzog, pp. 151–66. Orlando: Academic Press.

Hoy, M. A. (1986). Use of genetic improvement in biological control. *Agric. Ecosyst. Environ.* **15**: 109–19.

Hoy, M. A., G. L. Cunningham & L. Knutson (Eds). (1983). *Biological Control of Pests by Mites*. Division of Agricultural Sciences, University of California, Publ. **3304**.

Hoy, M. A. & D. C. Herzog (Eds). (1985). *Biological Control in Agricultural IPM Systems*. Orlando: Academic Press.

Hoy, M. A. & J. J. McKelvey (Eds). (1979). *Genetics in Relation to Insect Management*. New York: Rockefeller Foundation.

Huffaker, C. B. (1964). Fundamentals of biological weed control. In: *Biological Control of Insect Pests and Weeds*, ed. P. DeBach, pp. 631–49. New York: Reinhold.

Huffaker, C. B. (1967). A comparison of the status of biological control of St. Johnswort in California and Australia. *Mushi*, **39** (Suppl.): 51–73.

Huffaker, C. B. (1971a). The ecology of pesticide interference with insect populations. (Upsets and resurgences in insect populations.) In: *Agricultural Chemicals – Harmony or Discord*, ed. J. E. Swift, pp. 92–104. Division of Agricultural Sciences, University of California.

Huffaker, C. B. (Ed.). (1971b). *Biological Control*. New York: Plenum Press.

Huffaker, C. B. (1977). Augmentation of natural enemies in the People's Republic of China. In: *Biological Control by Augmentation of Natural Enemies*, ed. R. L. Ridgway & S. B. Vinson, pp. 329–39. New York: Plenum Press.

Huffaker, C. B. (Ed.). (1980). *New Technology of Pest Control*. New York: John Wiley & Sons.

Huffaker, C. B. (1985). Biological control in integrated pest management: an entomological perspective. In: *Biological Control in Agricultural IPM Systems*, ed. M. A. Hoy & D. C. Herzog, pp. 13–23. Orlando: Academic Press.

Huffaker, C. B. & L. E. Caltagirone. (1986). The impact of biological control on the development of the Pacific. *Agric. Ecosyst. Environ*. **15**: 95–107.

Huffaker, C. B. & A. P. Gutierrez. (1990). Evaluation of efficiency of natural enemies in biological control. In: *Armored Scale Insects, Their Biology, Natural Enemies and Control*, ed. D. Rosen, pp. 473–96. World Crop Pests, Vol. 4B. Amsterdam: Elsevier Science Publishers.

Huffaker, C. B. & C. E. Kennett. (1956). Experimental studies on predation: predation and cyclamen mite populations on strawberries in California. *Hilgardia*, **26**: 191–222.

Huffaker, C. B., C. E. Kennet & G. L. Finney. (1962). Biological control of the olive scale, *Parlatoria oleae* (Colvée), in California by imported *Aphytis maculicornis* (Masi) (Hymenoptera: Aphelinidae). *Hilgardia*, **32**: 541–636.

Huffaker, C. B., C. E. Kennett & R. L. Tassan. (1986). Comparisons of parasitism and desities of *Parlatoria oleae* (1952–1982) in relation to ecological theory. *Amer. Nat*. **128**: 379–93.

Huffaker, C. B., R. F. Luck & P. S. Messenger. (1977). The ecological basis of biological control. *Proc. XV Int. Congr. Entomol*. (Washington, DC, 1976), pp. 560–86.

Huffaker, C. B. & P. S. Messenger. (1964a). Population ecology – historical development. In: *Biological Control of Insect Pests and Weeds*, ed. P. DeBach, pp. 45–73. New York: Reinhold.

Huffaker, C. B. & P. S. Messenger. (1964b). The concept and significance of natural control. In: *Biological Control of Insect Pests and Weeds*, ed. P. DeBach, pp. 74–117. New York: Reinhold.

Huffaker, C. B. & P. S. Messenger (Eds). (1976). *Theory and Practice of Biological Control*. New York: Academic Press.

Huffaker, C. B., P. S. Messenger & P. DeBach. (1971). The natural enemy component in natural control and the theory of biological control. In: *Biological Control*, ed. C. B. Huffaker, pp. 16–67. New York: Plenum Press.

Huffaker, C. B. & R. L. Rabb (Eds). (1984). *Ecological Entomology*. New York: John Wiley & Sons.

Huffaker, C. B., F. J. Simmonds & J. E. Laing. (1976). The theoretical and empirical basis of biological control. In: *Theory and Practice of Biological Control*, ed. C. B. Huffaker & P. S. Messenger, pp. 41–78. New York: Academic Press.

Huffaker, C. B. & C. H. Spitzer, Jr. (1951). Data on the natural control of the cyclamen mite on strawberries. *J. Econ. Entomol.* **44**: 519–22.

Huffaker, C. B., M. van den Vrie & J. A. McMurtry. (1969). The ecology of tetranychid mites and their natural control. *Ann. Rev. Entomol.* **14**: 125–74.

Huffaker, C. B., M. van den Vrie & J. A. McMurtry. (1970). Ecology of tetranychid mites and their natural enemies: a review. II. Tetranychid populations and their possible control by predators: an evaluation. *Hilgardia*, **40**: 391–458.

Hull, L. A. & E. H. Beers. (1985). Ecological selectivity: modifying chemical control practices to preserve natural enemies. In: *Biological Control in Agricultural IPM Systems*, ed. M. A. Hoy & D. C. Herzog, pp. 103–21. Orlando: Academic Press.

Hummel, H. E. & T. A. Miller (Eds). (1984). *Techniques in Pheromone Research*. New York: Springer-Verlag.

Hussey, N. W. & L. Bravenboer. (1971). Control of pests in glasshouse culture by the introduction of natural enemies. In: *Biological Control*, ed. C. B. Huffaker, pp. 195–216. New York: Plenum Press.

Hussey, N. W. & C. B. Huffakker. (1976). Spider mites. In: *Studies in Biological Control*, ed. V. L. Delucchi, pp. 179–228. Cambridge: Cambridge University Press.

Hussey, N. W. & N. E. A. Scopes (Eds). (1985). *Biological Pest Control: the Glasshouse Experience*. Ithaca: Cornell University Press.

Ignoffo, C. M. (1985). Manipulating enzootic-epizootic diseases of arthropods. In: *Biological Control in Agricultural IPM Systems*, ed. M. A. Hoy & D. C. Herzog, pp. 243–62. Orlando: Academic Press.

International Atomic Energy Agency. (1982). *Sterile Insect Technique and Radiation in Insect Control*. IAEA, Vienna, STI/PUB/595.

Jepson, P. C. (Ed.). (1989). *Pesticides and Non-target Invertebrates*. Andover: Intercept.

Jervis, M. A. & N. A. Kidd. (1986). Host-feeding strategies in hymenopteran parasitoids. *Biol. Rev.* **61**: 395–434.

Jiménez A., J. G. & J. L. Carrillo S. (1978). Fauna insectil benefica en algodonero con maiz intercalado, comparada con algodonero solo. *Agric. Téc. Méx.* **4**: 143–56.

Jones, V. P., N. C. Toscano, M. W. Johnson, S. C. Welter & R. R. Youngman (1986). Pesticide effects on plant physiology: integration into a pest management program. *BioScience*, **32**: 103–9.

Julien, M. H. (Ed.). (1987). *Biological Control of Weeds: a World Catalogue of Agents and their Target Weeds*, 2nd edition. Wallingford: CAB International.

Julien, M. H., J. D. Kerr & R. R. Chan. (1984). Biological control of weeds: an evaluation. *Protec. Ecol.* **7**: 3–25.

Jutsum, A. R. (1988). Commercial application of biological control: status and prospects. *Phil. Trans. R. Soc. Lond. B* **318**: 357–73.

Jutsum, A. R. & R. F. S. Gordon (Eds). (1989). *Insect Pheromones in Plant Protection*. Chichester: John Wiley & Sons.

Kenmore, P. E., J. A. Listinger, J. P. Bandong, A. C. Santiago & M. M. Salac. (1987). Philippine rice farmers and insecticides: thirty years of growing dependency and new options for change. In: *Management of Pests and Pesticides. Farmers' Perceptions and Practices*, ed. J. Tait & B. Napompeth, pp. 98–108. Boulder: Westview Press.

Kennedy, G. G., F. Gould, O. M. B. dePonti & R. E. Stinner. (1987). Ecological, agricultural, genetic and commercial considerations in the deployment of insect-resistant germplasm. *Environ. Entomol.* **16**: 327–38.

Kerrich, G. J. (1960). The state of our knowledge of the systematics of the Hymenoptera Parasitica, with particular reference to the British fauna. *Trans. Soc. Brit. Entomol.* **14**: 1–18.

Khachatourians, G. G. (1986). Production and use of biological pest control agents. *Trends Biotechnol.* **4**: 120–4.

Kilgore, W. W. & R. L. Doutt (Eds). (1967). *Pest Control – Biological, Physical and Selected Chemical Methods.* New York: Academic Press.

Kirby, W. & W. Spence. (1815). *An Introduction to Entomology.* London: Longman, Brown, Green and Longmans.

Kirby, W. & W. Spence. (1826). Diseases of insects. Letter (Chapter) XLIV. In: *An Introduction to Entomology: or Elements of the Natural History of Insects*, vol. 4. London: Longman, Brown, Green and Longmans.

Knipling, E. F. (1979). *The Basic Principles of Insect Population Suppression and Management.* Agric. Handbook No. 512. Washington, DC: US Department of Agriculture.

Knutson, L. & R. D. Gordon. (1982). Status of insect taxonomy in China with notes on biological control of pests. In: *Biological Control of Pests in China*, pp. 216–65. Washington, DC: US Department of Agriculture.

Koebele, A. (1890). Report of a trip to Australia to investigate the natural enemies of the fluted scale. *US Dept. Agric. Bur. Entomol. Bull.* **21**.

Kogan, M. (1982). Plant resistance in pest management. In: *Introduction to Insect Pest Management*, 2nd edn, ed. R. L. Metcalf & W. H. Luckmann, pp. 93–134. New York: John Wiley & Sons.

Kurstak, E. (Ed.). (1982). *Microbial and Viral Pesticides.* New York: Marcel Dekker.

Kuwana, I. (1934). Notes on a newly imported parasite of the spiny whitefly attacking citrus in Japan. *Proc. 5th Pac. Sci. Congr.* (5): 3521–3 (1933).

Kydonieus, A. F. & M. Beroza (Eds). (1982). *Insect Suppression with Controlled Release Pheromone Systems.* 2 vols. Boca Raton: CRC Press.

Laing, J. E. & J. Hamai. (1976). Biological control of insect pests and weeds by imported parasites, predators, and pathogens. In: *Theory and Practice of Biological Control*, ed. C. B. Huffaker & P. S. Messenger, pp. 685–743. New York: Academic Press.

Lambur, M. T., R. F. Kazmierczak & E. G. Rajotte. (1989). Analysis of private consulting firms in integrated pest management. *Bull. Entomol. Soc. Amer.* **35**: 5–11.

Le Baron, W. (1870). The chalcideous parasite of the apple-tree bark-louse (*Chalcis* (*Aphelinus*) *mytilaspidis*, n. sp.). *Amer. Entomol. Bot.* **2**: 360–2.

Legner, E. F. (1972). Observations on hybridization and heterosis in parasitoids of synanthropic flies. *Ann. Entomol. Soc. Amer.* **65**: 254–63.

Legner, E. F. (1983). Broadened view of *Muscidifurax* parasites associated with endophilous synanthropic flies and sibling species in the *Spalangia endius* complex. *Proc. Calif. Mosquito and Vector Control Assoc.* **51**: 47–48.

Legner, E. F. (1986). Trapping efficiency of exophilous synanthropic Diptera increased through considerations of behavior. *Proc. Calif. Mosquito and Vector Control Assoc.* **54**: 160–2.

Legner, E. F. (1988). Quantitation of heterotic behavior in parasitic Hymenoptera. *Ann. Entomol. Soc. Amer.* **81**: 657–81.

Legner, E. F., R. D. Sjorgen & I. M. Hall. (1974). The biological control of medically important arthropods. *CRC Crit. Rev. Envir. Contr.* **4**: 85–113.

LePelley, R. H. (1943). The biological control of a mealy bug on coffee and other crops in Kenya. *Emp. J. Exp. Agric.* **11** (42): 78–88.

LeRoux, E. J. (1960). Effects of 'modified' and 'commercial' spray programs on the fauna of apple orchards in Quebec. *Ann. Soc. Entomol. Quebec,* **76**: 87–119.

Liu, Z., Z. Wang, Y. Sun, J. Liu & W. Yang. (1988). Studies on culturing *Anastatus* sp., a parasitoid of litchi stink bug, with artificial host eggs. In: *Trichogramma and Other Egg Parasites,* 2nd Int. Symp. (Guangzhou, China, 1986), ed. J. Vogele, J. Waage & J. van Lenteren, pp. 353–60. Paris: INRA.

Luck, R. F. (1981). Parasitic insects introduced as biological control agents for arthropod pests. In: *CRC Handbook of Pest Management in Agriculture,* vol. II, ed. D. Pimentel, pp. 125–284. Boca Raton: CRC Press.

Luck, R. F. & H. Podoler. (1985). Competitive exclusion of *Aphytis lingnanensis* by *A. melinus*: potential role of host size. *Ecology,* **66**: 904–13.

Luck, R. F., B. M. Shepard & P. E. Kenmore. (1988). Experimental methods for evaluating arthropod natural enemies. *Ann. Rev. Entomol.* **33**: 367–91.

Luck, R. F., R. van den Bosch & R. Garcia. (1977). Chemical insect control – a troubled pest management strategy. *BioScience,* **27**: 606–11.

Luginbill, P. (1969). Developing resistant plants – the ideal method of controlling insects. *USDA, ARS Production Res. Rept.* **111**.

Lüthy, P. (1986). Insect pathogenic bacteria as pest control agents. In: *Biological Plant and Health Protection,* ed. J. M. Franz, pp. 201–16. Stuttgart: Gustav Fischer Verlag.

MacArthur, R. H. & E. O. Wilson. (1967). *The Theory of Island Biogeography.* Princeton: Princeton University Press.

Mackauer, M., L. E. Ehler & J. Roland (Eds.) (1990). *Critical Issues in Biological Control.* Andover: Intercept.

MacPhee, A. W. & C. R. MacLellan. (1971). Cases of naturally-occurring biological control in Canada. In: *Biological Control,* ed. C. B. Huffaker, pp. 312–28. New York: Plenum Press.

MacPhee, A. W. & C. R. MacLellan. (1972). Ecology of apple orchard fauna and development of integrated pest control in Nova Scotia. *Proc. Tall Timbers Conf. Ecol. Anim. Contr. Habit. Mgmt,* **3**: 197–208.

Maple, J. D. (1947). The eggs and first instar larvae of Encyrtidae and their morphological adaptations for respiration. *Univ. Calif. Publ. Entomol.* **8**: 25–122.

Maramorosch, K. (1981). Spiroplasmas: agents of animal and plant diseases. *BioScience,* **31**: 374–80.

Maramorosch, K. & K. E. Sherman (Eds.). (1985). *Viral Insecticides for Biological Control.* Orlando: Academic Press.

Massee, A. M. (1954). Problems arising from the use of insecticides: effect on the balance of animal populations. *Rep. 6th Commonwealth Entomol. Conf. London,* pp. 53–7.

Matthews, R. (1974). Biology of Braconidae. *Ann. Rev. Entomol.* **19**: 15–32.

Maxwell, F. G. & F. A. Harris (Eds.). (1974). *Proceedings of the Summer Institute on Biological Control of Plant Insects and Diseases.* Jackson: University Press of Mississippi.

Maxwell, F. G. & P. R. Jennings (Eds). (1980). *Breeding Plants Resistant to Insects.* New York: John Wiley & Sons.

McMurtry, J. A. (1982). The use of phytoseiids for biological control: progress and future prospects. In: *Recent Advances in Knowledge of the Phytoseiidae,* ed. M. A. Hoy, pp. 23–48. Division of Agricultural Sciences, University of California, Publ. **3284**.

McMurtry, J. A., C. B. Huffaker & M. van den Vrie. (1970). Ecology of tetranychid mites and their natural enemies: a review. I. Tetranychid enemies: their biological characters and the impact of spray practices. *Hilgardia,* **40**: 331–90.

Mellanby, K. (1967). *Pesticides and Pollution.* New Naturalist No. 50. London: Collins.

Metcalf, R. L. (1980). Changing role of pesticides in crop protection. *Ann. Rev. Entomol.* **25**: 219–56.

Metcalf, R. L. & W. H. Luckmann (Eds). (1982). *Introduction to Insect Pest Management*, 2nd edn. New York: John Wiley & Sons.

Michael, A. D. (1903). *British Tyroglyphidae*, Vol. II. London: The Ray Society.

Mitchell, E. R. (Ed.). (1981). *Management of Insect Pests with Semiochemicals: Concepts and Practice.* New York: Plenum Press.

Morrison, D. E., J. R. Bradley, Jr. & J. W. Van Duyn. (1979). Populations of corn earworm and associated predators after applications of certain soil-applied pesticides to soybeans. *J. Econ. Entomol.* **72**: 97–100.

Muir, F. (1931). Introduction. In: *Handbook of the Insects and Other Invertebrates of Hawaiian Sugar Cane Fields*, by Francis X. Williams. Hawaiian Sugar Planters Association Experiment Station.

Mukerji, K. G. & K. L. Garg (Eds). (1988). *Biocontrol of Plant diseases.* 2 vols. Boca Raton: CRC Press.

Myers, J. G. (1935). Second report on an investigation into the biological control of West Indian pests. *Bull. Entomol. Res.* **26**: 181–252.

National Academy of Sciences. (1969). *Principles of Plant and Animal Pest Control.* Vol. 3, *Insect-Pest Management and Control.* NAS, Washington, DC, Publ. **1695**.

National Academy of Sciences. (1975). *Pest Control: an Assessment of Present and Alternative Technologies.* Vol. 1, *Contemporary Pest Control Practices and Prospects: the Report of the Executive Committee.* Washington, DC: NAS.

National Research Council. (1986). *Pesticide Resistance: Strategies and Tactics for Management.* Washington, DC: National Academy Press.

National Research Council. (1987). *Regulating Pesticides in Food: The Delaney Paradox.* Washington, DC: National Academy Press.

Neuenschwander, P. & W. N. O. Hammond. (1988). Natural enemy activity following the introduction of *Epidinocarsis lopezi* (Hymenoptera: Encyrtidae) against the cassava mealybug, *Phenacoccus manihoti* (Homoptera: Pseudococcidae), in southwestern Nigeria. *Environ. Entomol.* **17**: 894–902.

Neuenschwander, P. & H. R. Herren. (1988). Biological control of the cassava mealybug, *Phenacoccus manihoti*, by the exotic parasitoid *Epidinocarsis lopezi* in Africa. *Phil. Trans. R. Soc. Lond.* B **318**: 319–33.

Newsom, L. D. (1967). Consequences of insecticide use on nontarget organisms. *Ann. Rev. Entomol.* **12**: 257–86.

Nickle, W. R. (Ed.). (1984). *Plant and Insect Nematodes.* New York: Marcel Dekker.

Nordlund, D. A., R. L. Jones & W. J. Lewis (Ed.). (1981). *Semiochemicals, Their Role in Pest Control.* New York: John Wiley & Sons.

O'Connor, B. A. (1960). Decade of biological control work in Fiji. *Agric. J. Fiji*, **30** (2): 44–54.

Oka, I. N. (1983). The potential for the integration of plant resistance, agronomic, biological, physical/mechanical techniques, and pesticides for pest control in farming systems. In: *Chemistry and World Food Supplies: The New Frontiers*, ed. L. W. Shemilt, pp. 173–84. Oxford: Pergamon Press.

Olkowski, H. (1971). *Common Sense Pest Control.* Richmond: Consumers Coop. of Berkeley, Inc.

Ooi, P. A. C., L. G. Chan, K. C. Khoo, C. H. Teoh, M. M. Jusoh, C. T. Ho & G. S. Lim (Eds). (1987). *Management of the Cocoa Pod Borer.* Kuala Lumpur: Malayan Plant Protection Society.

Ordish, G. (1967). *Biological Methods in Crop Pest Control.* London: Constable.

Painter, R. H. (1951). *Insect Resistance in Crop Plants.* New York: Macmillan.

Painter, R. H. (1958). Resistance of plants to insects. *Ann. Rev. Entomol.* **3**: 267–90.

Painter, R. H. (1960). Breeding plants for resistance to insect pests. In: *Biological and Chemical Control of Plant and Animal Pests*, ed. L. P. Retz, pp. 245–66. American Association for the Advancement of Science, Washington, DC, Publ. **61**.

Painter, R. H. (1968). Crops that resist insects provide a way to increase world food supply. *Kansas State Agric. Expt. Sta. Bull.* **520**.

Palti, J. & R. Auscher (Eds). (1986). *Advisory Work in Crop Pest and Disease Management*. Berlin: Springer-Verlag.

Papavizas, G. C. (Ed.). (1981). *Biological Control in Crop Production*. Beltsville Symposia in Agricultural Research 5, Totowa: Allanheld, Osmun.

Parker, F. D. (1971). Management of pest populations by manipulating densities of both hosts and parasites through periodic releases. In: *Biological Control*, ed. C. B. Huffaker, pp. 365–76. New York: Plenum Press.

Parker, F. D. & R. E. Pinnell. (1972). Further studies of the biological control of *Pieris rapae* using supplemental host and parasite releases. *Environ. Entomol.* **1**: 150–7.

Payne, C. C. (1988). Pathogens for the control of insects: where next? *Phil. Trans. R. Soc. Lond.* B **318**: 225–48.

Pedigo, L. P., S. H. Hutchins & L. G. Higley. (1986). Economic injury levels in theory and practice. *Ann. Rev. Entomol.* **31**: 341–68.

Pemberton, C. E. (1948). History of the entomology department experiment station, H.S.P.A., 1904–1945. *Hawaiian Planters' Record*, **52**: 53–90.

Perkins, J. H. (1982). *Insects, Experts, and the Insecticide Crisis*. New York: Plenum Press.

Peterson, A. (1936). Biological control of insects at the University. In: *Conference on Biological Methods of Controlling Insect Pests*. Department of Agriculture Entomology Branch, Ottawa at Belleville, Ontario, June 25–26, 1936, pp. 1–86, mimeo.

Philipps, J. R., A. P. Gutierrez & P. L. Adkisson. (1980). General accomplishments toward better insect control in cotton. In: *New Technology for Pest Control*, ed. C. B. Huffaker, pp. 123–53. New York: John Wiley & Sons.

Pickett, A. D. & N. A. Patterson. (1953). The influence of spray programs on the fauna of apple orchards in Nova Scotia. IV. A review. *Canad. Entomol.* **85**: 472–8.

Pimentel, D. (Ed.). (1981). *CRC Handbook of Pest Management in Agriculture*, 3 vols. Boca Raton: CRC Press.

Pimentel, D. (1986). Population dynamics and the importance of evolution in successful biological control. In: *Biological Plant and Health Protection*, ed. J. M. Franz, pp. 3–18. Stuttgart: Gustav Fischer Verlag.

Pimentel, D., D. Andow, D. Gallahan, I. Schreiner, T. E. Thompson, R. Dyson-Hudson, S. N. Jacobson, M. A. Irish, S. F. Kroop, A. M. Moss, M. D. Shepard & B. G. Vinzant. (1980). Pesticides: environmental and social costs. In: *Pest Control: Cultural and Environmental Aspects*, ed. D. Pimentel & J. H. Perkins, pp. 99–158. AAAS Selected Symposium 43. Boulder: Westview Press.

Pimentel, D., J. Krummel, D. Gallahan, J. Hough, A. Merrill, I. Schreiner, P. Vittum, F. Koziol, E. Back, D. Yen & S. Fiance. (1981). A cost-benefit analysis of pesticide use in U.S. food production. In: *CRC Handbook of Pest Management in Agriculture*, ed. D. Pimentel, vol. 2, pp. 27–54. Boca Raton: CRC Press.

Pimentel, D. & J. H. Perkins (Eds). (1980). *Pest Control: Cultural and Environmental Aspects*. AAAS Selected Symposium 43, Boulder: Westview Press.

Plaut, H. N. & F. A. Mansour. (1981). Effects of photostable pyrethroids on spider mites (Tetranychidae). *Phytoparasitica*, **9**: 218.

Poinar, G. O. (1979). *Nematodes for Biological Control of Insects*. Boca Raton: CRC Press.

Poinar, G. O., Jr. & G. M. Thomas. (1984). *Laboratory Guide to Insect Pathogens and Parasites*. New York: Plenum Press.

Postel, S. (1987). *Defusing the Toxics Threat: Controlling Pesticides and Industrial Waste*. Worldwatch Paper 79. Washington, DC: Worldwatch Inst.

Quezada, J. R. & P. DeBach. (1973). Bioecological and population studies of the cottony-cushion scale, *Icerya purchasi* Mask., and its natural enemies, *Rodolia cardinalis* Mul. and *Cryptochaetum iceryae* Will., in southern California. *Hilgardia*, 41: 631–88.

Rabb, R. L. (1969). Environmental manipulations as influencing populations of tobacco hornworms. *Proc. Tall Timbers Conf. Ecol. Anim. Contr. Habit. Mgmt*, 1: 175–91.

Rabb, R. L. (1971). Naturally-occurring biological control in the eastern United States, with particular reference to tobacco insects. In: *Biological Control*, ed. C. B. Huffaker, pp. 294–311. New York: Plenum Press.

Rabb, R. L., R. E. Stinner & R. van den Bosch. (1976). Conservation and augmentation of natural enemies. In: *Theory and Practice of Biological Control*, ed. C. B. Huffaker & P. S. Messenger, pp. 233–54. New York: Academic Press.

Ragsdale, N. N. & R. J. Kuhr (Eds). (1987). *Pesticides: Minimizing the Risks*. ACS Symposium Series 336. Washington, DC: American Chemical Society.

Rao, S. V. & P. DeBach. (1969). Experimental studies on hybridization and sexual isolation between some *Aphytis* species (Hymenoptera: Aphelinidae) I. Experimental hybridization and an interpretation of evolutionary relationships among the species. *Hilgardia*, 39: 515–53.

Ratcliffe, F. N. (1965). Biological Control. *Austr. J. Sci.* 28: 237–40.

Report of the Environmental Pollution Panel, President's Science Advisory Committee. (1965). *Restoring the Quality of our Environment*. Washington, DC: The White House.

Report of the Secretary's Commission on Pesticides and their Relationship to Environmental Health, parts I and II. (1969). Washington, DC: US Department of Health, Education and Welfare.

Retnakaran, A., J. Granett & T. Ennis. (1985). Insect growth regulators. In: *Comprehensive Insect Physiology, Biochemistry and Pharmacology*, ed. G. A. Kerkut & L. I. Gilbert, vol. 12, *Insect Control*, pp. 529–601. Oxford: Pergamon Press.

Reynolds, H. T. (1971). A world review of the problem of insect population upsets and resurgences caused by pesticide chemicals. In: *Agricultural Chemicals – Harmony or Discord*, ed. J. E. Swift, pp. 108–112. University of California: Division of Agricultural Sciences.

Reynolds, H. T., P. L. Adkisson, R. F. Smith & R. E. Frisby. (1982). Cotton insect pest management. In: *Introduction to Insect Pest Management*, 2nd edn, ed. R. L. Metcalf & W. H. Luckmann, pp. 375–441. New York: John Wiley & Sons.

Ridgway, R. L., J. R. Ables, C. Goodpasture & A. W. Hartstack. (1981). *Trichogramma* and its utilization for crop protection in the U.S.A. In: *Proceeding of the Joint American–Soviet Conference on Use of Beneficial Organisms in the Control of Crop Pests* (Washington, DC, 1979), ed. J. R. Coulson, pp. 41–8. Entomological Society of America.

Ridgway, R. L. & S. L. Jones. (1969). Inundative releases of *Chrysopa carnea* for control of *Heliothis* on cotton. *J. Econ. Entomol.* 62: 177–80.

Ridgway, R. L., E. G. King & J. L. Carrillo. (1977). Augmentation of natural enemies for control of plant pests in the Western Hemisphere. In: *Biological Control by Augmentation of Natural Enemies*, ed. R. L. Ridgway & S. B. Vinson, pp. 379–416. New York: Plenum Press.

Ridgway, R. L. & S. B. Vinson (Eds). (1977a). *Biological Control by Augmentation of*

Natural Enemies. New York: Plenum Press.

Ridgway, R. L. & S. B. Vinson. (1977*b*). Commercial sources of natural enemies in the U.S. and Canada. In: *Biological Control by Augmentation of Natural Enemies*, ed. R. L. Ridgway & S. B. Vinson, pp. 451–3. New York: Plenum Press.

Riechert, S. E. & T. Lockley. (1984). Spiders as biological control agents. *Ann. Rev. Entomol.* **29**: 299–320.

Riley, C. V. (1893). Parasitic and predaceous insects in applied entomology. *Insect Life*, **6**: 130–41.

Ripper, W. E. (1956). Effect of pesticides on balance of arthropod populations. *Ann. Rev. Entomol.* **1**: 403–38.

Rivnay, E. (1942). *Clausenia purpurea* Ishii, a parasite of *Pseudococcus comstocki* Kuw. introduced into Palestine. *Bull. Soc. Fouad 1^{et} Entomol. d'Egypte*, **26**: 1–19.

Rivnay, E. (1968). Biological control of pests in Israel (a review 1905–1965). *Israel J. Entomol.* **3** (1): 1–156.

Robinson, R. R., J. H. Young & R. D. Morrisson. (1972). Strip-cropping effects on abundance of predatory and harmful cotton insects in Oklahoma. *Environ. Entomol.* **1** (2): 145–9.

Rose, M. (1990). Periodic colonization of natural enemies. In: *Armored Scale Insects, Their Biology, Natural Enemies and Control*, ed. D. Rosen, pp. 433–40. World Crop Pests, Vol. 4B. Amsterdam: Elsevier Science Publishers.

Rosen, D. (1980). Integrated control of citrus pests in Israel. *Proc. Int. Symp. IOBC/ WPRS on Integrated Control in Agriculture and Forestry* (Vienna, 1979), pp. 289–92.

Rosen, D. (Ed.). (1981). *The Role of Hyperparasitism in Biological Control: a Symposium.* Division of Agricultural Sciences, University of California, Publ. **4103**.

Rosen, D. (1985). Biological control. In: *Comprehensive Insect Physiology, Biochemistry and Pharmacology*, ed. G. A. Kerkut & L. I. Gilbert, vol. 12, *Insect Control*, pp. 413–64. Oxford: Pergamon Press.

Rosen, D. (1986). The role of taxonomy in effective biological control programs. *Agric. Ecosyst. Environ.* **15**: 121-9.

Rosen, D. & P. DeBach. (1976). Biosystematic studies on the species of *Aphytis* (Hymenoptera: Aphelinidae). *Mushi*, **49**: 1–17.

Rosen, D. & P. DeBach. (1978). Diaspididae. In: *Introduced Parasites and Predators of Arthropod Pests and Weeds: a World Review*, ed. C. P. Clausen, pp. 78–128. Agric. Handbook 480. Washington, DC: US Department of Agriculture.

Rosen, D. & P. DeBach. (1979). *Species of* Aphytis *of the World* (*Hymenoptera: Aphelinidae*). Jersusalem: Israel Universities Press, and The Hague: W. Junk.

Rosen, D. & C. B. Huffaker. (1983). An overview of desired attributes of effective biological control agents, with particular emphasis on mites. In: *Biological Control of Pests by Mites*, ed. M. A. Hoy, G. L. Cunningham & L. Knutson, pp. 2–11. Division of Agricultural Sciences, University of California, Publ. **3304**.

Rosenthal, S. S., D. M. Maddox & K. Brunetti. (1985). Biological control methods. In: *Principles of Weed Control in California*, sponsored by California Weed Conference, pp. 65–94. Fresno: Thomson.

Rössler, Y. & D. Rosen. (1990). A case history: IPM on citrus in Israel. In: *Armored Scale Insects, Their Biology, Natural Enemies and Control*, ed. D. Rosen, pp. 519–26. World Crop Pests, Vol. 4B. Amsterdam: Elsevier Science Publishers.

Roush, R. T. & M. A. Hoy. (1980). Selection improves sevin resistance in spider mite predator. *Calif. Agric.* **34** (2/3): 11–14.

Roush, R. T. & J. A. McKenzie. (1987). Ecological genetics of insecticide and acaricide resistance. *Ann. Rev. Entomol.* **32**: 361–80.

Rudd, R. L. (1964). *Pesticides and the Living Landscape*. Madison: The University of Wisconsin Press.

Rudd, R. L. (1971). Pesticides. In: *Environment – Resources, Pollution and Society*, ed. W. W. Murdoch, pp. 279–301. Stamford: Sinauer Assoc.

Sailer, R. I. (1978). Our immigrant insect fauna. *Bull. Entomol. Soc. Amer.* **24**: 3–11.

Sailer, R. I. (1983). History of insect introductions. In: *Exotic Plant Pests and North American Agriculture*, ed. C. L. Wilson & C. L. Graham, pp. 15–38. New York: Academic Press.

Sailer, R. I., R. E. Brown, B. Munir & J. C. E. Nickerson. (1984). Dissemination of the citrus whitefly (Homoptera: Aleyrodidae) parasite *Encarsia lahorensis* (Howard) (Hymenoptera: Aphelinidae) and its effectiveness as a control agent in Florida. *Bull. Entomol. Soc. Amer.* **30**: 36–9.

Samways, M. J. (1981). *Biological Control of Pests and Weeds*. Studies in Biology No. **132**. London: Edward Arnold.

Ščepetilnikova, V. A. (1970). Perspektiven der Kenntnis und Anwendung von Eiparasiten der Gattung *Trichogramma* zur Bekämpfung land- und forstwirtschaftlicher Schädlinge. (The prospects of the study and use of egg parasites of the genus *Trichogramma* to control agricultural and forest pests.) In: *Biologische Bekämpfungsmethoden von Forstschädlingen*, pp. 117–36. Berlin: Tagungsbericht der Deutschen Akademie der Landwirtschaftswissenschaften zu Berlin.

Schroder, R. F. W. (1981). Biological control of the Mexican bean beetle, *Epilachna varivestis* Mulsant, in the United States. In: *Biological Control in Crop Production*, ed. G. C. Papavizas, pp. 351–60. Beltsville Symposia in Agricultural Research **5**. Totowa: Allanheld, Osmun.

Schroeder, D. & R. D. Goeden. (1986). The search for arthropod natural enemies of introduced weeds for biological control – in theory and practice. *Biocontrol News and Information*, **7**: 147–55.

Schuster, M. F., J. C. Boling & J. J. Marony, Jr. (1971). Biological control of rhodesgrass scale by airplane releases of an introduced parasite of limited dispersing ability. In: *Biological Control*, ed. C. B. Huffaker, pp. 227–50. New York: Plenum Press.

Sheets, T. J. & D. Pimentel (Eds). (1979). *Pesticides: Contemporary Roles in Agriculture, Energy, and the Environment*. Clifton: The Humana Press.

Sheppard, C. (1983). House fly and lesser fly control utilizing the black soldier fly in manure management systems for caged laying hens. *Envir. Entomol.* **12**: 1439–42.

Shorey, H. H. & J. J. McKelvey, Jr. (Eds). (1977). *Chemical Control of Insect Behavior*. New York: John Wiley & Sons.

Shumakov, E. M., G. V. Gusev & N. S. Fedorinchik (Eds). (1974). *Biological Agents for Plant Protection*. (Kolos Publishing House, Moscow, in Russian). English Translation. Washington, DC: US Department of Agriculture.

Silvestri, F. (1909). Consideration of the existing condition of agricultural entomology in the United States of North America, and suggestions which can be gained from it for the benefit of Italian agriculture. *Bull. Soc. Ital. Agric.* **14**: 305–67. (Translated and reprinted in: *Hawaiian Forester and Agric.* **6**: 287–336, 1909.)

Silvestri, F. (1937). Insect polyembryony and its general biological aspects. *Bull. Mus. Compar. Zool. Harvard Coll.* **81**: 469–98.

Silvestri, F. & G. Martelli. (1908). La cocciniglia del fico (*Ceroplastes rusci L.*). *Boll. Lab. Zool. Gen. Agrar. Portici*, **2**: 297–358.

Smirnoff, W. A. (1972). Promoting virus epizootics in populations of the Swaine jack pine sawfly by infected adults. *BioScience*, **22**: 662–3.

Smith, H. D., H. L. Maltby & E. Jiménez-Jiménez. (1964). Biological control of the citrus blackfly in Mexico. *US Dept. Agric. Tech. Bull.* **1311**.

Smith, H. S. (1919). On some phases of insect control by the biological method. *J. Econ. Entomol.* **12**: 288–92.

Smith, R. F. (1971). What is being done by the universities? In: *Agricultural Chemicals – Harmony or Discord*, ed. J. E. Swift, pp. 138–45. Division of Agricultural Sciences, University of California.

Smith, R. F. (1972). Economic aspects of pest control. *Proc. Tall Timbers Conf. Ecol. Anim. Contr. Habit. Mgmt*, **3**: 53–83.

Smith, R. F. & H. T. Reynolds, (1972). Effects of manipulation of cotton agroecosystems on insect pest populations. In: *The Careless Technology: Ecology and International Development*, ed. M. T. Farvar & J. P. Milton, pp. 373–406. Garden City, NY: The Natural History Press.

Smith, R. F. & R. van den Bosch. (1967). Integrated control. In: *Pest Control: Biological, Physical and Selected Chemical Methods*, ed. W. W. Kilgore & R. L. Doutt, pp. 295–340. New York: Academic Press.

Sparks, A. N., J. R. Ables & R. L. Jones. (1982). Notes on biological control of stem bores in corn, sugarcane, and rice in the People's Republic of China. In: *Biological Control of Pests in China*, pp. 193–215. Washington, DC: US Department of Agriculture.

Starý, P. (1970). *Biology of Aphid Parasites*. Series Entomologica 6. The Hague: W. Junk.

Steinberg, S., H. Podoler & D. Rosen. (1986). Biological control of the Florida red scale, *Chrysomphalus aonidum*, in Israel by two parasite species: current status in the coastal plain. *Phytoparasitica*, **14**: 199–204.

Steinhaus, E. A. (1946). *Insect Microbiology*. Ithaca: Comstock Publishing Co.

Steinhaus, E. A. (1949). *Principles of Insect Pathology*. New York: McGraw-Hill.

Steinhaus. E. A. (1956). Microbial control – the emergence of an idea. A brief history of insect pathology through the nineteenth century. *Hilgardia*, **26**: 107–60.

Steinhaus, E. A., (Ed.). (1963). *Insect Pathology, an Advanced Treatise*, 2 vols. New York: Academic Press.

Stern, V. M. (1969). Interplanting alfalfa in cotton to control lygus bugs and other insect pests. *Proc. Tall Timbers Conf. Ecol. Anim. Contr. Habit. Mgmt*, **1**: 55–69.

Stern, V. M. (1973). Economic thresholds. *Ann. Rev. Entomol.* **18**: 259–80.

Stern, V. M., P. L. Adkisson, O. Beingolea G. & G. A. Viktorov. (1976). Cultural controls. In: *Theory and Practice of Biological Control*, ed. C. B. Huffaker & P. S. Messenger, pp. 593–613. New York: Academic Press.

Stern, V. M., R. F. Smith, R. van den Bosch & K. S. Hagen. (1959). The integrated control concept. *Hilgardia*, **29**: 81–101.

Stevens, L. M., A. L. Steinhauer & J. R. Coulson. (1975). Suppression of Mexican bean beetle on soybeans with annual inoculative releases of *Pediobius foveolatus*. *Environ. Entomol.* **4**: 947–52.

Stinner, R. E. (1977). Efficacy of inundative releases. *Ann. Rev. Entomol.* **22**: 515–31.

Sullivan, D. J. (1987). Insect hyperparasitism. *Ann. Rev. Entomol.* **32**: 49–70.

Summy, K. R., F. E. Gilstrap, W. G. Hart, J. M. Caballero & I. Saens. (1983). Biological control of citrus blackfly (Homoptera: Aleyrodidae) in Texas. *Environ. Entomol.* **12**: 782–6.

Surtees, G. (1971). Epidemiology of microbial control of insect pest populations. *Int. J. Environ. Studies*, **2**: 195–201.

Swan, L. A. (1964). *Beneficial Insects*. New York: Harper and Row.

Swirski, E., Y. Izhar, M. Wysoki, E. Gurevitz & S. Greenberg. (1980). Biological control of the long-tailed mealybug *Pseudococcus longispinus* (Coccoidea, Pseudococcidae) in the avocado plantations of Israel. *Entomophaga*, **25**: 415–26.

Tallian, L. (1975). *Politics and Pesticides*. Los Angeles: People's Lobby Press.

Tanaka, M. (1982). Biological control of arrowhead scale, *Unaspis yanonensis* (Kuwana), in the implementation of IPM programs of citrus orchards in Japan. *Proc. Int. Soc. Citriculture* (Tokyo, 1981), pp. 636–40.

Tanaka, M. (1989). Proven feasibility of biological control in Japanese citrus orchards and future prospects. *Proc. Int. Soc. Citriculture* (Tel Aviv, 1988), **3**: 1201–7.

Taylor, T. H. C. (1935). The campaign against *Aspidiotus destructor* Sign. in Fiji. *Bull. Entomol. Res.* **26**: 1–102.

Taylor, T. H. C. (1936). The biological control of the coconut leaf miner (*Promecotheca reichei*, Baly) in Fiji. *Agric. J. Fiji*, **8**: 17–21.

Taylor, T. H. C. (1937). *The Biological Control of an Insect in Fiji. An Account of the Coconut Leaf-Mining Beetle and its Parasite Complex*. London: Imperial Institute of Entomology (now CAB International Institute of Entomology).

Thompson, S. N. (1986). Nutrition and in vitro culture of insect parasitoids. *Ann. Rev. Entomol.* **31**: 197–219.

Tingey, W. M. (1981). The environmental control of insects using plant resistance. In: *CRC Handbook of Pest Management in Agriculture*, ed. D. Pimentel, vol. 1, pp. 175–97. Boca Raton: CRC Press.

Tooke, F. G. C. (1953). The eucalyputs snout-beetle *Gonipterus scutellatus* Gyll. A study of its ecology and control by biological means. *Union S. Africa, Dept. Agric. Entomol. Mem.* **3**.

Tothill, J. D., T. H. C. Taylor and R. W. Paine. (1930). *The Coconut Moth in Fiji. A History of its Control by Means of Parasites*. London: Imperial Bureaux of Entomology (now CAB International Institute of Entomology).

Townes, H. (1971). Ichneumonidae as biological control agents. *Tall Timbers Conf. Ecol. Anim. Contr. Habit. Mgmt*, **3**: 235–48.

Turnock, W. J., K. L. Taylor, D. Schroder & D. L. Dahlsten. (1976). Biological control of pests of coniferous forests. In: *Theory and Practice of Biological Control*, ed. C. B. Huffaker & P. S. Messenger, pp. 289–311. New York: Academic Press.

van den Bosch, R. (1971*a*). Experimental field studies on upsets and resurgences of pest populations associated with agricultural chemicals. In: *Agricultural Chemicals – Harmony or Discord*, ed. J. E. Swift, pp. 104–7. Division of Agricultural Sciences, University of California.

van den Bosch, R. (1971*b*). The Melancholy Addiction of Ol' King Cotton. *Nat. Hist.* **80** (10): 86–90.

van den Bosch, R. (1978). *The Pesticide Conspiracy*. Garden City, NY: Doubleday.

van den Bosch, R., B. D. Frazer, C. S. Davis, P. S. Messenger and R. Hom. (1970). *Trioxys pallidus* – an effective new walnut aphid parasite from Iran. *Calif. Agric.* **24** (11): 8–10.

van den Bosch, R. & K. S. Hagen. (1966). Predaceous and parasitic arthropods in California cotton fields. *Calif. Agric. Exp. Sta. Bull.* **820**.

van den Bosch, R., R. Hom, P. Matteson, B. D. Fraser, P. S. Messenger & C. S. Davis. (1979). Biological control of the walnut aphid in California: impact of the parasite, *Trioxys pallidus*. *Hilgardia*, **47**: 1–13.

van den Bosch, R., T. F. Leigh, L. H. Falcon, V. M. Stern, D. Gonzalez & K. S. Hagen. (1971). The developing program of integrated control of cotton pests in California. In: *Biological Control*, ed. C. B. Huffaker, pp. 377–94. New York: Plenum Press.

van den Bosch, R., P. S. Messenger & A. P. Gutierrez. (1982). *An Introduction to Biological Control.* New York: Plenum Press.

van den Bosch, R. & V. M. Stern. (1969). The effect of harvesting practices on insect populations in alfalfa. *Proc. Tall Timbers Conf. Ecol. Anim. Contr. Habit. Mgmt,* **1**: 47–54.

van den Bosch, R. & A. D. Telford. (1964). Environmental modification and biological control. In: *Biological Control of Insect Pests and Weeds,* ed. P. DeBach, pp. 459–88. New York: Reinhold.

van Emden, H. F. (1987). Cultural methods: the plant. In: *Integrated Pest Management,* ed. A. J. Burn, T. H. Coaker & P. C. Jepson, pp. 27–68. London: Academic Press.

van Lenteren, J. C. (1987). Environmental manipulation advantageous to natural enemies of pests. In: *Integrated Pest Management: Quo Vadis?,* ed. V. Delucchi, pp. 123–63. Geneva: Parasitis.

van Lenteren, J. C. (1988). Biological and integrated pest control in greenhouses. *Ann. Rev. Entomol.* **33**: 239–69.

van Lenteren, J. C. & K. Bakker. (1975). Discrimination between parasitised and unparasitised hosts in the parasitic wasp *Pseudeucoila bochei*: a matter of learning. *Nature,* **254**: 417–19.

Viggiani, G. (1981). Hyperparasitism and sex differentiation in the Aphelinidae. In: *The Role of Hyperparasitism in Biological Control: a Symposium,* ed. D. Rosen, pp. 19–26. Division of Agricultural Sciences, University of California, Publ. **4103**.

Viggiani, G. (1984). Bionomics of the Aphelinidae. *Ann. Rev. Entomol.* **29**: 257–76.

Vinson, S. B. (1977). Behavioral chemicals in the augmentation of natural enemies. In: *Biological Control by Augmentation of Natural Enemies,* R. L. Ridgway & S. B. Vinson, pp. 237–79. New York: Plenum Press.

Vinson, S. B. (1985). The behavior of parasitoids. In: *Comprehensive Insect Physiology, Biochemistry and Pharmacology,* ed. G. A. Kerkut & L. I. Gilbert, vol. 9, *Behavior,* pp. 417–69. Oxford: Pergamon Press.

Vinson, S. B. (1986). New prospects for the augmentation of natural enemies of pests of the Pacific region. *Agric. Ecosyst. Environ.* **15**: 167–73.

Vogele, J. 1988. Reflections upon the last ten years of research concerning *Trichogramma* (Hym. Trichogrammatidae). In: *Trichogramma and Other Egg Parasites,* 2nd Int. Symp. (Guangzhou, China, 1986), ed. J. Vogele, J. Waage & J. van Lenteren, pp. 17–29. Paris: INRA.

Waage, J. & D. Greathead (Eds). (1986). *Insect Parasitoids.* London: Academic Press.

Waage, J. K. & D. J. Greathead. (1988). Biological control: challenges and opportunities. *Phil. Trans. R. Soc. Lond.* B **318**: 111–28.

Walter, G. H. (1983). 'Divergent male ontogenies' in Aphelinidae (Hymenoptera: Chalcidoidea): a simplified classification and a suggested evolutionary sequence. *Biol. J. Linn. Soc.* **19**: 63–82.

Waterhouse, D. F. (1974). The biological control of dung. *Sci. Amer.* **230**: 101–9.

Waterhouse, E. F., L. E. LaChance & M. J. Whitten. (1976). Use of autocidal methods. In: *Theory and Practice of Biological Control,* ed. C. B. Huffaker & P. S. Messenger, pp. 637–59. New York: Academic Press.

Waterhouse, D. F. & K. R. Norris. (1987). *Biological Control: Pacific Prospects.* Melbourne: Inkata Press.

Whalon, M. E. & B. A. Croft. (1984). Apple IPM implementation in North America. *Ann. Rev. Entomol.* **29**: 435–70.

Wharton, R. A. (1990). Classical biological control of fruit-infesting Tephritidae. In: *Fruit Flies, Their Biology, Natural Enemies and Control,* ed. A. S. Robinson and

G. Hooper, pp. 303–13. World Crop Pests, Vol. 3B. Amsterdam: Elsevier Science Publishers.

White, E. B., P. DeBach & M. J. Garber. (1970). Artificial selection for genetic adaptation to temperature extremes in *Aphytis lingnanensis* Compere (Hymenoptera: Aphelinidae). *Hilgardia*, **40**: 161–92.

White, G. G. (1981). Current status of prickly pear control by *Cactoblastis cactorum* in Queensland. *Proc. Fifth Int. Symp. Biol. Contr. Weeds* (Brisbane, 1980), pp. 609–16.

Whitten, M. J. (1985). The conceptual basis for genetic control. In: *Comprehensive Insect Physiology, Biochemistry and Pharmacology*, ed. G. A. Kerkut & L. I. Gilbert, vol. 12, *Insect Control*, pp. 465–528. Oxford: Pergamon Press.

Whitten, M. J. & G. G. Foster. (1975). Genetical methods of pest control. *Ann. Rev. Entomol.* **20**: 461–76.

Wilson, L. T., C. H. Pickett, D. L. Flaherty & T. A. Bates. (1989). French prune trees: refuge for grape leafhopper parasite. *Calif. Agric.* **43** (2): 7–8.

Woglum, R. S. (1913). Report of a trip to India and the Orient in search of the natural enemies of the citrus white fly. *US Dept. Agric. Bur. Entomol. Bull.* **120**.

Wolf, R., (Ed.). (1977). *Organic Farming: Yesterday's and Tomorrow's Agriculture*. Emmaus: Rodale Press.

Wood, B. J. (1971). Development of integrated control programs for pests of tropical perennial crops in Malaysia. In: *Biological Control*, ed. C. B. Huffaker, pp. 422–57. New York: Plenum Press.

Wood, B. J. (1987). Economic aspects and chemical control. In: *Slug and Nettle Caterpillars: the Biology, Taxonomy and Control of the Limacodidae of Economic Importance on Palms in South-east Asia*, ed. M. J. W. Cock, C. J. H. Godfray & J. D. Holloway, pp. 223–35. Farnham Royal (now Wallingford): CAB International.

Wood, B. J. (1988). Overview of IPM infrastructure and implementation on estate crops in Malaysia. In: *Pesticide Management and Integrated Pest Management in Southeast Asia*, ed. P. S. Teng & K. L. Heong, pp. 31–41. Proceedings, Southeast Asia Pesticide Management Workshop (Pattaya, Thailand, 1987). College Park: Consortium for International Crop Protection.

Wood, B. J. & G. F. Chung. (1989). Integrated management of insect pests of cocoa in Malaysia. *The Planter*, **65**: 389–418.

Wood, K. R. S. & M. J. Way (Eds). (1988). *Biological Control of Pests, Pathogens and Weeds*. London: The Royal Society. (*Phil. Trans. Roy. Soc. Lond.* B **318**: 109–376).

Woodworth, C. W. (1908). The theory of the parasitic control of insect pests. *Science*, **28**: 227–30.

Xie, Z. N., W. C. Nettles, Jr., R. K. Morrison, K. Irie & S. B. Vinson. (1986). Three methods for the *in vitro* culture of *Trichogramma pretiosum*. *J. Entomol. Sci.* **21**: 133–8.

Yasumatsu, K. (1958). An interesting case of biological control of *Ceroplastes rubens* Maskell in Japan. *Proc. 10th Int. Congr. Entomol.* (Montreal, 1956), **4**: 771–5.

Yespen, R. B., Jr. (Ed.). (1984). *The Encyclopedia of Natural Insect and Disease Control*. Emmaus: Rodale Press.

Zelazny, B., L. Chiarappa & P. Kenmore. (1985). Integrated pest control in developing countries. *FAO Plant Prot. Bull.* **33**: 147–58.

Zwölfer, H., M. A. Ghani & V. P. Rao. (1976). Foreign exploration and importation of natural enemies. In: *Theory and Practice of Biological Control*, ed. C. B. Huffaker & P. S. Messenger, pp. 189–207. New York: Academic Press.

INDEX